Springer Series in Optical Sciences   Volume 59

Editor: Theodor Tamir

Springer Series in Optical Sciences

Volumes 1–41 are listed on the back inside cover

M.P. Petrov S.I. Stepanov
A.V. Khomenko

Photorefractive Crystals in Coherent Optical Systems

With 102 Figures

Springer-Verlag
Berlin Heidelberg GmbH

Professor Mikhail P. Petrov
Dr. Sergei I. Stepanov
Dr. Anatoly V. Khomenko

A. F. Ioffe Physical Technical Institute
Academy of Sciences of the USSR
194021 Leningrad, USSR

ISBN 978-3-662-13805-2

Library of Congress Cataloging-in-Publication Data.
Petrov, M. P. (Mikhail Petrovich) Photorefractive crystals in coherent optical systems / M. P. Petrov, S. I. Stepanov, A. V. Khomenko. p. cm. – (Springer series in optical sciences ; v. 59) Includes bibliographical references and index.
ISBN 978-3-662-13805-2   ISBN 978-3-540-47056-4 (eBook)
DOI 10.1007/978-3-540-47056-4
1. Holography. 2. Optical data processing. 3. Photorefractive materials. I. Stepanov, S. I. (Sergeĭ Ivanovich) II. Khomenko, A. V. (Anatoliĭ Vasil'evich) III. Title. IV. Series.
TA1542.P48 1991   621.36'7 – dc20   91-11650

This text was prepared using the PS™ Technical Word Processor
54/3140-543210 – Printed on acid-free paper

Preface

The phenomenon of photorefraction was discovered in 1966 in studies of propagation of a fairly powerful laser beam through the electro-optic crystals $LiNbO_3$, $LiTaO_3$, and some other compounds. The laser beam illuminating part of the sample was found to cause a local change in the refractive index of the crystal, thereby leading to distortion of the beam's wave front. The light had deteriorated the initially high optical quality of the crystal, in other words, it had given rise to a nonuniform distribution of the refractive index in the illuminated region. The effect was first called "optical damage".

The practical significance of the phenomenon was soon appreciated, applications were proposed, and a vast amount of activity began. In the years that followed, the phenomenon was termed the "photorefractive effect". Because of the reversible behavior of the refractive index variations due to photorefraction, photorefractive crystals have been regarded as recyclable photosensitive media. This became a valuable finding for optical engineers engaged in holography and optical information processing. On the other hand, the research into the nature of the photorefractive effect proved to be of considerable interest to physicists working in the fields of solid-state physics, semiconductors, and coherent optics.

This monograph covers, though to differing extents, all of the most important properties and characteristics of photorefractive crystals and their applications. In writing the book we attempted to give an impartial state-of-the-art review of the field. Nonetheless, we are fully aware that our scientific tastes and interests have influenced the choice of priorities and the emphasis on certain problems.

To make the subject understandable for readers who are engaged in other fields, the first three chapters give a qualitative and fairly general treatment of the major problems that are discussed later in detail, and present definitions of a number of parameters and important terms used throughout the book.

We wish to express our sincere gratitude to the members of the Department for Quantum Electronics of the A.F. Ioffe Physical Technical Institute of the Academy of Sciences of the USSR who took part in the investigations discussed in the book. We also thank Professor L.I. Korovin and Professor V.V. Bryksin for many helpful and stimulating discussions, and Mrs. N. Nazina for translating and typing the manuscript.

Leningrad, January 1991

M.P. Petrov
S.I. Stepanov
A.V. Khomenko

V

Contents

X

1. The Phenomenon of Photorefraction

A qualitative model of the photorefractive effect discovered by *Ashkin* et al. [1.1] is as follows. The photoelectrons excited in the illuminated area of a crystal by incident light leave the region through diffusion or drift due to an externally applied electric field (or via the photovoltaic effect). They are trapped in nonilluminated regions to give rise to a nonuniform space charge and, hence, to an electric field distribution within the crystal. The electric field, in turn, causes a nonuniform distribution of the refractive index through the electro-optic effect. This principle of image recording suggests immediately the main aspects of the physics of photorefractive media [1.2]:

1) The nature of photosensitive centers that absorb light and produce mobile charge carriers, mostly electrons.
2) The formation of a nonuniform space charge (diffusion and drift of carriers, relaxation effects, influence of the space-charge field itself on its formation process).
3) Electro-optic effects in a nonuniform electric field. Analysis of spatial variations of the refractive index that are directly related to the charge distribution pattern within the crystal.
4) Propagation and diffraction of light in the crystals with a spatially nonuniform refractive index.

This chapter provides a qualitative discussion of the first three aspects to reveal the basic physics associated with photorefraction. It points out specific features of their analysis in a number of cases.

1.1 Photosensitive Centers

Despite the vital importance of charge-carrier photoexcitation and trapping (by shallow and deep traps) for the information recording mechanism in PhotoRefractive Crystals (PRCs), the mechanism is still understood incompletely. Only in certain situations can we speak of generally accepted ideas. It is quite apparent that the important condition for light-induced space charge to be formed is the presence of impurity centers in the band gap that serve both as the sources of excited photoelectrons (donors) and as electron trapping centers (deep traps or acceptors). In the simplest, though fairly often encountered case, both the donors and traps are impurities of ions of one and the same type of atom but in different valence states.

Let us consider Fe^{2+} and Fe^{3+} ions as examples. They give rise to impurities that are thought to be most important for information recording in such crystals as $LiNbO_3$, $KNbO_3$, $BaTiO_3$, and others [1.3-6]. In oxide crystals Fe ions can be present either because of Fe impurities in the starting material or as a result of doping. Where Fe ions are located is often unknown (whether they substitute for certain cations in the crystal, are interstitial ions, or form other types of defects). For instance, in $LiNbO_3$ and $KNbO_3$, Fe ions substitute for Nb^{5+}, the local electroneutrality being preserved in this case by creation of an oxygen vacancy near the Fe^{2+}, that is, an $Fe^{2+}-V_O$ center is formed. Along with Fe^{2+}, Fe^{3+} is present, too.

Upon illumination of the crystal, the light is absorbed by Fe^{2+} impurities, further ion ionization with the formation of $Fe^{3+}+e^-$ occurs, and the excited electron migrates from the illuminated region until it is captured by a deep trap, for instance another Fe^{3+} ion in the nonilluminated region of the crystal.

Typical energies for photoexcitation of Fe^{2+} ions in $LiNbO_3$ and $KNbO_3$ crystals are 3.1-3.2 eV.

The concentrations of Fe^{3+} and Fe^{2+} ions can vary in a wide range ($10^{16}-10^{19} cm^{-3}$) with crystal doping. The proportion between Fe^{2+} and Fe^{3+} can also greatly differ depending on the subsequent treatment (annealing in reducing or oxidizing atmosphere). The presence of Fe impurities changes the crystal's conductivity noticeably, and affects the drift and diffusion lengths of electrons. These values vary from tens of Ångstroms to tens of micrometers from crystal to crystal.

As mentioned above, the photosensitive centers in the form of Fe impurities are typical of $LiNbO_3$, $KNbO_3$, and $BaTiO_3$. The most important type of photoactive centers in the highly popular and practically significant crystals $Bi_{12}SiO_{20}$, $Bi_{12}GeO_{20}$, and $Bi_{12}TiO_{20}$ have not yet been determined unambiguously. The models given in the literature assume, e.g., Si (or Ge) vacancies, a complex BiO_7 ion bonded to a silicon or oxygen vacancy, the presence of chromium impurities, and so on [1.7-10].

Not only electrons, but often also holes, can participate in the photoexcited charge transport. However, because of a much lower mobility of holes as compared with that of electrons - and a shorter lifetime - their contribution to photorefraction is typically small. Nonetheless, several researchers have found that holes play an important role in the space-charge formation [1.11-15].

For ferroelectric crystals (or, generally speaking, for polar crystals, including the pyroelectric ones), an essential mechanism that controls the electron transport in the absence of the external field is the photovoltaic effect [1.3,16]. The essence of the phenomenon is readily understood if we consider the situation when the probability of the excited-electron motion in one or another direction (impulse direction) is anisotropic, and a preferred electron motion generates a resulting EMF. Necessary conditions for the photovoltaic effect to arise are that the photoactive center itself be spatially asymmetric, that it has a dipole moment, and that such centers be preferentially oriented to prevent the total averaging of the directions of the

2

photoexcited electron transfer. Equally important contributions to the photorefractive effect come from the anisotropy of carrier recombinations, scattering anisotropy, and so on. Such necessary conditions for the photovoltaic effect are encountered in polar crystals. The observed photovoltaic fields in dielectric crystals of the $LiNbO_3$ type can be as high as 10^4 to 10^5 V/cm.

Generally speaking, nonpolar crystals also exhibit the phenomenon of photo-EMF generation. Here we mean the crystals that lack a center of inversion where the linear and circular photovoltaic effects are observed [1.17]. The preferred direction in space is then governed by the direction of linear or circular light polarization. This generated voltage is, however, typically small, though exceptions are possible, for instance, if the resistivity of the crystal is sufficiently high.

1.2 Mechanism for Optical Information Recording

1.2.1 Linear Approximation of Information Recording

The photorefractive crystals offer the capability for recording both images and holograms. For either of these cases we shall use the term "information recording". Depending on the task to be solved, recording is accomplished using either ordinary incoherent light or laser radiation. For research into the properties of crystals, however, recording of simple sinusoidal gratings formed by interference of two coherent beams is, as a rule, preferred (though not always). The assumption of linearity of information recording in photorefractive crystals is mainly responsible for the wide popularity of such a research technique.

A fact is that no matter how complex the three-dimensional recording intensity pattern $I(x, y, z)$ may be, it can be represented by a superposition of cosinusoidal and sinusoidal patterns of the type $I(\mathbf{K})\cos(\mathbf{K} \cdot \mathbf{r})$, $I(\mathbf{K}) \times \sin(\mathbf{K} \cdot \mathbf{r})$, or, in exponential form, by $I(\mathbf{K})\exp(i\mathbf{K} \cdot \mathbf{r})$, where $I(\mathbf{K})$ is the Fourier amplitude of the light intensity, $\mathbf{K} \cdot \mathbf{r} = 2\pi(\nu x + \xi y + \gamma z)$, and $\mathbf{K}$ is the wave vector of the pattern with the projections $K_x = 2\pi\nu$, $K_y = 2\pi\xi$, and $K_z = 2\pi\gamma$. The values ν, ξ, γ are called the spatial frequencies; $\nu = 1/\Lambda_x$, $\xi = 1/\Lambda_y$, $\gamma = 1/\Lambda_z$, where Λ_x, Λ_y, Λ_z are the grating spacings in the x, y, and z directions, respectively. The tradition has been established in the literature on photorefractive media to refer to the wave vector projections K_x, K_y, and K_z as spatial frequencies as well; no confusion has arisen so far.

Thus, in the most general case, the light intensity to be recorded can be given by the Fourier integral

$$I(x, y, z) = I_0 \iiint m(\nu, \xi, \gamma)\exp[-i2\pi(\nu x + \xi y + \gamma z)]d\nu d\xi d\gamma , \qquad (1.1)$$

where I_0 is the average intensity, and $m(\nu, \xi, \gamma)$ is the relative spectral (spatial) density of the light intensity. In specific cases, expansion in a Fourier series of sines and cosines functions is possible. Then, with the

3

same notations, $m(\nu,\xi,\gamma)$ implies a relative spectral amplitude or coefficient of light-intensity spatial modulation for given values of ν, ξ, and γ.

In a similar fashion, the crystal response, i.e., the spatial distribution of the refractive-index change $\Delta n(x,y,z)$, arising in the crystal, can be expanded in a Fourier series or represented by a Fourier integral. (The reasons for the use of the refractive index to describe the PRC response are discussed in Sect. 1.3)

If we assume that the recording mechanism is linear, in other words, the principle of superposition is applicable, recording of each of the sinusoidal patterns making up the complex image will occur, regardless of the presence of other sinusoidal gratings, with one sinusoidal light-intensity pattern producing only one refractive-index grating. Consequently, to understand properly the crystal parameters which determine the recording of complex patterns, it suffices to know how the recording of individual sinusoidal patterns, that differ from one another by the spatial frequency alone, proceeds. Sinusoidal patterns can readily be produced by interfering two coherent plane waves. Therefore we shall quite justifiably confine the analysis of all recording processes in the following discussion to that of simple sinusoidal gratings, with the aim of establishing the dependence of the amplitude and phase of the recorded refractive-index grating on spatial frequencies.

An alternative approach, entirely equivalent to that discussed above, in the framework of linear space-invariant systems theory is in general possible, too. It involves the investigation of the crystal response to an input delta function, i.e. the impulse response. From the standpoint of linear systems theory, the two methods are formally equivalent because they are related by the Fourier transformation. Experimentally, however, the study of light diffraction from the refractive index grating presents a more convenient research tool than the tiresome analysis of details of the impulse response shape.

Therefore, in what follows, we perform the analysis in terms of "elementary gratings", with the term "grating" used to describe the sinusoidal distribution of charge, electric field, refractive index, etc.. We emphasize that, though the linear approximation is a very useful research tool, it is not always applicable to photorefractive crystals (Chaps. 4, 8).

1.2.2 Diffusion and Drift Models

We turn now to information recording mechanisms in photorefractive crystals and restrict ourselves, in this section, to the formation of a nonuniform space charge and electric field. A relation between the electric-field distribution within the crystal and that of the refractive index will be treated in Sect. 1.3.

There exist two basic recording mechanisms, namely, diffusion and drift [1.18-21]. In the diffusion model, the electrons excited by light migrate from the illuminated region, where their concentration is higher, toward the nonilluminated regions, where the carrier concentration is lower.

There they become trapped. Let the crystal be exposed to an interference pattern of the type

$$I(x) = I_0[1 + m\cos(K_x x)] . \tag{1.2}$$

The light intensity is uniform along the y and z axes. In the diffusion model, the redistribution of the photoexcited electrons gives rise to three charge gratings (Fig.1.1). The first is the grating of positively charged donors, with the charge density given by

$$\rho^+(x) = \rho\cos(K_x x) + \rho_0 , \tag{1.3}$$

where

$$\rho = emI_0 t_{ex} \beta\alpha/(\hbar\omega) .$$

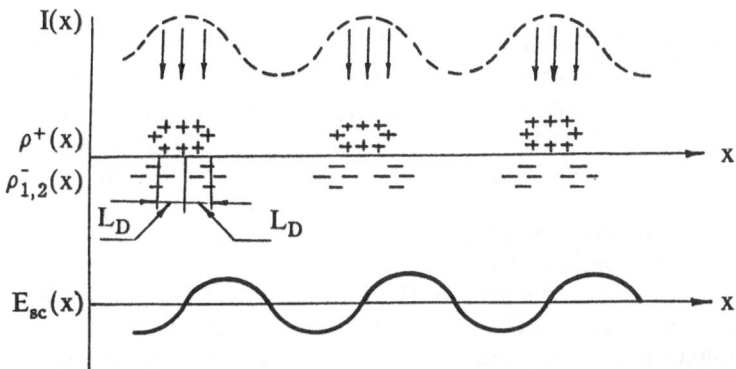

Fig.1.1. Formation of charge and electric-field gratings through diffusion on illumination by the recording light

Here, ρ_0 is the mean density of positive charge, α is the absorption coefficient, e is the electronic charge, t_{ex} is the exposure time, β is the quantum efficiency of electron excitation, and $\hbar\omega$ is the energy per recording light photon. The two other gratings are the gratings of negatively charged traps with the charge density for short exposures (the initial stage of recording), i.e.,

$$\rho_1^-(x) = - \frac{1}{2}\rho\cos[K_x(x + L_d)] - \frac{1}{2}\rho_0$$

and $\tag{1.4}$

$$\rho_2^-(x) = - \frac{1}{2}\rho\cos[K_x(x - L_d)] - \frac{1}{2}\rho_0 .$$

These gratings arise because of electron diffusion to the right and left from the site of the excitation. Here, L_d is the diffusion length, i.e., a typical

5

mean distance the electron moves from the excitation site to the trapping location. According to Poisson's equation

$$\text{div}E(x) = \frac{\rho(x)}{\epsilon\epsilon_0} , \qquad (1.5)$$

where $\rho(x) = \rho^+(x) + \rho_1^-(x) + \rho_2^-(x)$, the resultant field $E_{sc}(x)$ formed by the three gratings is

$$E_{sc}(x) = \frac{\rho}{\epsilon\epsilon_0 K_x}[1 - \cos(K_x L_d)]\sin(K_x x) , \qquad (1.6)$$

where ϵ_0 is the free-space electric permeability, and ϵ is the relative dielectric constant of the crystal. The grating field is along the x axis. Other components, along the y or z axes, are zero. For a short diffusion length L_d such that $K_x L_d \ll 1$ (this situation is fairly common in photorefractive media),

$$E_{sc}(x) = \frac{\rho L_d^2}{2\epsilon\epsilon_0} K_x \sin(K_x x) . \qquad (1.7)$$

Note that the field grating is shifted by $\pi/2$ with respect to the interference pattern [$\cos(K_x x)$ has been transformed into $\sin(K_x x)$] and the grating amplitude is proportional to the spatial frequency K_x. Moreover, comparison of (1.6 and 3) also reveals that the buildup rate of the field grating grows with increasing L_d and decreasing ϵ, with the recording light intensity being the same.

The other limiting case, when $L_d K_x \gg 1$, needs a more thorough treatment; we shall dwell upon it in Sect.4.2.

As the exposure grows, the grating field increases and begins to inhibit diffusion of electrons. At last, the amplitudes of the charge and field gratings grow no longer and steady state is reached; in other words, despite the presence of light that excites electrons, no further growth of the grating-field amplitude occurs. This situation arises when the grating field fully compensates for the so-called effective diffusion field E_D. The fact is that diffusion of electrons from the regions of higher concentration to those of lower concentration, which occurs exclusively at the expense of thermal motion, can be represented as electron motion in an effective electric field whose magnitude is [1.22]

$$E_D(x) = \frac{k_B T}{e} \frac{1}{n(x)} \frac{dn(x)}{dx} , \qquad (1.8)$$

where k_B is the Boltzmann constant, T is the temperature, $n(x)$ is the electron concentration at point x, and $dn(x)/dx$ is the concentration gradient. The grating field that is a real electric field is opposite to $E_D(x)$. Steady state is reached when $|E_{sc}(x)| = |E_D(x)|$. The grating field E_{sc} at steady state is independent of such crystal parameters as the dielectric permeability or the diffusion length.

Since $n(x) \propto I(x)$, $E_D(x)$ is shifted by $\pi/2$ with respect to the interference pattern and is proportional to the spatial frequency. The diffusion field for a sinusoidal grating with m<<1 can be written as

$$E_D(x) = mE_D \sin(K_x x) ,$$ (1.9)

where

$$E_D = K_x \frac{k_B T}{e} .$$ (1.10)

The condition m<<1 implies that there is a fairly strong uniform illumination of the crystal and, hence, a high level of photoconductivity. In further discussion of the diffusion mechanism we shall refer to the amplitude E_D as the diffusion field. For T = 300K and $K_x/2\pi \simeq 10^2 - 10^3$ mm^{-1}, E_D is 0.1-1 kV/cm.

We now turn to the drift recording mechanism in which, as distinguished from the diffusion mechanism, the motion of the photoexcited charge occurs in an external electric field (Fig.1.2). The photoexcited electrons move in one direction and pass, on the average, a certain typical distance L_0 before being trapped. Here L_0 denotes the drift length and is related to the external field E_0 by $L_0 = \mu \tau E_0$, μ being the mobility and τ the lifetime of a photoexcited electron.

The essential features of the drift mechanism are readily understood from the following model. We assume, as before, that the crystal is illuminated by light with $I(x) = I_0[1+m\cos(K_x x)]$. The external field is along the x axis. In the initial stage of recording, when the electron transport is still unaffected by the grating field, the charge distribution for short drift lengths can be represented as a sum of two gratings $\rho^+(x) = \rho\cos(K_x x)+\rho_0$ and $\rho^-(x) = -\rho\cos[K_x(x+L_0)]-\rho_0$, which are formed by positively charged donors and negatively charged traps; ρ is determined from (1.3). The net charge density is

$$\rho(x) = \rho \Big[\cos(K_x x) - \cos[K_x(x+L_0)] \Big].$$ (1.11)

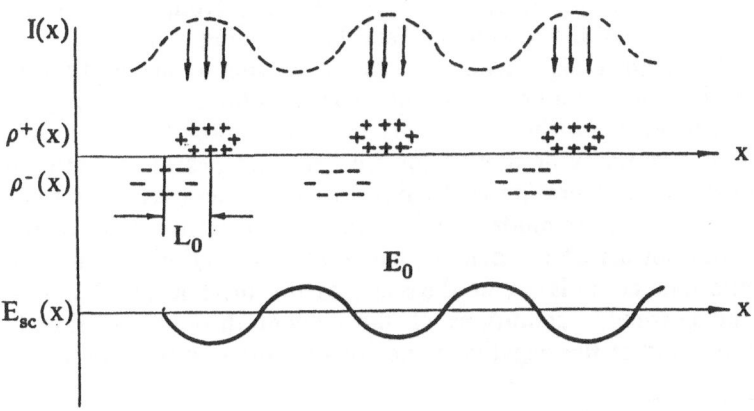

Fig.1.2. Formation of charge and electric-field gratings in a photorefractive crystal through drift

For short drift lengths $K_x L_0 \ll 1$, the field of such a grating is, according to (1.5),

$$E_{sc}(x) = \frac{-\rho}{\epsilon\epsilon_0} L_0 \cos[K_x(x + \tfrac{1}{2}L_0)] . \tag{1.12}$$

As seen from (1.12), in this case the grating field is almost in phase (unshifted grating) with the recording light grating, and its amplitude is independent of K_x, but depends on the drift length. In the other limiting case $(L_0 K_x \gg 1)$, a negative charge grating is not formed since electrons are uniformly "spread" throughout the crystal. Only the grating $\rho^+(x) + \rho_0$ is left, which, according to (1.5), gives the field grating shifted by $\pi/2$ with respect to the light intensity grating, and the field amplitude is $\propto 1/K_x$.[1]

For long exposures, when the recording exhibits a steady-state behavior, the grating-field amplitude attains its highest possible value. More sophisticated consideration (Sect.4.2) is needed to analyze how the grating reaches this regime. Here we note only that in the drift model the maximum grating field is limited either by the external field E_0, or (as in the diffusion mechanism) by depletion of donors and traps (the typical field value is here denoted by E_q). We define E_q as the maximum amplitude of a sinusoidal grating field when depletion of either donors or traps occurs. Thus, the field E_q is limited by the lower of the two values, i.e., the donor concentration N_D or the trap concentration N_A. If we denote this lower concentration by N_{min}, then

$$E_q = \frac{eN_{min}}{\epsilon\epsilon_0 K_x} . \tag{1.13}$$

Since E_0 and E_q can be well above E_D for moderate K_x, the efficiency of the drift mechanism can be appreciably higher.

The drift mechanism provides a great variety of recording conditions. The external field can be applied perpendicular to the grating fringes or, alternately, an AC or DC field can be used, the crystal's electrode contacts may be ohmic or rectifying, the crystal can be insulated from electrodes by a dielectric layer, etc.. All these factors affect the recording efficiency.

Of utmost practical significance is the recording configuration in which the electric field is applied along the z axis and the recording is performed by light incident on the crystal surface that is normal to the z axis. This geometry is commonly employed for recording images in spatial light modulators. In this case, a grating $\rho(x)$ having a finite thickness along the z axis is formed. In the linear mode of operation (as long as the field of the grating itself does not affect the drift of electrons) the grating thickness in practically important cases is about the same as the drift length L_0. Depending on the particular conditions of the current through the sample upon illumination, either the negative or positive charge grating may domi-

---

[1] It should be remembered that a *shifted grating* is one whose maxima are shifted with respect to the maxima of the interference pattern. A *unshifted grating* is one whose maxima coincide with those of the intereference pattern.

nate the process. Because of the finite thickness of the charge grating, non-zero values of the field are observed not only along the x axis, but also along the z axis. The field grating $E_x(x)$ is always shifted by $\pi/2$ from the interference pattern when the external field is along the z axis, and the grating $E_z(x)$ is in phase with it. The spatial frequency dependences of $E_x(x)$ and $E_z(x)$ also turn out to be different. These and other specific features of the drift mechanism will be discussed in more detail in Sects. 3.1 and 7.5.

The drift mechanism can take place not only in an external electric field, but also because of the photovoltaic effect mentioned above. From the formal point of view, the electron transport caused by the photovoltaic EMF for m << 1 can be represented as electron drift in an effective external field. The drift mechanism in information recording by means of the photovoltaic effects is therefore nearly equivalent to the process taking place in an external field.

1.3 Electro-Optic Effects

1.3.1 Longitudinal and Transverse Electro-Optic Effects

A nonuniform charge - and thus electric field - distribution induced in a photorefractive crystal cause variation of the index of refraction through the electro-optic effect.

In this section, electro-optic phenomena will be discussed from a general point of view [1.22-24], with particular attention given to specific features for photorefractive crystals.

The PRCs may originally be either isotropic or anisotropic. In an isotropic crystal, the dielectric permeability ϵ^ω is a scalar.[2] The refractive index ($n = \sqrt{\epsilon^\omega}$) and the light-propagation velocity depend neither on the direction of light-wave propagation nor on the direction and state of its polarization. Optical activity is ignored here. In an anisotropic crystal, $\hat{\epsilon}^\omega$ is a tensor, and the refractive index will generally depend on both the propagation direction and polarization of this wave. Nonetheless, one or two directions can exist in a spatially homogeneous crystal, the refractive index of the light traveling along which is independent of the wave polarization. These directions are called optical axes.

The electro-optic effect consists of a change of the refractive index of the crystal and a change of the orientation of its optical axes on application of an external electric field. In terms of crystal optics, the electro-optic effect rests on the dependence of components of the dielectric impermeability tensor α_{ij}^ω, $\hat{\alpha}^\omega = (\hat{\epsilon}^\omega)^{-1}$, on the electric field E. If α_{ij}^ω and E are linearly related, the effect is termed the linear electro-optic (Pockels) effect. If the relation is quadratic, we have the Kerr effect. In this monograph we deal exclusively with the Pockels effect.

---

[2] The superscript ω indicates that this parameter is considered for optical frequencies.

A necessary "existence" condition for the Pockels effect is the absence of an inversion center in the crystal. The linear relation between α_{ij}^{ω} and E is typically given by

$$\Delta\alpha_{ij}^{\omega} = \sum_n r_{ijn} E_n = r_{ijn} E_n , \qquad (1.14)[3]$$

where E_n is the electric-field projection (n = x, y, z), and the third-rank tensor r_{ijn} is the electro-optic tensor. Since $\Delta\alpha_{ij}^{\omega}/\alpha_{ij}^{\omega} \ll 1$, the component of the dielectric permeability tensor for a cubic crystal acquires the form

$$\epsilon_{ij}^{\omega}(E) \simeq n^2 \delta_{ij} - n^4 r_{ijn} E_n , \qquad (1.15)$$

where $\delta_{ij} = 1$ for i = j, and $\delta_{ij} = 0$ for i ≠ j. Equation (1.15) reveals that, in the most general case, the orientation of the coordinate system where tensor $\hat{\epsilon}^{\omega}$ is diagonal, will depend on the electric-field E direction, and the refractive indices in this coordinate system ($n_1 = \sqrt{\epsilon_{11}^{\omega}}$, $n_2 = \sqrt{\epsilon_{22}^{\omega}}$, $n_3 = \sqrt{\epsilon_{33}^{\omega}}$) depend on the magnitude of the field. When field E is applied, the originally optically isotropic crystal ($n_1 = n_2 = n_3$) can become uniaxial ($n_1 = n_2 \neq n_3$) or biaxial ($n_1 \neq n_2 \neq n_3$).

Let us consider some features of the electro-optic effect, taking the originally uniaxial and isotropic crystals as illustrations. A plane light wave with an arbitrary propagation direction and an arbitrary linear polarization direction in the crystal can be represented as a superposition of two so-called eigenmodes. These modes are identified as waves with polarizations oriented perpendicular to each other and traveling through the crystal with their own refractive indices. One of the eigenmodes is the wave with the polarization normal to the optical axis and the direction of propagation. This wave is referred to as the "ordinary" wave, and the "ordinary" refractive index n_o corresponds to it. Once the ordinary wave is defined, the second mode is determined unambiguously and is called the "extraordinary" wave, with the corresponding "extraordinary" refractive index n_e. Note that n_o is the same for all ordinary waves in the crystal, and n_e depends on the propagation direction of the extraordinary wave.[4] For an arbitrarily incident wave, the ordinary and extraordinary beams will refract at different angles - because $n_o \neq n_e$ - to give rise to birefringence.

If the original wave is incident perpendicular to the crystal surface and the optical axis is parallel to the crystal surface, the ordinary and extraordinary waves travel in the same direction within the crystal, but, because of different propagation velocities (generally n_o is not equal to n_e), the waves

---

[3] Henceforth, we shall follow the familiar convention that repeated subscripts denote summation and omit the summation sign.

[4] In the literature n_e often denotes the maximum (if $n_e > n_o$) or the minimum (if $n_e < n_o$) value of the refractive index for an extraordinary wave.

have different phase delays. The difference between phase delays is

$$\phi_{oe} = \frac{2\pi d(n_o - n_e)}{\lambda}, \tag{1.16}$$

where d is the crystal thickness. Application of the electric field can cause variations in n_o and n_e, and a change of ϕ_{oe}, i.e., $\Delta\phi_{oe}$ arises. Variation in phase retardation induced by the applied field can easily be demonstrated experimentally. This is precisely the effect used in polarization techniques for studying electro-optic properties of crystals, and also in the design of electro-optic light modulators.

The electro-optic effects can be classified depending on the orientation of the applied field E and the light-propagation direction. If E‖K, the longitudinal effect is possible; if E⊥K, the transverse effect can arise (Fig. 1.3). Apparently the values of $\Delta\phi_{oe}$ for both the longitudinal and transverse effects will depend on how the crystal is cut, in other words, on the direction in which the light travels with respect to the crystallographic axes and electric-field vector E.

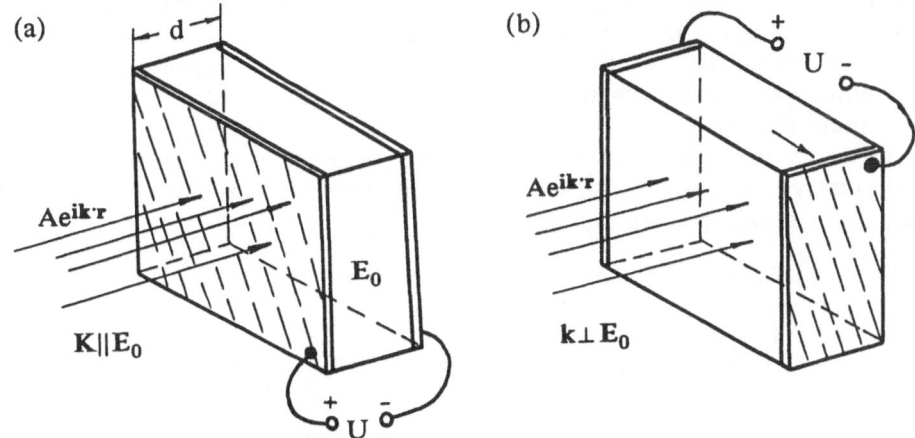

Fig.1.3. Mutual orientations of the applied electric field E_0 and light propagation direction (direction of **K**) for (a) the longitudinal electro-optic effect, and (b) the transverse electro-optic effect

The simplest relations are obtainable for an originally isotropic (cubic) crystal. With no applied field, $n_o = n_e = n$ and $\phi_{oe} = 0$. If the field E_ℓ is applied along one of the cubic crystallographic axes - for instance, the [001] direction - the crystal becomes a biaxial one with $n_z = n$, $n_{x'} = n+\Delta n$, $n_{y'} = n-\Delta n$, where $\Delta n = \frac{1}{2}n^3 r_{41} E_\ell d$. We assume here that the z axis ‖ [001], and the two other axes x' and y' are at ±45° to the [100] axis. For light travelling along the z axis (the longitudinal effect), the eigenmodes are the waves polarized along x' and y'. Because calculations of phase relations for these modes are similar to those for ordinary and extraordinary waves in uniaxial crystals, we retain here the notation $\Delta\phi_{oe}$ and, to make it clear, assume that the wave which is polarized along x' is similar to the ordinary beam, and that polarized along y' is similar to the extraordinary beam. Then for the longitudinal electro-optic effect

$$\Delta\phi_{oe} = \frac{2\pi d n^3 r_{41} E_\ell}{\lambda} = \frac{2\pi n^3 r_{41} U}{\lambda}, \qquad (1.17)$$

where U is the voltage applied to the crystal plate. r_{41} is the respective electro-optic coefficient with the subscript 41 being used instead of ijn, when ijn = 321 = 231

A specific characteristic of the magnitude of the electro-optic effect is the so-called half-wave voltage $U_{\lambda/2}$. It is defined as the value of the applied voltage U at which $\Delta\phi_{oe}$ becomes equal to π. The half-wave voltage is given by

$$U_{\lambda/2} = \frac{\lambda}{2 n^3 r_{41}}. \qquad (1.18)$$

The half-wave voltage differs for different crystal cuts. However, handbooks typically specify $U_{\lambda/2}$ for such a cut where the half-wave voltage is at the minimum. When $\Delta\phi_{oe} = \pi$ is reached, the polarization of the resultant wave (superposition of the ordinary and extraordinary beams or corresponding eigenmodes) changes by 90° at the crystal exit (the direction of the input light polarization is shown in Fig.1.4). Then the crystal is equivalent to the so-called half-wave plate. The vertically polarized input beam will emerge from the crystal with a horizontal polarization. For an arbitrary $\Delta\phi_{oe}$, the output polarization is elliptic.

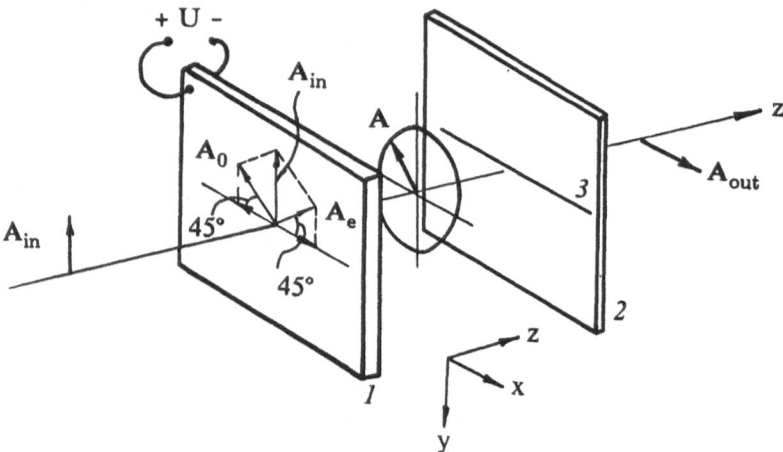

Fig.1.4. Propagation of linearly polarized light through an electro-optic crystal: (1) crystal, (2) analyzer, (3) orientation of the analyzer axis. A_{in} denotes the incident light with vertical linear polarization; A_o and A_e the ordinary and extraordinary components into which the incident light is decomposed; A is the elliptically polarized beam emerging from the crystal, and A_{out} the horizontally polarized light wave resulting from the elliptically polarized wave passing through the analyzer

If the transverse electro-optic effect is employed, then, for a cubic crystal,

$$\Delta\phi_{oe} = \frac{2\pi d n^3 r_{41} E_t p}{\lambda} , \tag{1.19}$$

where E_t is the field applied in the plane perpendicular to the light-propagation direction, and p is a coefficient ($p \leq 1$) that depends on the crystal orientation. Now the effect is governed rather by the magnitude of the transverse field than by the potential difference, as with the longitudinal effect.

A change in the light-polarization state at the crystal exit, resulting from different phase delays experienced by the ordinary and extraordinary beams, can be converted into the light intensity variations by a polarization analyzer. For instance (Fig.1.4), if light is polarized vertically at the crystal input, and an analyzer with a horizontally directed transmission axis is placed behind the crystal, then, in a general case for an arbitrary $\Delta\phi_{oe}$, the light intensity behind the analyzer ($I_{\rightarrow}$) is given by

$$I_{\rightarrow}(\Delta\phi_{oe}) = I_{\uparrow} \sin^2(\tfrac{1}{2}\Delta\phi_{oe}) , \tag{1.20}$$

where $I_{\uparrow}$ is the light intensity at the input.

1.3.2 Electro-Optic Effects in PRCs

As pointed out above, the notions of the transverse and longitudinal electro-optic effects, and also that of the half-wave voltage, are widely used in polarization techniques for investigating electro-optic media and also for analyzing electro-optic light modulators. Generally speaking, these notions are used in studies of photorefractive crystals as well, however here they acquire somewhat different meanings. These differences are attributable first to the fact that the electro-optic effect arises not in externally applied fields, but in internal fields induced by an exposure of the crystal to recording light. Secondly, we have to deal with a nonuniform (both in magnitude and direction) field within the crystal. Several examples follow.

1) The notions of the longitudinal and transverse effects are traditionally associated with the mutual orientation of the applied electric field and the light-propagation direction. In photorefractive crystals, however, the electro-optic effect of interest arises because of internal fields that are caused by a nonuniform space charge, i.e., field E_{sc}. Thus, the notions of the longitudinal and transverse effects are now used to describe the orientation of the light-propagation direction with respect to the space-charge field direction within the crystal. The role of the external field may be unimportant here, since it serves merely to induce the drift of charge carriers. For the diffusion mechanism, the external field may not be applied at all.

2) In photorefractive crystals, especially when they are employed in coherent optical systems, extensive use is made of the phenomenon of light

diffraction from the refractive index grating. In this diffraction experiment the index variations should be treated separately for the ordinary and extra-ordinary beams, since light can diffract in an entirely independent manner for the two beams. Of primary concern here is a change of each refractive index n_o and n_e by itself, not of their difference. For instance, sometimes n_o and n_e exhibit similar dependences on the electric field, but induce no changes in birefringence ($\Delta\phi_{oe} = 0$), and are not detected by the polariza-tion technique, whereas the diffraction experiments allow measurement of the individual variations (n_o and n_e). Introducing here also the notion of the half-wave voltage is sometimes convenient, but its meaning and mag-nitude prove to be somewhat different from (1.18). The half-wave voltage can be specified in this case as the voltage under which the phase retarda-tion for an individual beam alters by π. This parameter is therefore dif-ferent for different beams (ordinary and extraordinary) and its magnitude typically differs from that in the definition of $U_{\lambda/2}$ given previously.

We should comment specifically on the use of the refractive index as a performance parameter related to recording of a two- or three-dimensional light intensity distribution $I(x, y, z)$. Take recording of a two-dimensional image as an example. In particular, difficulties are encountered for the transverse electro-optic effect. The fact is that on recording a two-dimen-sional pattern $I(x, y)$ the nonuniform internal electric field within the crystal changes from point to point not only in magnitude, but also in direction. However, the direction of the transverse electric field in a given region of the crystal is what determines a local coordinate system $x'y'$ in this region (x', y' are the directions of the optical axes). A change in the transverse field direction (in the xy plane) leads to a change in the local coordinate system ($x'y'$) that establishes the polarization direction of the ordinary and extraordinary beams. This consideration raises doubts as to whether it is reasonable to analyze information recording in PRCs using the description in terms of the refractive index variations, because decomposition of the recording light intensity or the space-charge electric field into spatial fre-quencies is carried out in a fixed coordinate system, and the refractive in-dices require a local coordinate system that depends on the field direction and, what is more, changes from one image to another.

In this case, the convention of "eigenmodes" (normal modes) of light waves and refractive indices are meaningless for the crystal as a whole. Thus, to find the spatial two-dimensional dependence of the output light amplitude $A_{out}(x, y)$, we should calculate the light wave field at the crystal output using a more general approach. The general solution of this problem for thin plates was given in [1.25]. The input-output relation of the complex light amplitude in a thin plate is determined by a tensor that is a linear function of the recording intensity $I(x, y)$. No difficulties arise for the long-itudinal electro-optic effect, since the field direction $E_z(x, y)$ is uniquely specified for the entire crystal plane. The situation is also simplified for the transverse effect if spatial frequencies differ markedly in magnitude (either $\nu/\xi \ll 1$ or $\nu/\xi \gg 1$), and also if the initial birefringence is fairly large $(n_o - n_e) \gg \Delta n(x, y)$. In these cases, a description in terms of a nonuniform distribution of the refractive index $\Delta n(x, y)$ can often be used quite justifiably.

The analysis of the electro-optic effect associated with a nonuniform field in PRCs can give birth to a new branch of the optics of solids, i.e. to electro-optics in nonuniform fields.

2. Holography, and Optical Information Processing Systems

This chapter is concerned with the use of photorefractive crystals as reusable photosensitive media in holography and optical information processing systems. The underlying principles are outlined and discused in connection with specific applications.

2.1 Holographic Recording of Wave Fronts

Holographic recording exploits the well-known phenomenon of interference between wave fields [2.1-3]. The interference pattern formed by two coherent beams and recorded in a photosensitive material, i.e., an interferogram, contains information on the amplitudes and the relative phases of these beams. Thus, in principle, the pattern carries all necessary data for reconstruction of all the parameters of one of the beams, provided that the characteristics of the other one are known. In practice, it is possible to reconstruct one of the original beams used to form the interferogram by illuminating it with the second beam.

We turn our attention for the moment to the geometries used for recording and reconstructing holograms. Let two coherent plane waves, one of which is called the reference wave (A_r) and the other one the object wave (A_{ob}), be incident on a thin photographic plate (Fig.2.1), i.e.,

$$A_{ob}(x, y, z, t) = A_{ob}\cos(\omega t - \mathbf{K}_{ob}\cdot\mathbf{r} + \psi_{ob}) \, ,$$

and $\hspace{8cm}$ (2.1)

$$A_r(x, y, z, t) = A_r\cos(\omega t - \mathbf{K}_r\cdot\mathbf{r} + \psi_r) \, ,$$

where $\mathbf{K}_{ob}$ and $\mathbf{K}_r$ are the wave vectors, and ψ_{ob} and ψ_r are the initial phases of the light waves. These waves produce an interference pattern of the light intensity in the $z = 0$ plane

$$I(x, y) = A_{ob}^2 + A_r^2 + A_{ob}(x, y, 0)A_r^*(x, y, 0) + A_{ob}^*(x, y, 0)A_r(x, y, 0)$$

$$= I_1 + I_2 + 2A_{ob}A_r\cos[(\mathbf{K}_{ob}-\mathbf{K}_r)\cdot\mathbf{r} + \psi_{ob} - \psi_r] \, . \hspace{2cm} (2.2)$$

The asterisk (*) denotes complex conjugation. Here, we use the definition for the light intensity $I(x, y)$ in vacuum employed in the literature on holography. It differs from the conventional definition by a multiplicative

16

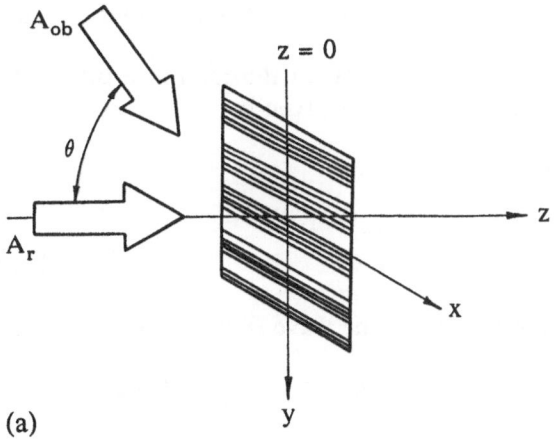

(a)

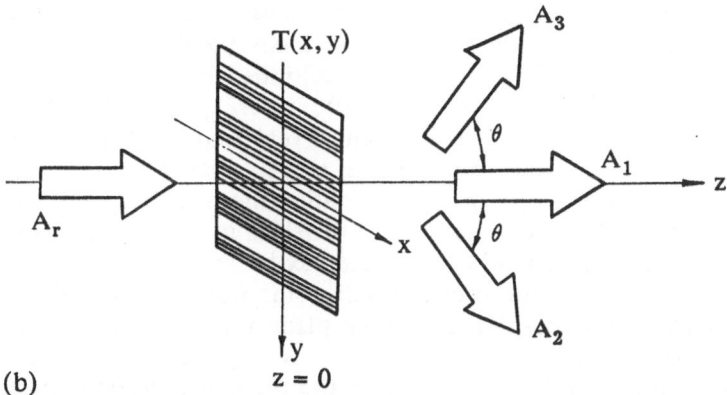

(b)

Fig.2.1a,b. Schematic of the simplest hologram recording. (a) Sinusoidal grating, and (b) schematic of hologram readout

$$A_1 = T_0 A_r(x,y,0)e^{-iK_r z} ,$$
$$A_2 = T_1 A_r{}^2 A_{ob}(x,y,0)e^{-iK_{ob} z} ,$$
$$A_3 = T_1 A_r{}^2 A_{ob}^*(x,y,0)e^{-iK_{ob} z}$$

factor $\frac{1}{2}c\epsilon_0$, c being the velocity of light. In addition, $A_{ob,r}(x,y,z)$ is the complex amplitude of the respective wave:

$$A_{ob,r}(x,y,z) = A_{ob,r}\exp[-i(\mathbf{K}_{ob,r}\cdot\mathbf{r} - \psi_{ob,r})] . \qquad (2.3)$$

We further assume that the amplitude transmittance $T(x,y)$ of a developed photosensitive plate is proportional to $I(x,y)$. Then we can write

$$T(x,y) = T_0 + T_1 A_{ob}(x,y,0)A_r^*(x,y,0) + T_1 A_{ob}^*(x,y,0)A_r(x,y,0), \qquad (2.4)$$

17

where T_0 and T_1 are coefficients that depend on the sensitivity of the photographic material.

If the interferogram is again illuminated with the reference beam, the light amplitude immediately behind the plate is given by

$$A_{out}(x,y,0) = A_r(x,y,0)T(x,y)$$
$$= T_0 A_r(x,y,0) + T_1 A_r^2 A_{ob}(x,y,0)$$
$$+ T_1 A_r^2(x,y,0) A_{ob}^*(x,y,0) . \qquad (2.5a)$$

Far away (to the right) from the hologram, the wave field has the form

$$A_{out}(x,y,z) = T_0 A_r(x,y,0)e^{-iK_{rz}z} + T_1 A_r^2 A_{ob}(x,y,0)e^{-iK_{obz}z}$$
$$+ T_1 A_r^2(x,y,0) A_{ob}^*(x,y,0)e^{-iK_{obz}z} , \qquad (2.5b)$$

where $K_{obz} = (K^2 - K_{obx}^2 - K_{oby}^2)^{1/2}$. The expressions (2.5a, b) reveal that three beams are present behind the plate. One of them (the first term) is the readout beam with the amplitude $A_r(x,y,z)$ multiplied by coefficient T_0. The second beam given by $T_1 A_r^2 A_{ob}(x,y,0)\exp(-iK_{obz}z)$ is the object beam times the coefficient $T_1 A_r^2$. This result is very important, for it reveals the essence of the holographic technique, namely that it involves reconstruction of the beam $A_{ob}(x,y,z)$ by illuminating the interferogram with the beam $A_r(x,y,z)$. The interferogram $T(x,y)$ itself is in this case the simplest hologram or holographic grating. If a complex beam scattered from an object is used for recording instead of a simple plane wave $A_{ob}(x,y,z)$, we can represent it by a superposition of simple plane waves and repeat the manipulation described above for each of the waves separately. Then illumination by the beam $A_r(x,y,z)$ reconstructs the original superposition of waves, i.e., the light beam scattered from the object, and the observer will be able to see the image stored in the hologram.

Let us continue our analysis of (2.5). For the sake of simplicity, we can take $\psi_r = \psi_{ob} = 0$ and assume, according to Fig.2.1, that $K_{rx} = K_{ry} = 0$. Now the third term contains the multiplier $A_{ob}^*(x,y,0)\exp(-iK_{obz}z)$, i.e., it describes the wave with phase $K_{obx}x + K_{oby}y$ that has the sign opposite to that of $A_{ob}(x,y,z)$ and travels in the positive z direction. Referring to the scheme in Fig.2.1, where it is assumed that $K_{obx} = 0$, this simply means that the wave of interest propagates at an angle $(\theta'=-\theta)$ symmetric to that of the reconstructed object wave.

Regarding the complex wave scattered from the object as $A_{ob}(x,y,z)$, the last term in (2.5b) gives rise to the so-called pseudoscopic image, which has inverted depth relations, i.e., elevations look like cavities, and vice versa.

An interesting phenomenon arises if the hologram is reconstructed by the complex conjugate $A_r^*(x,y,z)$ to $A_r(x,y,z)$. For the plane wave with $K_{rx} = K_{ry} = 0$, this is simply a beam propagating in the negative z direction. Then the third term of (2.5a) will produce a beam whose complex

18

amplitude $A_{ob}^*(x, y, 0)\exp(iK_{obz}z) = A_{ob}^*(x, y, z)$. This implies that the complex amplitude is an ideal conjugate of that of the original object beam $A_{ob}(x, y, z)$; in other words, a phase-conjugate wave is reconstructed. The conjugate wave exhibits a curious property; namely, it retraces the path which the object wave has traveled before arriving at the hologram plane. In this situation, the diverging wave becomes a converging one, the wavefront distortions suffered during the first pass through the nonhomogeneous medium are "healed", etc. Formally, such properties could be exhibited by a wave if time is reversed - naturally, in a nonabsorbing medium (Fig.2.2).

Phase conjugation attracts a great deal of attention both scientifically and practically.

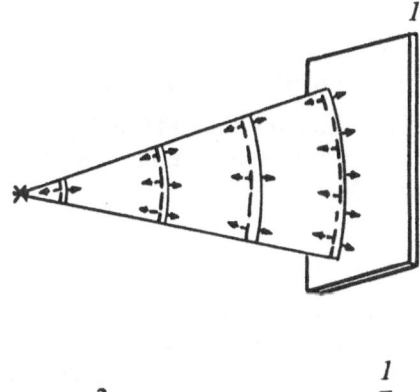

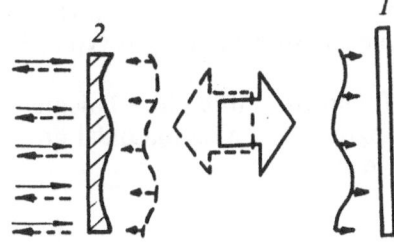

Fig.2.2. Phase conjugation. (*1*) Phase-conjugate mirror (the medium providing phase conjugation); (*2*) distorting medium. (Original wave: solid line; phase-conjugate wave: dashed line)

2.2 Classification and Properties of Holograms

2.2.1 Thin and Thick Holograms

In this section we discuss only those properties and types of holograms that are directly related to the PRCs.

Since the information is recorded in the crystals as a spatial modulation of the refractive index, the holograms so formed can be classified as phase holograms. During readout, the readout-beam phase, not the amplitude, is modulated. The holograms involved may be thin or thick. The term "thin" means that the hologram thickness is small enough so that we can ignore the diffraction effects inside the hologram. In a thick (volume) hologram, the diffraction within it plays a dominating role. In the latter case, the effect of light diffraction in the medium is termed Bragg diffraction. The parameter [2.4]

$$Q' = \frac{d\lambda}{n\Lambda^2} ,$$ (2.6)

where d is the hologram thickness and Λ is the grating spacing, can serve as a quantitative criterion for classification of "thin" and "volume" holograms. For $Q' \ll 1$, the grating is thin, and for $Q \gg 1$, the grating is thick. According to this criterion, the hologram is said to be thick if the light beam (incident or diffracted) passing through the hologram intersects a plane of equal refractive index many times (i.e., intersects many grating periods). The hologram is thin if the beam displacement in the xy plane across the hologram thickness is far smaller than the grating spacing.

2.2.2 Diffraction Efficiency and Selectivity of Phase Holograms

The parameters of utmost importance here include the diffraction efficiency, and the angular and wavelength selectivity of holograms. The diffraction efficiency is expressed as the ratio [2.5]

$$\eta = I_1/I_{in} ,$$ (2.7)

where I_{in} is the readout intensity incident on the hologram, and I_1 is the light intensity diffracted into the first diffraction order. In order to compare diffraction efficiencies of different photosensitive materials, diffraction from a hologram in the form of a simple sinusoidal grating is typically considered.

The diffraction efficiency η is readily found for a thin phase hologram $\psi(r) = \psi_1 \cos K \cdot r$. For the simplest phase hologram, i.e., a cosinusoidal grating with $K = K_x$, the amplitude transmittance is

$$T(x) = T_0 \exp(i\psi_1 \cos Kx) .$$ (2.8)

Neglecting the effects due to reflection, light absorption, and uniform phase shift, we obtain $T_0 = 1$. Expansion of the exponential factor in a Fourier series yields

$$T(x) = \sum_{p=-\infty}^{\infty} i^p J_p(\psi_1) e^{ipKx} ,$$ (2.9)

where $J_p(\psi_1)$ is the Bessel function of the first kind and order p.

To observe the diffraction orders separately, measurements should be made at a sufficient distance from the grating. In the Fraunhofer-diffraction region, the light amplitude is proportional to the Fourier transform of T(x). Therefore, each Fourier component (2.9) produces a beam in the far field whose complex amplitude is proportional to $i^p J_p(\psi_1)$. The angle into which it is diffracted, is defined by $\sin\theta_p = pK\lambda/2\pi$, where p is the diffraction order. Thus the diffracted light intensity is

$$I_1 = I_{in}J_1{}^2(\psi_1) \quad \text{and} \quad \eta = J_1{}^2(\psi_1) \,. \tag{2.10}$$

For a holographic grating recorded as the refractive index distribution $\Delta n(x) = \Delta n \cos Kx$,

$$\psi_1 = \frac{2\pi\Delta n d}{\lambda} \,. \tag{2.11}$$

For a thin sinusoidal phase grating the maximum possible η is $\eta_{max} = 33.9\%$.

Calculating the diffraction efficiency for a volume holographic grating is far more difficult. This is attributable to the following: because of the Bragg nature of diffraction, the wave-field distribution behind the volume hologram can be found only by solving the problem of light propagation in a three-dimensional medium with a periodic variation of the refractive index. For a thin hologram, on the other hand, it is easily obtained by multiplying the incident light amplitude by the transmittance T(x). We do not give a detailed derivation (Sect.5.1.2) and only present the final results [2.5].

For a transmission volume grating (the readout and reconstructed beams travel in the +z direction)

$$\eta = \sin^2 \frac{\pi\Delta n d}{\lambda \cos\theta_B} \,, \tag{2.12}$$

where θ_B is the Bragg angle inside the medium.

For a reflection grating, where the diffracted beam travels opposite to the readout beam (in the -z direction),

$$\eta = \text{cth}^{-2}\left[\frac{\pi\Delta n d}{\lambda \sin\theta_B}\right] \,. \tag{2.13}$$

One of the characteristic features of volume phase holograms is their high diffraction efficiency, which can, in principle, approach 100%. Another unique property of volume holograms is their high angular and wavelength selectivity in certain recording and reconstruction geometries. Because of the Bragg nature of diffraction from volume holograms, there is a definite relation between the wavelength λ, the spatial frequency of the grating $1/\Lambda$, and the angle of incidence (θ_B) at which the efficient diffraction is observed. For a hologram of infinite thickness,

$$\sin\theta_B = \frac{2\lambda}{n\Lambda} \,. \tag{2.14}$$

Condition (2.14) becomes less strict for a hologram of finite thickness d. According to theory [2.6], it can be shown [2.5-7] that the incidence angle and wavelength of the readout beam may deviate from the values given by (2.14) for a transmission volume phase hologram by

$$\Delta\theta = \Lambda/d \tag{2.15}$$

and

$$\frac{\Delta\lambda}{\lambda} = \frac{\Lambda}{d} \text{ctan}\theta_B .$$
(2.16)

The relative deviation of the readout-beam wavelength for the reflection hologram is

$$\Delta\lambda/\lambda = \Lambda/d .$$
(2.17)

In practice, the angular selectivity is $\sim 10^{-3}$ and the wavelength selectivity of the volume holograms discussed is $\sim 0.1\%$ at $\Lambda^{-1} = 500$ mm^{-1} and d = 2 mm. Equations (2.15-17) imply that if the incidence angle and wavelength of the readout beam differ by $\Delta\theta$ and $\Delta\lambda$, respectively, from the values given by (2.14), the diffraction efficiency approaches its first zero.

Thin holograms do not exhibit such a selective property. They can be read out at different incidence angles and wavelengths. The image will be reconstructed; however, its location and magnification will depend on the angle and the wavelength of the readout light. Geometrical distortions may arise.

2.2.3 Specific Features of Holograms in PRCs

So far we have discussed salient features of holograms irrespective of the photosensitive media in which they are recorded. At least three additional properties that the hologram acquires at recording in photorefractive media should be noted here (to be discussed in more detail in Chaps.5 and 6). The first is the possibility of anisotropic diffraction, which is a consequence of the birefringent properties of photorefractive crystals and the anisotropy of the refractive-index grating itself. The anisotropic diffraction can give rise to such a phenomenon as light diffraction with the polarization plane rotated, a change in the Bragg condition on hologram readout, and a suppression of wavelength selectivity of volume holograms.

The second property is the dynamic nature of holograms. The fact is that, simultaneously with a hologram's recording, erasure, and diffraction of the incident beams from the hologram recorded previously (the so-called self-diffraction) will occur. As a result of self-diffraction, the waves entering from outside interfere not only with one another, but also with the diffracted waves within the crystal volume. One of the natural consequences is that the conditions of recording can change markedly throughout the crystal depth to give rise to a strong nonhomogeneity of the hologram across the crystal thickness. A more interesting consequence is the possibility of automatic "adjustment" of the hologram to the corresponding recording conditions.

The third property is an appreciably nonlocal nature of the response of the photorefractive medium to the recording light. This is primarily attributable to the electrostatic mechanisms; in other words, the presence of the charge at a particular point gives rise to the electric field and, hence, to the refractive-index variations at distances much larger than the light wavelength. The nonlocal response results not only in a limited resolution of the photorefractive media, but, in many cases, causes a shift between the refractive-index grating and the interference pattern formed by the recording light. This problem was already discussed briefly in Sect.1.2, when we considered the diffusion recording mechanism.

The grating shift produces a curious phenomenon called energy coupling. For a definite spatial shift between grating and interference pattern, the diffracted ray of one wave can add in phase or subtract with the zero order (undiffracted ray) of the other wave. As a result, energy transfer from one wave to the other occurs. Energy coupling is most efficient when the interference pattern is shifted by a quarter of the grating spacing. In photorefractive crystals, a $\Lambda/4$ shift can be preset (for instance, through the diffusion recording mechanism) and automatically maintained because of the dynamic nature of the hologram recording. Note that this procedure can be applied not only to simple gratings, but also to any complex hologram. No special measures are needed to ensure stability of the hologram or the position of the interference pattern. If an occasional misalignment occurs (a change in shift other than $\Lambda/4$), the existing hologram is erased and, simultaneously, a new hologram is automatically recorded in such a way that the shift required for energy coupling is retained. Thus it is ensured by the recording mechanism itself. Therefore, though energy coupling can take place not only in photorefractive media (for simple gratings it can occur in any phase hologram), the dynamic and nonlocal nature of recording that provides stability of the process of energy coupling is nonetheless an important specific feature of these media.

2.3 Coherent Optical Information Processing Systems

Typical examples for the processing of optical information by means of coherent light are Fourier processors, spatial-filtering devices, and correlators [2.8]. These systems are, in fact, analog computing devices that use light diffraction. Therefore, the problems they solve obey the equations used in describing diffraction processes. Generally speaking these problems can be more or less successfully solved for incoherent light as well; however, coherent optical radiation provides a simple and elegant tool (theoretically) for accomplishing the task. The coherent systems of interest are linear with respect to the readout-light amplitude - a significant feature that necessitates creation of optical elements to ensure linearity of the transformation of the wave amplitude.

The simplest, but an extremely important representative of coherent optical systems, is the Fourier processor [2.9]. It is of interest not only by itself, but also as a unit that can be used for constructing other coherent optical systems. Figure 2.3 shows one of the most typical schemes. In the input plane (the front focal plane of a lens) there is a transparency (a slide)

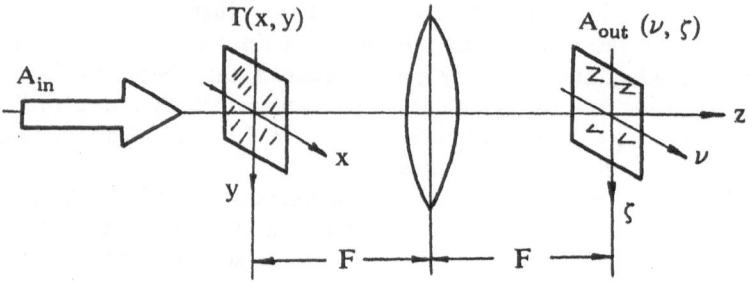

Fig.2.3. A coherent optical Fourier processor

whose amplitude transmittance $T(x, y)$ describes the input data to be processed. If the transparency is illuminated by a plane wave, the light amplitude distribution at the output $A_{out}(\nu, \xi)$ is given by a two-dimensional Fourier transform of $T(x, y)$:

$$A_{out}(\nu, \xi) = \frac{A_{in}}{i\lambda F} \iint T(x, y) \exp[i2\pi(\nu x + \xi y)] dx\, dy \; . \qquad (2.18)$$

Here, A_{in} is the readout light amplitude in the input plane, $\nu = x'/\lambda F$, and $\xi = y'/\lambda F$, $x'$and y' being the coordinates in the output plane, and F is the focal length of the lens.

If we place two Fourier processors in series, an imaging optical system is obtained [2.10]. The double Fourier transformation yields the original function, but in the inverted coordinate system, i.e., $T(-x,-y)$. If an appropriate holographic filter is placed in the input plane of the second processor, the result is the well-known Vander Lugt correlator [2.11, 12] used for pattern recognition.

The simplicity and the high potential of similar Fourier processors and correlators have attracted a great deal of attention. The practical realization of both systems is, however, difficult. In particular, high-quality components such as input devices, Fourier objectives, photodetectors, and lasers are required.

Photorefractive media are most appropriate for use in input devices and reversible holographic filters. If the input information is introduced by optical means - i.e., the transmittance $T(x, y)$ of the input transparency is formed by illuminating it with the recording light - such a transparency is called an optically addressed Spatial Light Modulator (SLM) or simply SLM.

Let us specify briefly some requirements an SLM must satisfy to make the optical information system competitive. We take the Fourier processor as an example, and its speed of operation as a criterion by which we can judge whether the system is competitive. A rather realistic measure of the input data is on the order of $N = 10^3 \times 10^3$ resolvable points (pixels) on the input transparency. To perform the Fourier transformation by an electronic computer using the algorithm of fast Fourier transformation (FFT) [2.13] $2N \cdot \log_2 N$ complex operations are required. If $N = 10^6$, then $2N \cdot \log_2 N \simeq$

$4 \cdot 10^7$. We assume that we can make 100 replacements of the input data (i.e., the image on the input transparency) per second. Then the equivalent speed of operation of the optical Fourier processor is on the order of $4 \cdot 10^9$ operations. Such a fast system can, in certain cases, compete with computers. What then are the requirements for SLMs?

For a modulator of moderate size (2–3cm) a resolution of 20 to 50 line pairs/mm is required to ensure recording of 10^6 pixels. A necessary condition to achieve linearity with respect to the readout-light amplitude is that the diffraction efficiency be well below its highest possible value, but at the same time it must be high enough to provide admissible losses of the readout light. So the desired diffraction efficiency is on the order of 1–5%. The speed is not less than 100 record-erase cycles per second. Then, allowing for the fact that to prevent overheating of the modulator, the absorption of the recording light power should not exceed 1 W, the sensitivity should be better than $10^{-3} - 10^{-4}$ J/cm^2. If ordinary light sources are used, the sensitivity should not be poorer than 10^{-7} to 10^{-8} J/cm^2.

The dynamic range of the modulator can be in the range of 40–60 dB for the recording light intensity. Besides the well-known parameters listed above, attention should be paid to two other important characteristics, i.e., the noise and the phase homogeneity of the modulator. The need for a low level of intrinsic noise is a consequence of the low diffraction efficiency, and the absence of phase distortions is a specific requirement to all coherent optical systems. Phase distortions of SLMs give rise to a spread of the light spot at the output of the Fourier processor and, hence, to a loss of its resolving ability and a drastic decrease of the signal-to-noise ratio.

Demands on the phase homogeneity are extremely severe. For instance, if we assume that the spurious modulation of the beam phase should not be more than 0.5 rad, the tolerance on the SLM thickness will be $\Delta d_{max} < 0.5\lambda(2\pi n)^{-1}$. At $\lambda = 0.6$ μm and n = 2.5, Δd_{max} should not exceed $2 \cdot 10^{-2}$ μm over the entire surface of the crystal.

As will be shown in Chap.7, the use of photorefractive crystals, in general, offers a solution to the problem of fabrication of SLMs, at least in terms of a reasonable compromise between the conflicting requirements (size and phase distortion, resolution and sensitivity, speed of operation and energy consumption).

3. Salient Features of Photorefractive Crystals for Holography

The mechanism of information recording determines such characteristics of the photorefractive crystals as diffraction efficiency, information capacity, spatial-frequency bandwidth, sensitivity, speed of operation, dynamic range, and storage time. The parameters listed are essential for the description of the media in terms of information theory, but they do not yield adequate estimates for other important features such as reliability, thermal stability, ease of fabrication, and the cost. In this chapter we shall be concerned with properties of photorefractive media that govern the application in coherent-optical information processing.

3.1 Diffraction Efficiency

Evaluating the diffraction efficiency involves calculation of the amplitude of refractive index variation Δn. Let us restrict our analysis to one eigenmode of the light wave in the crystal. Then Δn denotes the refractive index deviation for a given mode (for instance, ordinary or extraordinary beam). Because Δn is unambiguously related to the grating field amplitude through the electro-optic coefficient, the field in the crystal must be determined for a specified charge distribution.

In the linear regime of recording, i.e., when the field of the grating itself still does not affect the charge formation, the complex charge grating can be represented as a superposition of elementary gratings of the positive and negative charges linearly related to the intensity of the recording light. Now the field of each elementary grating can be found separately, and the resultant field is a sum of all fields of the different gratings. This approach reduces the problem to an analysis of the electrostatic factors for a single grating. The linear regime is typical of spatial light modulators and the initial stages of holographic recording (far from the steady-state regime). The amplitude of the elementary-charge grating is proportional to the recording light intensity, and is independent of the spatial frequency, though the amplitude of the resultant grating consisting of a number of elementary gratings can be frequency dependent. For simplicity, we shall use the model of an infinite, originally homogeneous medium [3.1], with the term "infinite" implying that the sample sizes in any direction are far greater than the grating spacing. Let us consider three examples: a volume grating, a thin grating, and a thin grating in a sample coated with electrodes.

3.1.1 Volume Grating

First we consider a volume grating with the charge density ρ, i.e.,

$$\rho(\mathbf{r}) = \rho \cos \mathbf{K} \mathbf{r}. \tag{3.1}$$

Accoring to Poisson's equation (1.5) the grating field is given by

$$\mathbf{E}(\mathbf{r}) = \frac{\mathbf{K}}{K^2} \frac{\rho}{\epsilon \epsilon_0} \sin \mathbf{K} \mathbf{r}. \tag{3.2}$$

It is common to consider the situation when the grating wave vector is along a selected axis of the coordinate system chosen. Take the grating directed along the x axis as an example, i.e., $\mathbf{K} = K_x = 2\pi\nu$. Then, instead of (3.2), we have

$$E_x(x) = \frac{\rho}{2\pi\epsilon\epsilon_0} \frac{1}{\nu} \sin(2\pi\nu x) , \quad \text{and} \quad E_y = E_z = 0 . \tag{3.3}$$

Equation (3.3) predicts that if the light propagates along the z or y axis, only the transverse electro-optic effect must be used; and if the light travels along the x axis, the longitudinal effect must be employed. The field grating and, hence, the refractive index grating are shifted by a quarter of the spacing with respect to the charge grating.

For a cubic crystal of the $\bar{4}$3m and 23 point group with the z axis oriented along the [110], with the y axis along the [1$\bar{1}$0], and the x axis along the [001] crystallographic axes, respectively, the transverse effect takes place for light polarized along the y axis. We then have

$$\Delta n(x) = \frac{1}{2} n^3 r_{41} E_x(x) = \Delta n \sin(2\pi\nu x) . \tag{3.4}$$

We assume that the hologram thickness, though large, is finite and equals d. Then, under the condition that $\pi \Delta n d / \lambda \cos \theta_B \ll 1$, Eqs.(3.4) and (2.12) yield for the diffraction efficiency

$$\eta = \left[\frac{1}{4} n^3 \frac{r_{41}}{\epsilon\epsilon_0} \frac{d}{\cos\theta_B} \frac{\rho}{\lambda} \frac{1}{\nu} \right]^2 . \tag{3.5}$$

Attention should be paid to the ν^{-2} frequency dependence of the diffraction efficiency, which is fundamental, for a single grating with $\mathbf{K}$ along a specified direction.

In a more general case, when the charge grating is arbitrarily oriented, i.e., $\mathbf{K} \neq K_x$, the diffraction efficiency is derived in a more complicated manner, and the frequency dependence can be different from ν^{-2}. To avoid misunderstanding, we emphasize that the frequency dependence of η for an arbitrary grating orientation will imply variations of η as a function of the individual spatial frequencies $\nu = K_x/2\pi$, $\xi = K_y/2\pi$, $\gamma = K_z/2\pi$, rather than of the overall modulus K - i.e., $\eta = \eta(\nu, \xi, \gamma)$ - because the recorded image I(x, y, z) is decomposed into harmonics with the frequencies ν, ξ, γ.

The complicated behavior of η as a function of spatial frequencies is attributable to the fact that the changes in the dielectric permeability or refractive index are governed by a superposition $\Sigma_j r_{inj} E_j$ of the field projections E_j rather than by the magnitude of modulus E. The different projections E_j have different frequency dependences, for instance

$$E_x(x) = \frac{\rho 2\pi\nu \sin Kr}{\epsilon\epsilon_0 K^2} = \frac{\rho\nu \sin Kr}{2\pi\epsilon\epsilon_0(\nu^2 + \xi^2 + \gamma^2)}, \tag{3.6}$$

while $E_y \propto \xi/K^2$ and $E_z \propto \gamma/K^2$ [3.1].

However, the condition $\Delta K_{max}/K_c \ll 1$ is typically satisfied in recording complex volume holograms (but not for SLMs). Here we assume that the individual gratings labeled j, constituting the complex hologram, have the frequency $K^j = K_c + \Delta K^j$, where K_c is the average (carrier) frequency, and ΔK_{max} is the maximum deviation of the spatial frequency from K_c. Therefore, the $\eta \propto 1/K^2$ dependence is a dominating one. Note that for the same reason ($\Delta K_{max}/K_c \ll 1$) recording of volume holograms can be adequately analyzed through the diffraction of eigenmodes and the refractive-index modulation.

In conclusion, we recall that the frequency dependences of the net field and the diffraction efficiency for the superposition of gratings of different charges with the same K can obey laws other than K^{-1} and K^{-2}, respectively. For instance, as it follows from the examples given in Sect.1.2, under certain conditions $E \propto K$ and $\eta \propto K^2$ for the diffusion mechanism, and η is independent of the spatial frequency for the drift mechanism and the short drift lengths.

3.1.2 Thin Grating

Now we consider examples for the frequency dependence of the diffraction efficiency for thin holograms or SLMs. That the hologram is thin means in this case that the charge grating thickness $h \ll \nu^{-1}$, but the grating spacing ν^{-1}, in turn, is far smaller than the crystal thickness d.

The first example is a thin charge grating in a volume crystal in the z = 0 plane (Fig.3.1), namely

$$\rho(x) = \rho \cos 2\pi\nu x .$$

Under the conditions mentioned above, we can introduce the surface charge density $\sigma = \rho h$, and the grating field is then given by

$$E_x(x,z) = \frac{\sigma}{2\epsilon\epsilon_0} e^{-|2\pi\nu z|} \sin 2\pi\nu x = E_x(z)\sin(2\pi\nu x) ,$$

$$E_y = 0 , \tag{3.7}$$

$$E_z(x,z) = \pm \frac{\sigma}{2\epsilon\epsilon_0} e^{-|2\pi\nu z|} \cos 2\pi\nu x = E_z(z)\cos(2\pi\nu x) .$$

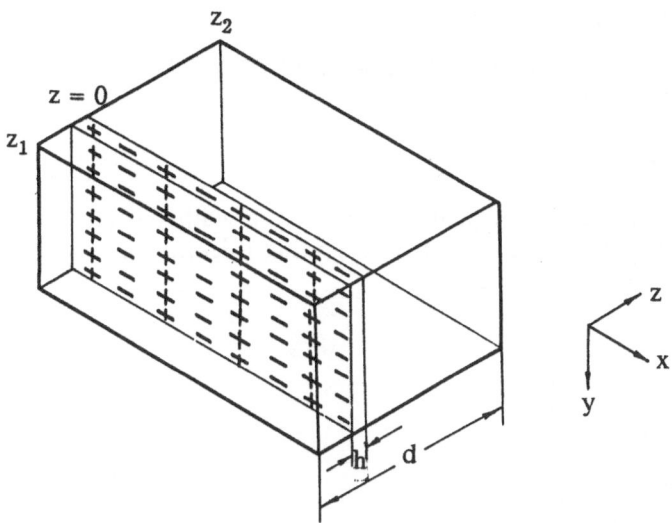

Fig.3.1. Thin charge grating in a crystal of thickness d (grating thickness: h<<d)

The sign + or − in the last expresion corresponds to z > 0 or z < 0, respectively.

For the transverse electro-optic effect where Δn arises from the field component E_x, the refractive index grating is shifted by $\pi/2$ from the charge grating. The amplitude modulation of the readout-beam phase taking account of (3.4) is

$$\psi_1 = \frac{\pi}{\lambda} n^3 r_{41} \int_{-\infty}^{+\infty} E_x(z) dz = \frac{1}{2\lambda} n^3 \frac{r_{41}}{\epsilon\epsilon_0} \frac{\sigma}{\nu}. \tag{3.8}$$

Since we are dealing with a thin hologram, we have in accordance with (2.10)

$$\eta = \left(\frac{n^3 r_{41} \sigma}{4\lambda\epsilon\epsilon_0 \nu} \right)^2. \tag{3.9}$$

Equation (3.7) reveals that due to the transverse effect the field grating is shifted with respect to the charge grating, and the diffraction efficiency $\eta \propto \nu^{-2}$ and is independent of the crystal thickness.

For the longitudinal effect in this geometry (i.e., when the grating is within the crystal volume), the field E_z has different signs according to z > 0 and z < 0. Therefore, $\psi_1 \propto \int E_z dz = 0$, and $\eta = 0$. If the charge grating is on the crystal surface, $\psi_1 \neq 0$ for the longitudinal effect. Solution of this problem for the longitudinal and transverse effects requires that the corresponding boundary conditions be taken into account, and the rigorous results will be given in Chaps.7 and 8.

3.1.3 Thin Grating with Electrodes

We consider one more example of a thin grating of particular concern for SLMs, i.e., a thin charge grating (Fig.3.2)

$$\rho(x) = \rho \cos(2\pi\nu x)$$

located in the center of a crystal plate of thickness d with the front and back faces coated with electrodes. The charge-grating thickness is h << ν^{-1}; the z axis is normal to the plate plane.

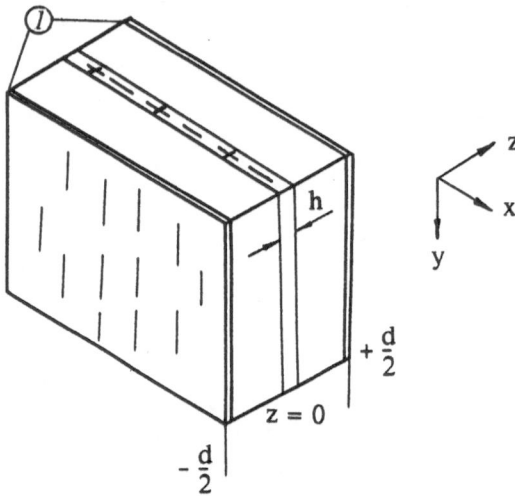

Fig.3.2. Thin charge grating in the center of a crystal. The front and back faces of the crystal are coated with electrodes

Then denoting $\rho h = \sigma$ we have

$$E_x(x,z) = \frac{\sigma\nu}{\pi\epsilon\epsilon_0} \frac{1}{d} \sin 2\pi\nu x \sum_{j=0}^{\infty} \frac{\cos[\pi(2j+1)/dz]}{\nu^2 + [(2j+1)/2d]^2} ,$$

$$E_y = 0, \quad E_z(x,z) \neq 0 . \tag{3.10}$$

Though $E_z \neq 0$, no phase modulation $\psi_1(x)$ through the longitudinal effect arises since $\psi_1(x)$ is proportional to the spatial modulation of the potential difference in the longitudinal effect. In our case, however, the crystal is between two equipotential surfaces and

$$\psi_1(x) \propto \int_{-d/2}^{+d/2} E_z(x,z)dz = 0 .$$

As in the previous case, the field grating, and hence the refractive index grating, is shifted by a quarter of the spacing for the transverse effect.

The field $E_x(x,z)$ in (3.10) is represented as a sum of harmonics, i.e., it can be regarded as a sum of fields from a set of gratings with j-dependent spacings. This is not surprising, because the presence of electrodes gives rise to an infinite number of mirror-charge layers, i.e., a periodic structure of the plane charges spaced by 2d. Some characteristic features of the field $E_x(x,z)$ may be analyzed by taking separate terms of (3.10) as examples. Let us assume j = 0. Then

$$E_x(x,z,j{=}0) = \frac{\sigma}{\pi\epsilon\epsilon_0 d}\frac{\nu}{\nu^2 + (2d)^{-2}}\sin(2\pi\nu x)\cos\left[\frac{2\pi z}{2d}\right]$$

$$= E_x(z)\sin(2\pi\nu x) . \tag{3.11}$$

For the transverse effect, taking into account (3.4),

$$\psi_1(j{=}0) = \frac{\pi}{\lambda}n^3 r_{41}\int_{-d/2}^{+d/2} E_x(z)dz = \frac{2\sigma}{\pi\lambda\epsilon\epsilon_0}n^3 r_{41}\frac{\nu}{\nu^2 + (2d)^{-2}} \tag{3.12}$$

and the diffraction efficiency

$$\eta\,(j{=}0) = \left[\frac{\sigma n^3 r_{41}}{\pi\lambda\epsilon\epsilon_0}\right]^2 \left[\frac{\nu}{\nu^2 + (2d)^{-2}}\right]^2 . \tag{3.13}$$

As seen from (3.11) the field grows for low spatial frequencies ($\nu <$ 1/2d) with the increasing frequency, rather than reducing as in other cases. Accordingly, for the transverse effect the diffraction efficiency tends to zero as $\nu \to 0$ and peaks in the frequency range $\nu \sim 1/2d$. Here the effect arises because of the electrodes present on the crystal plate. However, a similar dependence for the field E_x could be obtained for the grating given by (3.2) at $\xi = 0$. Equations (3.6, 11) bear a close resemblance to each other from the point of view of frequency dependence if we assume that $\gamma = (2d)^{-1}$, $\xi = 0$. Equation (3.11) for E_x can be regarded as resulting from the formation of two charge gratings with K_1 and K_2, where $K_{1x} = K_{2x} = 2\pi\nu$ and $K_{1z} = -K_{2z} = 2\pi\gamma$, and $\gamma = (2d)^{-1}$. Then

$$E_x(x,z,j{=}0) = \frac{\rho h}{2\pi\epsilon\epsilon_0 d}\frac{\nu}{\nu^2 + \gamma^2}(\sin K_1 r + \sin K_2 r) . \tag{3.14}$$

Note that, although we are discussing only one term of (3.10), Eq.(3.11) proves to give satisfactory results for the limiting values $\nu \ll 1/2d$ and $\nu \gg$ 1/2d since the term with j = 0 plays a dominating role.

Thus, deposition of electrodes on the crystal plate is qualitatively equivalent - with an accuracy on the order of unity - to the recording of two gratings in the crystal without electrodes, but with the wave vectors having nonzero components not only in the direction of x, but also of z.

3.2 Transfer Function

In the framework of linear space-invariant systems theory, spatial light modulators (or simply thin plates of PRCs) can be described by transfer functions or impulse responses. Evidently, because of the anisotropic properties of PRCs, the transfer functions are generally two-dimensional functions of spatial frequencies, and the impulse responses are the two-dimensional functions of coordinates. The impulse response and transfer function of the SLMs establish the relation between the input and output signals.

Let us first discuss what we mean by the input and output signals. For the optically addressed electro-optic SLMs discussed here, the input signal is the magnitude of exposure with recording light

$$W(x, y) = I(x, y)\tau_{ex} . \qquad (3.15)$$

The problem of interest in the analysis of coherent optical systems is how the complex amplitude of the readout light changes during its passage through the elements of the optical system. We assume here, with thin holograms, that the complex amplitude of the readout light at the output of a SLM $A_{out}(x, y)$ is related to the incident readout light amplitude A_{in} by $A_{out}(x, y) = A_{in} T(x, y)$, where $T(x, y)$ is the complex transmittance of the SLM. Therefore we regard a change of the complex transmittance $T(x, y)$ caused by the recording beam as the output signal. Note once more that $T(x, y)$ relates the complex amplitudes of the readout light, and not its intensities, at the input and output. Since A_{out} and A_{in} can differ by the polarization state, generally speaking, $T(x, y)$ is a tensor that can be written as a 2×2 matrix, because A_{in} and A_{out} are regarded as two-dimensional Maxwell vectors. However, we do not consider the tensorial properties of the transmittance now.

Assume that an image of an infinitesimally small point object with the coordinates x', y' is recorded on the SLM. With a corresponding normalization, the input signal is $\delta(x-x', y-y')$, i.e., a two-dimensional Dirac δ-function. Since the SLM, as any other recording medium, has a limited resolution, the changes of its transmittance, i.e., the output signal, will not be localized in an infinitesimally small region.

Let us introduce the function $h(x-x', y-y')$ which, with an accuracy to a normalizing factor, is the change of transmittance caused by a δ-function input signal. The function $h(x-x', y-y')$ is termed the impulse response of an SLM. If an arbitrary image $W(x'y')$ is recorded on the modulator, the output signal will be given by the superposition integral

$$T(x, y) = \iint W(x', y') h(x-x', y-y') dx' dy' . \qquad (3.16)$$

Thus, knowledge of the function $h(x-x', y-y')$ allows the SLM response to an arbitrary input signal $W(x', y')$ to be defined. In practical situations, however, it is common to use the transfer function $\chi(\nu,\xi)$ which is a Fourier transform of the impulse response at $x' = y' = 0$, i.e.,

$$\chi(\nu,\xi) = \iint h(x,y)\exp[2\pi i(\nu x + \xi y)]dxdy .\tag{3.17}$$

The transfer function relates the Fourier spectra of the recorded image and the output signal

$$\tilde{T}(\nu,\xi) = \chi(\nu,\xi)\tilde{W}(\nu,\xi) ,\tag{3.18}$$

where $\tilde{T}(\nu,\xi)$ and $\tilde{W}(\nu,\xi)$ are the Fourier spectra of $T(x,y)$ and $W(x,y)$, respectively.

3.2.1 Experimental Determination of the Transfer Function

Let us see how the transfer function can be defined using the diffraction technique. The image of a sinusoidal grating is recorded on the modulator

$$W(x,y) = W_0[1 + m\cos2\pi(\nu x + \xi y)] .\tag{3.19}$$

The grating may be recorded by the holographic method as a pattern of interference between two plane coherent beams or focused on the SLM from a transparency in incoherent light. If we do not take into account the limited sizes of the SLM's active surface, the Fourier spectrum of the recorded grating, (3.19), according to (3.18) yields the output signal spectrum density

$$\tilde{T}(\nu,\xi) = W_0\left[\chi(\nu',\xi')\delta(\nu',\xi') + \frac{m}{2}\chi(\nu',\xi')\delta(\nu-\nu',\xi-\xi')\right.$$
$$\left. + \frac{m}{2}\chi(\nu',\xi')\delta(\nu+\xi',\xi+\xi')\right] .\tag{3.20}$$

After the grating has been recorded, the diffraction of the readout light transmitted through the modulator is observed. The intensity of the first diffraction order is

$$I_1(\nu,\xi) = \left|\frac{A_{in}\chi(\nu,\xi)W_0 m}{2}\right|^2 .\tag{3.21}$$

Measurements of the diffraction intensity $I_1(\nu,\xi)$ yield the diffraction efficiency η. As (3.21) reveals, we can find the transfer function modulus by measuring η for gratings with different ν and ξ, but with the same amplitude mW_0

$$|\chi(\nu,\xi)| = \frac{2}{mW_0}\sqrt{\eta(\nu,\xi)} .\tag{3.22}$$

Next we must find how $\chi(\nu,\xi)$ is related to the electro-optic and photorefractive properties of the crystal. Generally, it is a fairly involved task if we allow for anisotropic properties of the materials used. The problem was solved for cubic crystals with an arbitrary electric field distribution within the crystal [3.2]. We take a thin charge grating (Fig.3.1) as an example, and assume that the readout light corresponds only to one eigenmode of the electro-optic crystal. Note that this is not a common situation, but a real problem of SLM can be reduced to the superposition of two eigenmodes.

Now, comparing (2.9) and (3.20), and using (3.4, 8) we obtain

$$\chi(\nu,0) = \frac{i\psi_1}{mW_0} = \frac{in^3 r_{41} \sigma}{2\lambda\epsilon\epsilon_0 \nu mW_0} . \tag{3.23}$$

In comparing (3.20) with (2.9) we took into account that, in order to apply a linear approximation, the condition $\psi \ll 1$ should be fullfilled and all terms with $|p| > 1$ should be neglected. The complete solution requires that the relation between σ and W_0 be known, but this relation depends on the particular structure of the device, and therefore this aspect of the problem will not be discussed in this section.

As seen from (3.23), $\chi(\nu,0) \propto 1/\nu$ for the example under discussion. If the modulator has electrodes on the front and back faces, then, similar to (3.12), we have $\chi(\nu,0) \propto \nu$ for $\nu \ll 1/2d$, and $\chi(\nu,0) \propto 1/\nu$ for $\nu \gg 1/2d$ for the transverse effect. As will be shown in Chaps.7 and 8, the $\chi(\nu,0)$ dependence of the longitudinal effect is close to $1/(\nu^2+(2d)^{-2})$.

We have given examples for gratings with $\mathbf{K}$ parallel to the axis. If $\mathbf{K}$ is arbitrarily oriented in the xy plane, χ is the function $\chi(\nu,\xi)$ of two spatial frequencies. Note that in experiments it is often more convenient to measure χ as a function of the grating wave-vector modulus K and the angle of the vector $\mathbf{K}$ with respect to a selected axis. Passing from one variable to the other is not difficult. For instance, as will be shown in Chaps.7 and 8 for one of the SLM modifications, the experimental dependence $\chi \propto (K\cos\theta)/[K^2+(2\pi/2d)^2]$, where θ is the angle between $\mathbf{K}$ and the direction of one of the crystal axes, corresponds to $\chi(\nu,\xi) \propto \nu[\nu^2+\xi^2+(2d)^{-2}]^{-1}$.

The question frequently arises as to whether it is possible to measure $T(x,y)$ and $\chi(\nu,\xi)$ - i.e., the parameters describing the system that is linear with respect to the readout amplitude - using the methods developed for the incoherent systems that are linear with respect to the readout intensity. It is interesting to consider the relationship between the SLM's coherent transfer function discussed here and the incoherent optical transfer function $\mathcal{H}(\nu,\xi)$. If we use the incoherent light, we suppose that the optical system is linear with respect to the readout intensity, and the modulator is described by means of the transmittance $\phi(x,y)$ such that

$$I_{out}(x,y) = \phi(x,y)I_{in} , \tag{3.24}$$

where $\phi(x,y) = |T(x,y)|^2$, and I_{out}, I_{in} are the readout light intensities at the front and back faces of the device, respectively.

The two conditions, (3.24) and $A_{out}(x,y) = T(x,y)A_{in}$, require that both $T(x,y)$ and $\phi(x,y) = |T(x,y)|^2$ be linear functions of the recording light intensity $I(x,y)$. Generally speaking, this is possible if there is a strong bias component of the readout light at the modulator output; that is, either

$|\chi(0,0)| >> \chi(\nu,\xi)$ at ν and $\xi \neq 0$ or there is a beam which is uniform over the modulator aperture and whose intensity is well above that of the diffracted beam. Then we can introduce the incoherent transfer function $\mathcal{H}(\nu,\xi)$, which is related to the coherent transfer function $\chi(\nu,\xi)$ by

$$\mathcal{H}(\nu,\xi) = \iint \chi^*(\nu+\nu',\xi+\xi')\chi(\nu',\xi')d\nu'd\xi' \ . \tag{3.25}$$

Since $\mathcal{H}(\nu,\xi)$ is the autocorrelation function of $\chi(\nu,\xi)$, it peaks at $\nu = \xi = 0$. Therefore normalizing it to $\mathcal{H}(0,0)$ is appropriate. The modulus of the relation

$$\left| \frac{\mathcal{H}(\nu,\xi)}{\mathcal{H}(0,0)} \right| = \mathrm{MTF} \tag{3.26}$$

is called the modulation transfer function. The MTF is typically measured as the contrast of the recorded grating versus frequency. This discussion clearly shows that the MTF can be unambiguously defined for a system that is linear with respect to both the amplitude and intensity of the readout light if $\chi(\nu,\xi)$ is known, but not vice versa. Also, $|\chi(\nu,\xi)|$ is readily found from the MTF measurements only at low contrasts and for $\chi(0,0) >> \chi(\nu,\xi)$.

We note in conclusion that, when used with coherent light, the electro-optic modulators are operated in such a mode that A_{out} and A_{in} have different polarization states. Therefore, in practical situations an analyzer of the polarization is commonly placed behind the SLM to suppress the undiffracted part of the light to reduce the noise. Transfer functions for such SLMs will be discussed in detail in Chaps. 7 and 8.

3.3 Dynamic Range and Information Capacity

The dynamic range of photorefractive media is still not understood fully despite the fact that the relatively low values of the dynamic range limit applicability of PRCs. Formally, the dynamic range D can be characterized by the ratio

$$D = 10\lg \frac{I_{max}}{I_{min}} \, , \tag{3.27}$$

where I_{min} is the minimum level of the usable optical signal that can be detected during readout of the information recorded in the PRC, and I_{max} is the maximum level of the readout signal within the linear region of the input-output signal relation (by a certain specified criterion). Generally, I_{min} is determined by the noise level due to scattered light and is related to the nonhomogeneities, i.e., defects of the photorefractive medium. Rather often the minimum level is determined by other optical elements incorporated in the optical scheme or by the sensitivity of the photodetector.

On the other hand, I_{max} is limited either by the nonlinearities of the recording mechanism or the number of photoactive centers in the crystal. For instance, as will be shown for SLMs, the field induced in the crystal during recording can strongly affect the recording process itself and give rise to higher-order spatial harmonics and nonlinear distortions. A limited concentration of donors and traps is essential for recording holograms at high spatial frequencies. A sufficiently accurate theoretical evaluation of the dynamic range cannot be made since the noise level can be measured only in experiments. However, situations are frequently encountered in which even the available experimental evidence does not provide an estimate of the dynamic range for an arbitrary form of the input signal.

The fact is that the experimental data depend on whether the dynamic range is measured in the image or Fourier plane and on how the spatial frequency bandwidth is controlled during measurements. For instance, if measurements are carried out in the Fourier plane and the sizes of the photodetector correspond to the diffraction-limited spot, then, if a sinusoidal grating is recorded at the input, the signal-to-noise ratio in the Fourier plane will approximately be N times as high as that in the input plane. This is due to the focusing of the light energy into a small spot. Here $N \simeq (2L\Delta\nu)^2$, where L is the size of the input aperture and $\Delta\nu$ is the transmitted frequency bandwidth. We assume that we have white noise and the signal-to-noise ratio in the input plane is measured at the spot of $(1/\Delta\nu)^2$ in size. If a noiselike signal is recorded and the conditions are those given above, the dynamic ranges in the image plane and the Fourier plane must be nearly the same.

Thus, the dynamic range of the measuring techniques based upon diffraction is a function of the input signal, the input aperture, and the bandwidth. Recalculating the data obtained for a single grating to apply to the case of a complex signal is possible in general. However, since certain measurement conditions are not known, it leads to fairly large errors in the estimate of D, and the definition of the information capacity of holograms and SLMs.

The information capacity is an extremely important parameter for an evaluation of the information properties of an individual optical element, and a comparison of the information processing system as a whole. In evaluating the information capacity we can encounter the same difficulties as those arising in the determination of the dynamic range; i.e., the result can differ noticeably depending on where the measurements are carried out (the image plane or the Fourier plane) and on the type of signal for which the analysis is performed. Moreover, confusion sometimes arises in experimental studies since the measured values are not always consistent with those required for calculation.

Let us estimate the information capacity of an SLM using the measurements in the Fourier plane as an example. From familiar theory [3.3], the information capacity of a signal recorded on a square-shaped modulator with the linear size L is given by

$$C = \frac{L^2}{2} \iint \log_2 \left[1 + \frac{P_s(\nu,\xi)}{P_n(\nu,\xi)} \right] d\nu d\xi , \qquad (3.28)$$

where $P_s(\nu,\xi)$ and $P_n(\nu,\xi)$ are the spectral densities of the signal and noise power, respectively. The factor $\frac{1}{2}$ takes into account the loss of information on phase when the intensity of the Fourier spectrum is measured in the Fourier plane. Note that the value of C depends on the type of the signal $P_s(\nu,\xi)$. Therefore, care is needed in inferring the highest possible information capacity of the device from the experimental data obtained for a specific type of the signal. For instance, for recording a single cosinusoidal grating $\cos[2\pi(\nu_0 x + \xi_0 y)]$ and measuring in the Fourier plane,

$$C \simeq \log_2 \left[1 + \frac{I_s(\nu_0,\xi_0)}{I_n(\nu_0,\xi_0)} \right] , \qquad (3.29)$$

where $I_s(\nu_0,\xi_0)$ is the magnitude of the signal, and $I_n(\nu_0,\xi_0)$ is the average noise level in the neighborhood of point ν_0,ξ_0. Note that the information capacity is not high and is determined essentially by the dynamic range of the modulator.

In another situation, if a signal with a rich spectrum, for instance, a noiselike signal, is recorded

$$C \simeq \frac{N}{2} \cdot \log_2 (1 + I_{sn}/I_n) , \qquad (3.30)$$

where I_{sn} and I_n are the mean values of the signal and noise in the Fourier plane, respectively. For simplicity, (3.30) assumes a constant transfer function of the modulator within the bandwidth and that the noise is white.

Equation (3.30) can be used to estimate the maximum information capacity of the modulator; i.e., we can state that

$$C_{max} \simeq N\log_2 (1 + I_{sn}/I_n) . \qquad (3.31)$$

However, experimental studies of the noiselike signal are rather complicated. The ratio I_{sn}/I_n in (3.30) is sometimes replaced by $I_s(\nu_0,\xi_0)/I_n(\nu_0,\xi_0)$ obtained from measurements of a sinusoidal signal, which gives the overestimated values of C_{max}. Since we can take $I_s(\nu_0,\xi_0) \simeq NI_{sn}$ for the sinusoidal and noiselike signals, the error can be very large and the estimate of C_{max} will be incorrect.

In practice, the inherent noises of the modulator or the detecting device can be so high that it is impossible to detect the noiselike signal with the spectrum equal to $\Delta\nu$,[1] i.e., $I_{sn}/I_n < 1$, in spite of the fact that on recording a single sinusoidal signal,

$$I_s(\nu_0,\xi_0)/I_n(\nu_0,\xi_0) > 1 .$$

---

[1] Here we assume that the bandwidth $\Delta\nu$ is defined in the conventional manner, i.e., it is the frequency band within which the transfer function decreases from its maximum to a specified level.

Here detection of the noiselike signal - but with a spectrum narrower than $\Delta\nu$ - is possible, and then the maximum information capacity can be estimated as

$$C_{max} \simeq (2L\Delta\nu_{eff})^2 \simeq \frac{I_s^{max}(\nu_0,\xi_0)}{I_n(\nu_0,\xi_0)} , \qquad (3.32)$$

where $\Delta\nu_{eff}$ is the maximum frequency spectrum ($\Delta\nu_{eff} < \Delta\nu$) in the initial noiselike signal that can be detected by the device, and $I_s^{max}(\nu_0,\xi_0)$ is the highest possible magnitude of the signal on recording a single sinusoidal grating. Thus, (3.32) shows that the dynamic range of the modulator is a deciding factor in the latter case.

Limitations imposed by the dynamic range are essential for the volume holograms as well. Though formally $C \sim \Delta\nu^3 L^2 d$ for a volume hologram, the noises of the crystal itself and the photodetecting system that limit the dynamic range prevent realization of the theoretically possible information capacity. In practice, $C_{max} \sim 10^5 - 10^7$ bit/cm^2 for thin holograms and SLMs with an area of 1 cm^2, while for volume holograms it is one to two orders of magnitude higher.

3.4 Sensitivity

The sensitivity characterizes the light energy needed to record information in a photosensitive material. Different researchers have defined sensitivity in different ways.

A parameter of great importance is the holographic sensitivity [3.4]

$$S_h = \frac{\sqrt{\eta}}{Wm} , \qquad (3.33)$$

where W is the total light energy (i.e., the sum of the energies of the reference and object beams) required for recording a single grating on an area of 1 cm^2, and m is the visibility (contrast) of the interference fringes. Holographic sensitivity is measured in cm^2/J. The diffraction efficiency of volume holograms depends on the sample thickness and the Bragg angle, i.e., the conditions of measurements. Therefore, the sensitivity of photosensitive material in volume holograms is sometimes defined by S_h per unit length [3.5]. Instead of S_h, the reciprocal value S_h^{-1} is often regarded as the holographic sensitivity.

The holographic media are as a rule insufficiently linear with respect to the recording light intensity, and S_h thus depends on W. The definition (3.33) is often used, but taken for a definite value of the diffraction efficiency, typically $\eta = 1\%$. Here sensitivity is the energy density W needed to achieve $\eta = 0.01$ at m = 1, i.e.,

$$1/S = W(1\%) . \qquad (3.34)$$

The sensitivity for linear media (in terms of S_h as a function of $\sqrt{\eta}$) is $1/S = 0.1/S_h$.

In the framework of information theory, the given definitions suffer from the limitation that they do not account for the reconstructed frequency bandwidth and the dynamic range of the medium. For comparing the estimates for the sensitivities of different media in this sense, it is more convenient to express sensitivity as the energy required to record a pixel or a bit of information. However, because of the difficulties in precise evaluation of the information capacity from the experimental data, this definition is seldom used. The sensitivity for SLMs is sometimes defined as the energy required to achieve phase modulation of $\psi_1 = \pi/2$.

Once the expressions for the diffraction efficiency on information recording are known, the equations describing the fundamental dependences of sensitivity are easily derived. Without examining specific cases, we give here the typical expressions for the sensitivity of photorefractive media in the generalized form for optimum charge-transport lengths and in the linear approximation (when $\sqrt{\eta} \propto W$)

$$1/S_h = \frac{\hbar\omega\epsilon\epsilon_0}{n^3 r_{ij}} \frac{1}{\alpha\beta} \frac{1}{m} F(\nu, d, \theta_B) . \tag{3.35}$$

Equation (3.35) shows that sensitivity becomes lower with the increasing energy of the photon quantum $\hbar\omega$ and increasing dielectric permeability, and becomes higher with increasing r_{ij}, n, absorption coefficient α, quantum efficiency β, and modulation depth m. The dependence on the absorption coefficient α is obtained under the assumption that $\alpha d < 1$, where d is the sample thickness. Function $F(\nu, d, \theta_B)$ describes the dependence on spatial frequencies. This dependence takes into account the specific features of the recording and readout geometries used, the role of sample thickness and the type of electro-optic effect.

Note the following curious fact. Equation (3.35) includes the ratio $\epsilon\epsilon_0/r_{ij}$. But it is known [3.6] from the data on the electro-optic properties of oxide crystals that this relation exhibits a weak dependence on both the type of crystal and the temperature. Thus in reality the sensitivity of photorefractive media is slightly affected by their electro-optic properties.

In practice, the observed sensitivities $1/S$ of the presently known crystals lie in the range 10 to 10^{-3} J/cm² at high spatial frequencies $\nu \sim 300$ to 1000 line pairs/mm. The sensitivity of SLMs is determined as in the case of thin holograms. For $\nu \sim 5$ to 10 line pairs/mm, the SLM sensitivity $1/S$ reaches 10^{-6} J/cm².

Estimates of sensitivity per pixel or bit of information yield more universal values that in favorable situations can amount to 10^{-10} to 10^{-11} J, irrespective of whether a volume or a thin grating is recorded.

3.5 Speed of Operation and Storage Time

The speed of operation of photosensitive elements fabricated from PRCs is determined by the rates of information recording and erasure. Two most typical cases can be singled out for holographic recording.

First typical case: The drift and diffusion lengths are much shorter than the grating spacing. The characteristic time for the charge grating formation is then governed by the time required for the photoexcited carriers to redistribute, i.e., Maxwell's relaxation time τ_M. The characteristic relaxation rate $1/\tau_M$ is determined by the dielectric permeability and conductivity of the crystal σ. With acceptable accuracy,

$$\frac{1}{\tau_M} = \frac{\sigma}{\epsilon\epsilon_0} = \frac{\sigma_D}{\epsilon\epsilon_0} + \frac{\sigma_I}{\epsilon\epsilon_0}, \tag{3.36}$$

where σ_D is the conductivity at dark, and $\sigma_I = e\beta\alpha I_0 \mu\tau/\hbar\omega$ is photoconductivity. The notation is the same as that used in Sect.1.2. The process of grating formation is here equivalent to recharging a capacitor with the time constant τ_M.

It is apparent from (3.36) that τ_M depends on the average light intensity incident on the crystal. Therefore, during recording, $\sigma_I \gg \sigma_D$ and $1/\tau_M \propto I_0$. After the recording light is switched off, erasure at a rate $1/\tau_M \propto \sigma_D$ occurs. The erase rate determines the information storage time of the crystal. To achieve a more rapid erasure, the crystal can be uniformly irradiated with the erase light I_{er}. The erase rate obeys the same relation (3.36), with I_0 being replaced by I_{er}. The process of erasure is equivalent to capacitor discharging. Note that it is convenient to short-circuit electrodes during accelerated erasure if the drift recording mechanism is used.

The storage time can sometimes be increased through special fixing techniques. In practice, storage times of PRCs at room temperature lie in a wide range, from microseconds to days.

Second typical case: The drift and diffusion lengths are well above the grating spacing. Here the charge-grating formation rate is primarily determined by the electron photoexcitation rate, since their subsequent redistribution scarcely affects the charge grating produced by the positively ionized donors. These processes will be analyzed in the chapter to follow.

The mechanisms governing the recording and erase rates during information recording on SLMs are basically the same. However, the most important parameter for SLMs is the allowable record-read-erase cycle rate. It can prove to be much less than $1/\tau_M$, since multiple recording and erasure processes give rise to heat release, causing overheating of the modulator. In practice, the cycle rate of such modulators as PROM and PRIZ, to be discussed in Chaps.7 and 8, is several tens of Hz, while τ_M can reach 10^{-3} to 10^{-6} s.

3.6 Resolution

Resolution is an important parameter of photosensitive materials, but, unfortunately, defined in different ways. In conventional optics, resolution is often determined as the size of the minimum element that can be detected by the photosensitive material on recording a test object, for instance, a resolution chart. This method is useful for noncoherent optical systems, but it is less attractive for coherent ones, since here more suitable diffraction methods can be used.

In holography (and, generally, coherent optical systems) resolution is typically defined as the bandwidth of spatial frequencies $\Delta \nu$ within which the transfer function $\chi(\nu)$ or diffraction efficiency $\eta(\nu)$ decreases by a certain factor. Although this is common practice, the definition is confusing, because, first, different researches use different criteria (a decrease of $\eta(\nu)$ by a factor of 2, 3, 10 etc.) and, second, the dynamic range or noise of the material itself are not taken into account. Therefore the resolution obtained cannot be used for estimation of the information capacity of the material. As mentioned in Sect.3.3, diffraction efficiencies of individual gratings as functions of ν can be measured in sequence in a wide range of spatial frequencies and, formally, the bandwidth $\Delta \nu$ can be fairly large. However, if a large number of gratings is recorded simultaneously (for a complicated image), not all these gratings can be simultaneously reconstructed because of a definite level of noise and a limited dynamic range of the photosensitive material. Thus, in practice, the effective bandwidth $\Delta \nu_{eff}$ can be much smaller than $\Delta \nu$. For instance, for thin holograms or SLMs using photorefractive materials, individual gratings with spatial frequencies up to hundreds of lines per millimeter can be recorded. However, $\Delta \nu_{eff}$ that provides a simultaneous reconstruction of a set of gratings with an appropriate signal-to-noise ratio is 15-20 lines/mm. The maximum spatial frequency of an individual grating and bandwidth $\Delta \nu_{eff}$ for volume holograms are several times higher than for thin holograms, because volume holograms have higher diffraction efficiencies. In spite of the fact that $\Delta \nu_{eff}$ is the most appropriate characteristic for information processing systems, literature rarely provides data on the effective bandwidth, and, therefore, in this book we give the magnitudes of the bandwidths for each case according to the definitions used in the papers referred to.

4. Formation of an Electric-Field Grating During Holographic Recording in PRCs

So far we have discussed the general problem of optical recording and readout. In this chapter we analyse in more detail one of the most important aspects, namely, the process of formation of a space-charge field $E_{sc}(r)$ in a layer of the PhotoRefractive Crystal (PRC) on its exposure to a sinusoidal interference pattern. This problem is considered mainly for the case of recording a volume hologram, but Sect.4.6 is devoted to Spatial Light Modulators (SMLs).

Up to now, several hundred publications have appeared, including comprehensive theoretical treatments by *Amodei* [4.1], *Deigen* et al. [4.2], *Alphonse* et al. [4.3], *Peltier* and *Micheron* [4.4], *Moharam* et al. [4.5], *Kuchtarev* et al. [4.6], *Günter* [4.7] and *Refregier* et al. [4.8], and others. Thus in further discussions we shall mainly refer to the original experimental studies or to theoretical papers covering problems beyond the scope of our analysis.

4.1 Main Assumptions and Equations

To facilitate understanding of basic features of hologram recording, a number of simplifying assumptions is made:

First, the light-intensity distribution to be recorded is the simple sinusoidal interference pattern (Fig.4.1)

$$I(r) = I(x) = I_0 [1 + m\cos(Kx)] , \qquad (4.1)$$

with the average light intensity I_0 being constant both across the sample thickness and in the incidence plane (along the x axis in Fig.4.1). This condition, as well as the requirements for ohmic contacts to which the external voltage U is applied, is necessary to avoid consideration of macroscopically inhomogeneous electric fields (i.e., slowly varying as compared with the grating spacing $\Lambda = 2\pi/K$) [4.9-11], to be analysed below in Sect. 4.6. This assumption allows consideration of a one-dimensional problem with cyclic boundary conditions when all basic parameters change only along the x axis.

Another essential assumption is to ignore self-diffraction of the recording light waves from the hologram being recorded (Chap.6). This means that the analysis is carried out in the approximation of a given light distribution in the interference pattern (4.1).

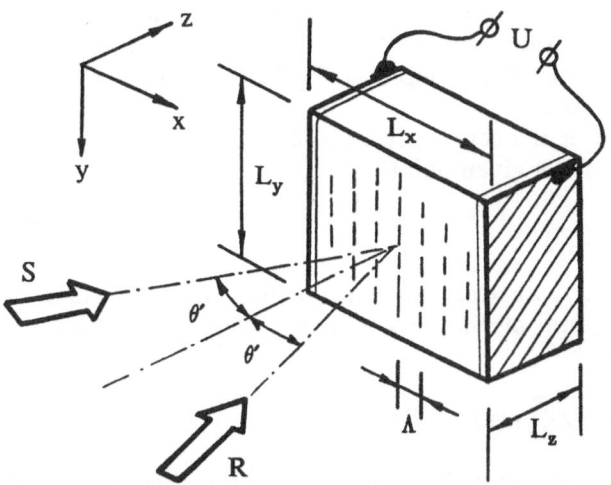

Fig.4.1. Recording of a sinusoidal grating in a photorefractive crystal (if the diffusion or photovoltaic recording mechanisms are used, the external field is not applied to the sample)

Finally, the theoretical analysis can be greatly simplified by assuming a low contrast of the recorded pattern (m << 1). It can indeed be shown [4.2, 3,5] that only as m→1 higher harmonics (with spatial frequencies 2K, 3K, ... pK) in the electric-field pattern $E_{sc}(x)$ begin to play an essential role. So, a low modulation depth of the original pattern (4.1) allows all higher harmonics beginning from p = 2 to be ignored.

It is convenient to use here the complex-function formalism for the light intensity

$$I(x) = I_0\left[1 + \frac{m}{2}\exp(iKx) + \frac{m}{2}\exp(-iKx)\right], \tag{4.2}$$

the electric field E(x)

$$E(x) = E_0 + E_{sc}(x) = E_0 + \frac{E_{sc}}{2}\exp(iKx) + \frac{E_{sc}^*}{2}\exp(-iKx), \tag{4.3}$$

and also other spatially periodic values relevant to the process, namely, charge density $\rho(x)$, current density $j(x)$, concentration of electrons in the conduction band $n(x)$, etc. [4.3,6,12]. As will be shown later, the equations that describe hologram recording are, in this case, reduced to a set of ordinary differential equations relating the time-varying complex amplitudes.

The basic equation that gives the time evolution of the complex amplitude of the electric field grating E_{sc} can be obtained through a self-consistent solution of a set of equations. It includes, in particular, Poisson's equation relating the spatially inhomogeneous component of the electric field and density of the net space-charge distribution $\rho(x)$:

$$\frac{\partial}{\partial x}E_{sc}(x) = \frac{1}{\epsilon\epsilon_0}\rho(x). \tag{4.4}$$

43

As (4.4) makes clear, we restrict our analysis to isotropic photorefractive crystals with a scalar dielectric constant ϵ. This allows consideration of only one component of the electric field amplitude $E_{sc}(x)$, namely, the component directed along the x axis.

The second equation is the continuity equation that describes a change of the charge density $\rho(x)$ at a given point

$$\frac{\partial}{\partial t}\rho(x) = - \frac{\partial}{\partial x}j(x) . \tag{4.5}$$

We consider mainly PRCs with electron photoconductivity, for which the current density $j(x)$ is a sum of three components, namely the drift, diffusion, and photovoltaic

$$j(x) = e\mu n(x)E(x) + eD\frac{\partial}{\partial x}n(x) + \alpha G I(x) , \tag{4.6}$$

where $n(x)$ is the density of mobile photoelectrons in the conduction band (the dark conductivity is ignored in the analysis), μ is the mobility, D is the diffusion coefficient [4.13-15], α is the optical absorption coefficient, and G is Glass' constant characterizing the photovoltaic effect [4.16].

The third equation is the balance equation for the density of photoelectrons

$$\frac{\partial}{\partial t}n(x) = g(x) - \frac{1}{\tau}n(x) + \frac{1}{e}\frac{\partial}{\partial x}j(x) , \tag{4.7}$$

where τ is the mean lifetime of a photoelectron in the conduction band and $g(x) = \beta\alpha I(x)/\hbar\omega$ is the photoelectron excitation rate. Here $\hbar\omega$ is the photon energy of the recording light, and β is the quantum efficiency of the photoconductivity. Equation (4.7) assumes linearity in the excitation and recombination of photoelectrons, i.e., these processes depend neither on the photoelectron concentration in the conduction band, nor on the concentration of the charge trapped on the impurity levels at a given point of the crystal.

To simplify our discussion further, we assume a quasi-stationary distribution of the electron concentration throughout the conduction band. This allows substitution of (4.7) by

$$0 = g(x) - \frac{1}{\tau}n(x) + \frac{1}{e}\frac{\partial}{\partial x}j(x) . \tag{4.8}$$

We can use such an assumption because the average electron lifetime in the conduction band is rather short $\tau \simeq 10^{-6} - 10^{-9}$ s [4.13-15]. That is why the time needed for the stationary distribution of $n(x)$ to build up for given $I(x)$ and $E(x)$ is far shorter than the hologram recording time, i.e., a typical time required for a noticeable change in $E_{sc}(x)$ to occur. One of the main results of this assumption is that we can neglect the contribution of mobile electrons into the net space-charge distribution $\rho(x)$. The latter is now determined almost entirely by the charges at donor and acceptor centers of the photorefractive crystal.

44

Eventually, the complete set of equations (4.4-8) is given by

$$iKE_{sc} = \frac{1}{\epsilon\epsilon_0}\rho \,, \quad \frac{\partial}{\partial t}\rho = -iKj_{sc} \,,$$

$$j_{sc} = e\mu n_0(E_{sc} + aE_0) + iKeDn_0 a + \alpha GmI_0 \,,$$

$$g_0 m = \frac{1}{\tau}n_0 a - i\frac{1}{e}Kj_{sc} \,. \tag{4.9}$$

In writing the equations we ignored terms quadratic with respect to m, and used the following complex representations

$$\rho(x) = \frac{\rho}{2}\exp(iKx) + \frac{\rho^*}{2}\exp(-iKx) \,,$$

$$j(x) = j_0 + \frac{j_{sc}}{2}\exp(iKx) + \frac{j_{sc}^*}{2}\exp(-iKx) \,,$$

$$n(x) = n_0\left[1 + \frac{a}{2}\exp(iKx) + \frac{a^*}{2}\exp(-iKx)\right] \,,$$

$$g(x) = g_0\left[1 + \frac{m}{2}\exp(iKx) + \frac{m^*}{2}\exp(-iKx)\right] \,. \tag{4.10}$$

Note that (4.9) does not include explicitly any parameters characterizing the structure and absolute concentration of impurity centers. Such apparent incompleteness of (4.9) is associated with the assumption of linear photoexcitation and recombination of photoelectrons, i.e., with the absence of saturation of impurity levels. It obviates the need for additional assumptions concerning the structure of impurity centers in photorefactive crystals [4.17].

4.2 Holographic Recording in a Monopolar Photoconductor Without Trap Saturation

4.2.1 Fundamental Equation for the Formation of an Electric-Field Grating

From (4.9) we derive the fundamental equation that describes the formation of holograms, namely the time evolution of the electric-field amplitude E_{sc}. Equations (4.9c, d) yield the following relation for the electron modulation depth in the conduction band:

$$a = \frac{m(1 + iK\mu\tau E_G) + iK\mu\tau E_{sc}}{1 + K^2 D\tau - iK\mu\tau E_0} \,. \tag{4.11}$$

Here $E_G = \alpha G/\sigma_0$ [4.16] is the so-called photovoltaic field established within the crystal by compensation of two opposite currents, namely, the photovoltaic current ($=\alpha GI_0$) and the photoconductivity current ($=E_G\sigma_0$ with $\sigma_0 = en_0\mu$ being an average photoconductivity of the crystal).

By inserting the first equation from (4.9) into the second one and taking into account the third relation and (4.11), we obtain the final expression describing the hologram recording process

$$iK \frac{\epsilon \epsilon_0}{e} \frac{\partial}{\partial t} E_{sc} = g_0(m - a) , \qquad (4.12)$$

which is equivalent [4.12] to

$$\frac{\partial}{\partial t} E_{sc} = - \frac{m(iE_D + E_0 + E_G) + E_{sc}}{\tau_M(1 + K^2 L_D^2 - iKL_0)} . \qquad (4.13)$$

Here $\tau_M = \epsilon \epsilon_0 / \sigma_0$ is the characteristic Maxwell dielectric relaxation time of the crystal corresponding to its average photoconductivity σ_0, $L_0 = \mu \tau E_0$ is the average drift length of photoelectrons in the external electric field E_0, $L_D = \sqrt{D\tau} = \sqrt{\mu \tau E_D / K}$ is the average diffusion length, and $E_D = KD/\mu = Kk_B T/e$ is the diffusion field.

The general equation (4.13) reveals that recording of the hologram in a photorefractive crystal is, in fact, nothing else but an exponential relaxation of the hologram amplitude from its initial value to a certain steady-state amplitude E_{sc}^{st}. Note, however, that the complex value of the characteristic time of the process implies that oscillating solutions are possible.

4.2.2 Steady-State Holograms

A steady-state regime of hologram recording is a state of the crystal that is continuously illuminated by the interference pattern (4.1) when the amplitude of the electric field grating remains constant ($\partial E_{sc} / \partial t = 0$). Equation (4.12) predicts that in this case the modulation depth of the electron density, a, coincides with the modulation depth of the interference pattern, m, regardless of the drift or diffusion length of the electrons (KL_D, $|KL_0|$ can even be far greater than 1).

From (4.13), the steady-state amplitude of the recorded grating is

$$E_{sc}^{st} = - m(iE_D + E_0 + E_G) . \qquad (4.14)$$

Let us analyse this equation. First, the steady-state hologram amplitude is proportional to the modulation depth of the interference pattern, i.e., the photorefractive crystals can be regarded as "linear" [4.18] holographic media. Second, E_{sc}^{st} proves to be independent of the recording light intensity. The latter, however, is valid only as long as we can ignore the dark conductivity of the crystal, as compared with its photoconductivity.

In the specific case of a crystal with a negligible photovoltaic field and with no external electric field applied (E_G, $E_0=0$), the first term in the parentheses on the right-hand side of (4.14) begins to play a dominating role. This is the so-called diffusion mechanism [4.1], where the holograms are formed by diffusion of mobile charge carriers (electrons in the case under discussion) from the bright interference fringes to the dark ones. The mechanism is characterized by a linear relationship between the steady-state grating amplitude and spatial frequency ($E_D = K k_B T/e$), and also by a $\Lambda/4$ shift (note the imaginary unity in front of E_D) of the electric field grating $E_{sc}(x)$ from the interference pattern $I(x)$ (a shifted hologram). It should be noted that the transmission grating recorded via the diffusion mechanism has a rather moderate amplitude E_{sc}^{st}, since $E_D = 1.6$ kV/cm for $\Lambda = 1$ μm (at T = 300K).

In the opposite case, when the photovoltaic field E_G exceeds the diffusion field ($E_G \gg E_D$), or a fairly high external field is applied to the crystal ($E_0 \gg E_D$), the imaginary term on the right-hand side of (4.14) can be neglected. This is the so-called drift mechanism of recording. The recording process consists here of a partial compensation of the respective field (E_0 or the effective field E_G, initially uniform throughout the sample volume) at the maxima of the recorded pattern $I(x)$. So the steady-state grating produced via the drift mechanism turns out to be unshifted.

The steady-state amplitude of the drift grating is K independent and can have a far greater magnitude limited by the breakdown field in the PRC, namely, 10-20 kV/cm on recording in the external field [4.19]; larger magnitudes of E_{sc}^{st} can be encountered in a ferroelectric PRCs with large E_G.

4.2.3 Characteristic Time and Rate of the Hologram Buildup

The characteristic time and rate of the hologram formation are important parameters which are directly related to the sensitivity of the PRC to recording, and they can also be obtained from the general equation (4.13). Specifically, the characteristic time τ_{sc} needed for the hologram to reach steady state is determined by the denominator of the right-hand side of (4.13). For the simplest case of $KL_D \ll 1$ and $|KL_0| \ll 1$, τ_{sc} is equal to the Maxwell relaxation time τ_M and is independent of either the grating spacing or the dominant mechanism of hologram recording (diffusion, drift in the external field E_0, or in an effective photovoltaic field E_G). The absolute rate of E_{sc} buildup in the initial stage of recording (and hence the sensitivity of the photorefractive crystal as a holographic medium) therefore proves to be proportional to $|E_{sc}^{st}|/\tau_M$. For the diffusion mechanism, a faster recording of higher spatial frequencies is implied, since here $|E_{sc}^{st}| \propto E_D \propto K$. For the drift mechanism, the sensitivity is independent of the grating's spatial frequency and the absolute value of the recording rate can be increased by enlarging the electric field E_0 (for $E_G \ll E_0$).

The record-erase time constant τ_{sc} of a crystal with a dominating diffusion mechanism turns out to be proportional to K^2 for $KL_D \gtrsim 1$. As a consequence, the absolute grating buildup rate in the initial stage of recording becomes inversely proportional to K. The physical meaning of this change is easily understood if we take into account that for $KL_D \gg 1$ the photoinduced electrons excited by the interference pattern (4.1), are captured by deep traps almost evenly throughout the crystal volume. Thus, of major importance for the initial stage of hologram recording is the free-electron excitation rate g_0 rather than the photoconductivity of the crystal

σ_0. The rate of the charge-grating buildup ρ turns out to be equal to emg_0. So the grating field E_{sc} is determined according to (4.9a) by

$$|E_{sc}| = \frac{t}{\epsilon\epsilon_0 K} \text{emg}_0 = \frac{t}{\epsilon\epsilon_0 K} \text{em} \frac{\beta\alpha I_0}{\hbar\omega}. \tag{4.15}$$

The spatially uniform retrapping of photoelectrons in the initial stage of hologram formation is also characteristic of the drift recording mechanism in the external electric field E_0, such that $|KL_0| \gg 1$. Therefore, the E_{sc} buildup rate (4.15) determined by the electron excitation rate is valid here too. Let us estimate the absolute value of the holographic sensitivity of PRCs in this approximation. Note that, since the charge-grating buildup rate is the highest in this case, the sensitivity to be obtained below is the highest possible for PRCs.

As will be shown in Sect. 5.4, the diffraction efficiency η of transmission volume phase holograms in a PRC can be estimated from the simple relation:

$$\eta = \sin^2\left[\frac{\pi d}{2} \frac{n^3 r}{\lambda} E_{sc}\right]. \tag{4.16}$$

Here r is the linear electro-optic coefficient relevant to the geometry of the experiment, n and d are the refractive index and crystal thickness, respectively, and λ is the readout wavelength. A direct substitution of (4.15) into (4.16) yields the expression that describes the initial stage of hologram formation:

$$\sqrt{\eta(t)} = \frac{\pi d}{2} \frac{n^3 r}{\lambda} \frac{1}{\epsilon\epsilon_0 K} \text{m} \frac{\beta\alpha I_0}{\hbar\omega} t. \tag{4.17}$$

Taking $\eta(t) = 0.01$ (1%), we obtain the holographic sensitivity of the photorefractive crystal (Sect. 3.4):

$$S = (I_0 tm)^{-1} = \frac{5}{2} \frac{rn^3 e\Lambda}{\lambda\epsilon\epsilon_0\hbar\omega}. \tag{4.18}$$

For simplicity, the quantum efficiency of the photoconductivity (β) and optical density of the sample at the recording wavelength (αd) are assumed here to be equal to unity. For the typical parameters $\lambda \simeq 0.5 \, \mu m$, $n^3 \simeq 10$, $r/\epsilon \simeq 10^{-10}$ cm/V [4.20], $\Lambda = 1 \, \mu m$, and $\hbar\omega \simeq 2$ eV, (4.18) gives $S \simeq 10^4$ cm^2/J approaching the sensitivity of conventional high-resolution silver halide photographic Kodak 649F plates [4.18]

Note that one of the highest holographic sensitivities in presently known photorefractive crystals was reported by *Huignard* and *Micheron* [4.21] for BSO. It amounts to $S = 5 \, 10^2$ cm^2/J at $\Lambda = 1 \, \mu m$ and, indeed, increases linearly with increasing fringe spacing Λ.

4.3 Trap Saturation (Violation of Quasi-Neutrality)

The linearity of photoexitation and recombination of charge carriers implies that the concentrations of donor impurities from which photoelectrons are excited or trapping centers on which electrons recombine are nearly constant during recording. The process of hologram recording by its very nature, however, involves redistribution of charges throughout the sample volume and, hence, varying degrees of population of donor and trapping impurity centers. According to Poisson's equation, the steady-state amplitude of the field grating E_{sc}^{st} and the corresponding amplitude of the sinusoidal distribution of the charge density ρ^{st} are related by (4.9a). Thus, the local concentration of electrons captured on deep impurity centers in the steady-state regime can vary by $\rho^{st}/e = \pm \epsilon\epsilon_0 K E_{sc}^{st}/e$.

That is why, for the condition of linear mobile carrier photoexcitation and recombination in the steady-state regime to be fulfilled, the initial concentrations of donor and trapping centers must distinctly exceed the above-given ρ^{st}/e value. If we assume [4.4, 12] that the concentration of donor centers is well above the concentration of traps ($N_D \gg N_A$), the condition of linear recombination will be satisfied only for

$$N_A \gg \rho^{st}/e = \epsilon\epsilon_0 K |E_{sc}^{st}|/e . \tag{4.19}$$

This inequality has also another fairly evident meaning. The maximum amplitude of the charge grating in a PRC with a limited trap concentration cannot also exceed eN_A. Therefore, in accordance with Poisson's equation, the maximum amplitude of the field grating is limited by

$$E_q = eN_A(\epsilon\epsilon_0 K)^{-1} . \tag{4.20}$$

For diffusion recording, when $|E_{sc}^{st}| \sim E_D$, the quasi-neutrality condition (4.19) is transformed into

$$K \sqrt{\frac{\epsilon\epsilon_0 k_B T}{e^2 N_A}} = K L_D' \ll 1 , \tag{4.21}$$

where L_D' is the so-called Debye screening length. The physical meaning of this parameter is well known [4.14]; L_D' is a typical steady-state depth of penetration of mobile charge carriers into the dark region of the crystal illuminated by a sharp light-to-dark transition.

Note that for $\epsilon \simeq 50$ typical of PRCs and a modest concentration $N_A \simeq 10^{16}$ cm^{-3}, we have $L_D' \simeq 0.1$ μm. The effect of trap saturation and an accompanying limitation of the steady-state grating amplitude, according to (4.20), are then observed for a grating spacing of $\Lambda \gtrsim 2\pi L_D' \simeq 0.5$ μm.

Similar effects lead to a limitation of the grating amplitude during hologram recording through the drift mechanism as well. Here $|E_{sc}^{st}| \sim E_0$ (at

$E_G = 0$), and (4.21) is replaced by

$$K \left[\frac{\epsilon \epsilon_0 E_0}{e N_A} \right] = K L_0' \ll 1 . \tag{4.22}$$

Note that the characteristic parameter L_0' (the so-called tightening length [4.14]) also determines the typical penetration depth of photocarriers under continuous illumination of the crystal (in the external field E_0) by a sharp light-to-dark transition. For the example given above ($\epsilon \simeq 50$, $N_A \simeq 10^{16}$ cm^{-3}) and $E_0 = 10$ kV/cm, $L_0' = 0.3$ μm. That is to say, the saturation effects begin to show up already for $\Lambda \stackrel{>}{\sim} 2$ μm.

A more thorough analysis of the steady-state regime of hologram recording with saturation of impurity centers (i.e., when the quasi-neutrality condition is violated) was given in [4.6]. Here we present only the final expression for the steady-state grating amplitude in the specific case of $E_G = 0$, which confirms the qualitative arguments given above:

$$E_{sc}^{st} = - m \frac{i E_D + E_0}{(1 + E_D / E_q) - i E_0 / E_q} . \tag{4.23}$$

A typical example of the experimental $\eta(K)$ dependence for BSO crystal where saturation of trapping centers is observed, was reported in [4.22] (Fig. 4.2).

The most important specific features of (4.23) should be pointed out. First, the saturation regime is reached in a particular PRC at certain fixed values of K and E_0, whatever the magnitude of m. Second, irrespective of the particular mechanism (be it drift, diffusion or a nonstationary mechanism) through which the regime of trapping centers' saturation is reached, a shifted hologram with the amplitude

$$E_{sc}^{st} = - i m E_q \tag{4.24}$$

is formed.

A major consequence of this limitation should be noted. Among basic parameters of PRCs as dynamic holographic media, the gain factor Γ is one of the most important ones (Sect. 6.2). Equation (4.24) sets an upper limit on the possible magnitude of its value

$$\Gamma^{max} = 2\pi \frac{n^3 r}{\lambda} E_q = \frac{\Lambda}{\lambda} n^3 \frac{e}{\epsilon_0} \frac{r}{\epsilon} N_A , \tag{4.25}$$

which, in fact, depends only on the trap concentration (N_A) in the PRC involved. For the parameters typical of presently known PRCs $r/\epsilon \simeq 10^{-10}$ cm/V, $n \simeq 2.5$, $N_A \simeq 10^{16}$ cm^{-3}, and for $\Lambda/\lambda \simeq 1$, we have $\Gamma^{max} \simeq 20$ cm^{-1}.

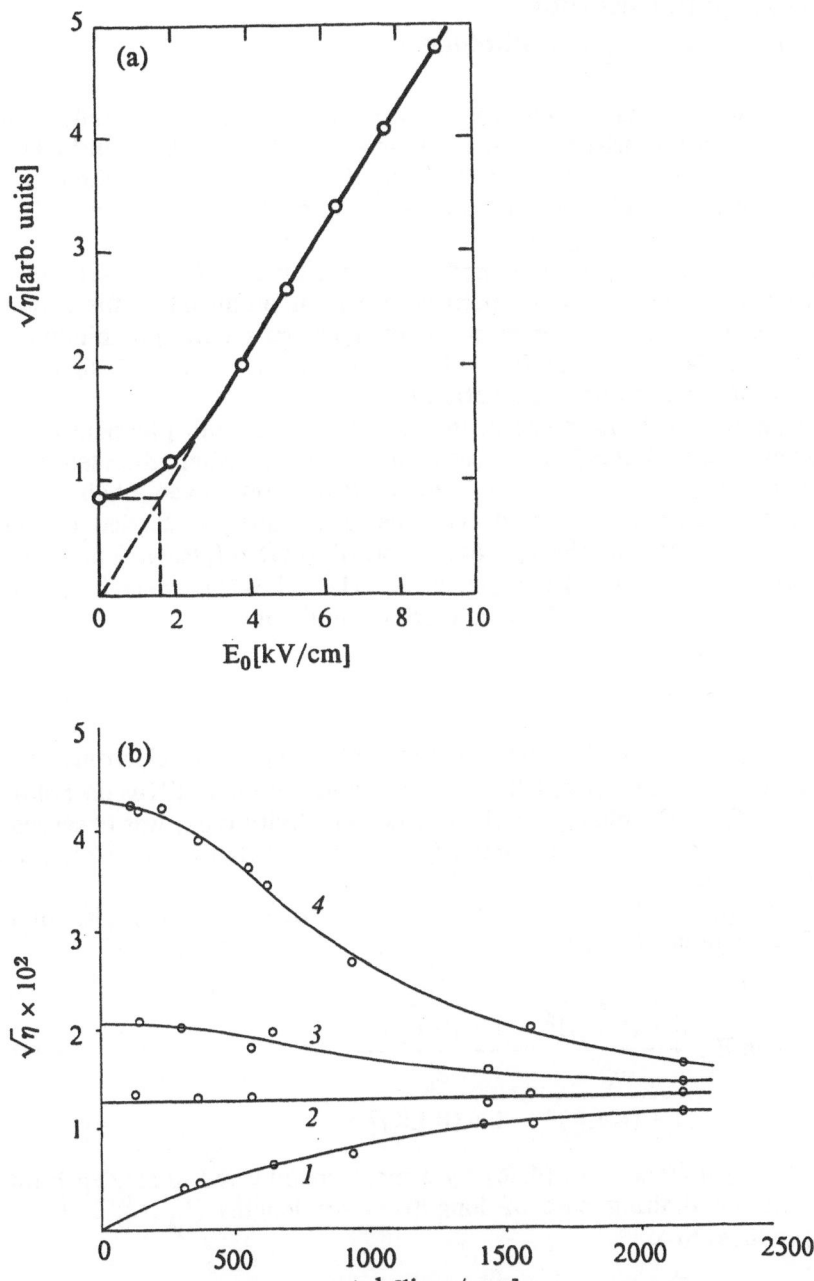

Fig.4.2. (a) Experimental curve of steady-state hologram amplitude ($\sqrt{\eta} \propto E_{sc}$) vs. DC electric field E_0 for a BSO sample with no trap saturation [4.12] ($\lambda = 514$nm, $\Lambda \simeq 1\mu$m, $E_D \simeq 1.5$kV/cm). (b) Experimental curves of steady-state hologram amplitude vs. spatial frequency for a BSO sample with trap saturation [4.22] ($\lambda = 514$nm, E_0 [kV/cm], *1*: 0, *2*: 2, *3*: 3, *4*: 6)

4.4 Holographic Recording in PRCs with Bipolar Photoconductivity

If the condition of quasi-neutrality is fulfilled, the steady-state hologram recorded through the drift mechanism in an external DC field is unshifted. The grating sign is independent of the type of dominant photoexcited carriers. Both for holes and electrons, compensation of the external field E_0 in the bright interference fringes is observed.

Another situation is encountered for diffusion recording ($E_0, E_G = 0$). Here an excess or deficiency of positive charge is produced in the bright fringes, depending on whether electrons or holes are mobile photoinduced charge carriers. That is, the shifted diffusion hologram reverses its sign depending on the type of photoconductivity.

In crystals with bipolar photoconductivity (where both photoelectrons and holes are excited) competition between the two oppositely directed holographic recording processes and, in certain cases, even their total compensation should be expected. The theoretical analysis carried out in [4.23] (see also [4.24]) in the approximation of short diffusion lengths of photoelectrons ($L_D^e \ll K^{-1}$) and photoholes ($L_D^h \ll K^{-1}$), neglecting trap saturation, yields for the steady-state grating amplitude:

$$E_{sc}^{st} = - imE_D \frac{\sigma_0^e - \sigma_0^h}{\sigma_0^e + \sigma_0^h} , \qquad (4.26)$$

where σ_0^e and σ_0^h are crystal photoconductivities due to the electrons and holes, respectively. Experimentally, the sign inversion of a diffusion hologram, arising from the change of the photoconductivity type, was observed in [4.25] in a series of $BaTiO_3$ samples, and also in $LiNbO_3:Fe$ [4.23] and $BaTiO_3$ [4.26] with a varying degree of reduction.

In [4.26-28] (4.26) was extended to the case of an arbitrary diffusion length of photoinduced carriers

$$E_{sc}^{st} = - imE_D \frac{\dfrac{\sigma_0^e}{1 + (KL_D^e)^2} - \dfrac{\sigma_0^h}{1 + (KL_D^h)^2}}{\dfrac{\sigma_0^e}{1 + (KL_D^e)^2} + \dfrac{\sigma_0^h}{1 + (KL_D^h)^2}} \qquad (4.27)$$

Equation (4.27) differs from (4.26) by a more complicated dependence on K. Also, in the limiting case of long transport lengths ($L_D^e, L_D^h \gtrsim K^{-1}$) Eq.(4.27) reduces to

$$E_{sc}^{st} = - imE_D \frac{g_0^e - g_0^h}{g_0^e + g_0^h} , \qquad (4.28)$$

where g_0^e and g_0^h are the average electron and hole excitation rates, respectively. In [4.28] the saturation of impurity centers in bipolar PRCs was also taken into account.

Note that as the spatial frequency K increases, the transformation of the grating amplitude from (4.26 to 28), which can differ not only in magnitude, but also in sign, is possible. A similar behavior was observed, for instance, in experiments with $BaTiO_3$ [4.25] and $KNbO_3$ [4.29]. A complicated dependence of the amplitude of a steady-state diffusion grating on K, that can be explained in terms of the bipolar photoconductivity model, was also observed in a cubic BSO [4.30]

4.5 Nonstationary Holographic Recording Mechanisms

Conventional techniques of holographic recording in PRCs employ stationary external conditions, such as a fixed interference pattern of constant intensity, application of a steady external field, etc. Recent experiments demonstrated that nonstationary recording conditions (specifically, a "moving" or an oscillating interference pattern and also recording in an external alternating electric field) can increase appreciably the efficiency of hologram recording in cubic crystals of the BSO and GaAs type.

Efficient nonstationary recording relies on long drift lengths of photoelectrons in an external electric field E_0, or to be more exact, on the following condition:

$$|KL_0| \gtrsim 1 + K^2 L_D^2 . \tag{4.29}$$

An immediate reason for the hologram enhancement on recording a moving interference pattern is a "moving" behavior of the hologram in a state of free relaxation under the external DC field [4.31]. This means that the hologram recorded through any of the mechanisms discussed above, moves along the external field direction during the uniform illumination in the crystal (m = 0). A similar phenomenon (the so-called "waves of spatial charge exchange") was also described for the classical semiconductor non-photorefractive crystals [4.32]. The phenomenon should be distinguished from the energy coupling enhancement observed during recording a drift grating by moving interference pattern [4.33].

4.5.1 Moving Holograms in PRCs Under an External DC Field

Let us consider free relaxation of a grating from its initial amplitude $E_{sc}(0) \neq 0$ at t = 0. For this purpose, it is necessary to set m = 0 in (4.13), which yields

$$E_{sc}(t) = E_{sc}(0)\exp\left[-\frac{t}{\tau_M}\,\frac{1}{(1 + K^2L_D{}^2) - iKL_0}\right]$$

$$= E_{sc}(0)\exp\left[-\frac{t}{\tau_M}\,\frac{(1 + K^2L_D{}^2)}{(1 + K^2L_D{}^2)^2 + K^2L_0{}^2}\right]$$

$$\times\exp\left[\frac{t}{\tau_M}\,\frac{-iKL_0}{(1 + K^2L_D{}^2)^2 + K^2L_D{}^2}\right]. \tag{4.30}$$

The grating decay is determined by the first exponential factor with a real index and has a characteristic time

$$\tau_{sc} = \tau_M\,\frac{(1 + K^2L_D{}^2)^2 + K^2L_0{}^2}{(1 + K^2L_D{}^2)} \tag{4.31}$$

longer than τ_M and proportional to the square of the external field ($L_0 \propto E_0$).

The second exponential factor in (4.30) with an imaginary argument implies that the grating moves in the direction of the external field E_0 with the characteristic velocity

$$v_{sc} = \frac{L_0}{\tau_M}\,\frac{1}{(1 + K^2L_D{}^2)^2 + K^2L_0{}^2}. \tag{4.32}$$

Indeed, the electric field distribution in this hologram is given by

$$E_{sc}(x,t) = Re\{E_{sc}(t)\exp(iKx)\} =$$

$$= \left|E_{sc}(0)\right|e^{-t/\tau_{sc}}\cos[K(x - tv_{sc}) + \psi(0)]. \tag{4.33}$$

The characteristic frequency of this moving grating is evidently $\Omega_{sc} = Kv_{sc}$. At $E_0 = 0$, the grating, which is erased, is at rest ($v_{sc}, \Omega_{sc} = 0$) and the characteristic erase time is

$$\tau_{sc} = \tau_M(1 + K^2L_D{}^2). \tag{4.34}$$

The physical mechanism of the processes can be explained as follows [4.31]. Let the hologram with an electric field distribution $E_{sc}(x)$ be initially recorded in the sample (Fig.4.3). With no external field applied ($E_0 = 0$), the hologram field causes the uniformly excited photoelectrons to concentrate at the left-hand slope of the distribution $E_{sc}(x)$. This gives rise to a new "secondary" hologram with the field distribution $E'_{sc}(x)$ out of phase with $E_{sc}(x)$ (Fig.4.3a). Thus, the resulting hologram amplitude decreases and this, in fact, means optical erasure of the hologram.

In the short drift length approximation ($|L_0| \ll K^{-1}$), application of the external dc field E_0 has no significant effect on the grouping of photoelectrons. In the other limiting case ($|L_0| \gg K^{-1}$) electrons move through

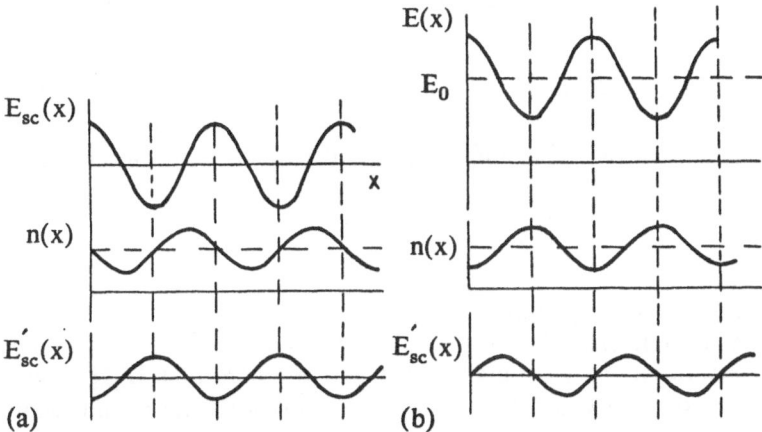

Fig.4.3. Spatial redistribution of uniformly excited photoelectrons in the hologram field $E_{sc}(x)$ (a) with no external field applied and (b) in an external DC electric field E_0 in the long drift length approximation ($|KL_0| \gg 1$)

several grating spacings Λ during their mean lifetime. The process of grouping is not as efficient as before, and the peak concentration of photoelectrons occurs at the minimums of the net field $E(x) = E_0 + E_{sc}(x)$. The average drift velocity of the photoelectrons is the lowest and the time of stay is the longest here. In this situation the secondary field $E'_{sc}(x)$ is shifted by $\Lambda/4$ relative to the initial pattern $E_{sc}(x)$. Thus, uniform illumination of the sample results in a shift of the initial hologram as a whole along the external field E_0 (or opposite to E_0 for hole conductivity) rather than in its relaxation.

4.5.2 Recording of Resonantly Moving Interference Patterns

The simplified analysis given above reveals that, if long drift lengths of the photoelectrons are involved, a stationary interference pattern is not optimum for recording. The peak steady-state hologram amplitude will be obtained if the condition of phase synchronism is satisfied, i.e., when the pattern moves with the hologram at the velocity v_{sc}.

Recording of a moving interference pattern in PRCs was theoretically analysed by many researchers [4.8, 31, 35, 36]. Here we restrict our analysis to a simplified derivation of the steady-state amplitude of the grating recorded by a resonantly-moving pattern [4.31]. The fact is that the steady-state amplitude E^{st}_{sc} of the hologram in this case is equal to its relaxation time τ_{sc} (4.31) multiplied by the rate of its recording. The recording rate is given by (4.13) and for $E_G = 0$ and for long drift lengths (4.29) is equal

$$\frac{\partial}{\partial t} E_{sc} \simeq - im E_0 \frac{1}{\tau_M KL_0} . \tag{4.35}$$

By multiplying the values mentioned above, we obtain the amplitude of the steady-state hologram:

$$E_{sc}{}^{st} \simeq - im E_0 \frac{KL_0}{1 + K^2 L_D{}^2} \cdot \tag{4.36}$$

The hologram is therefore of a shifted type and reaches its peak value (for a given E_0)

$$E_{sc}^{st} = - im E_0 \frac{L_0}{2L_D} \tag{4.37}$$

at the optimuml spatial frequency $K = L_D{}^{-1}$.

A more thorough analysis given in the papers mentioned above takes into account saturation of trapping centers and predicts that $|E_{sc}^{st}|$ is also limited by E_q. A direct consequence of this fact is that the maximum value (for a given E_0)

$$E_{sc}^{st} = - im E_q(K^{opt}) = - im E_0 \frac{L_D}{2(L'^2 + L_D'^2)^{1/2}} \tag{4.38}$$

is achieved at the optimum spatial frequency

$$K^{opt} = \frac{L'}{L_D(L'^2 + L_D'^2)^{1/2}} \cdot \tag{4.39}$$

Here the characteristic length $L' = k_B T/eE_0$ ($\simeq 0.25\ \mu m$ at $E_0 = 10$ kV/cm), and L_D' is the Debye screening length (4.21).

Thus the resonantly-moving shifted holograms with an amplitude higher than that of the conventional drift grating (mE_0) can be recorded in crystals with a fairly long diffusion length of the photoelectrons

$$L_D \gtrsim 2 \sqrt{L'^2 + L_D'^2} \ . \tag{4.40}$$

This does not, however, mean that a grating with an amplitude greater than the external field E_0 can be recorded. The theoretical analysis was, in fact, carried out in the linear approximation for m << 1.

The moving behavior of the holograms in BSO during their erasure in the external DC field was experimentally observed in [4.31, 37], slowing down of their erasure rate in [4.31, 38]. Recording of the resonantly moving interference pattern in this PRC was also investigated in [4.8, 31, 33, 35, 38]. Typical curves for slowing down of the erasure rate η^{st} as a function of velocity v of the moving pattern, and also the resonance velocity v_{sc} as a function of the light intensity I_0 and the field E_0 observed in BSO, are plotted in Fig.4.4. Recording of the moving interference pattern in an external DC electric field was also used recently in cubic GaAs [4.40].

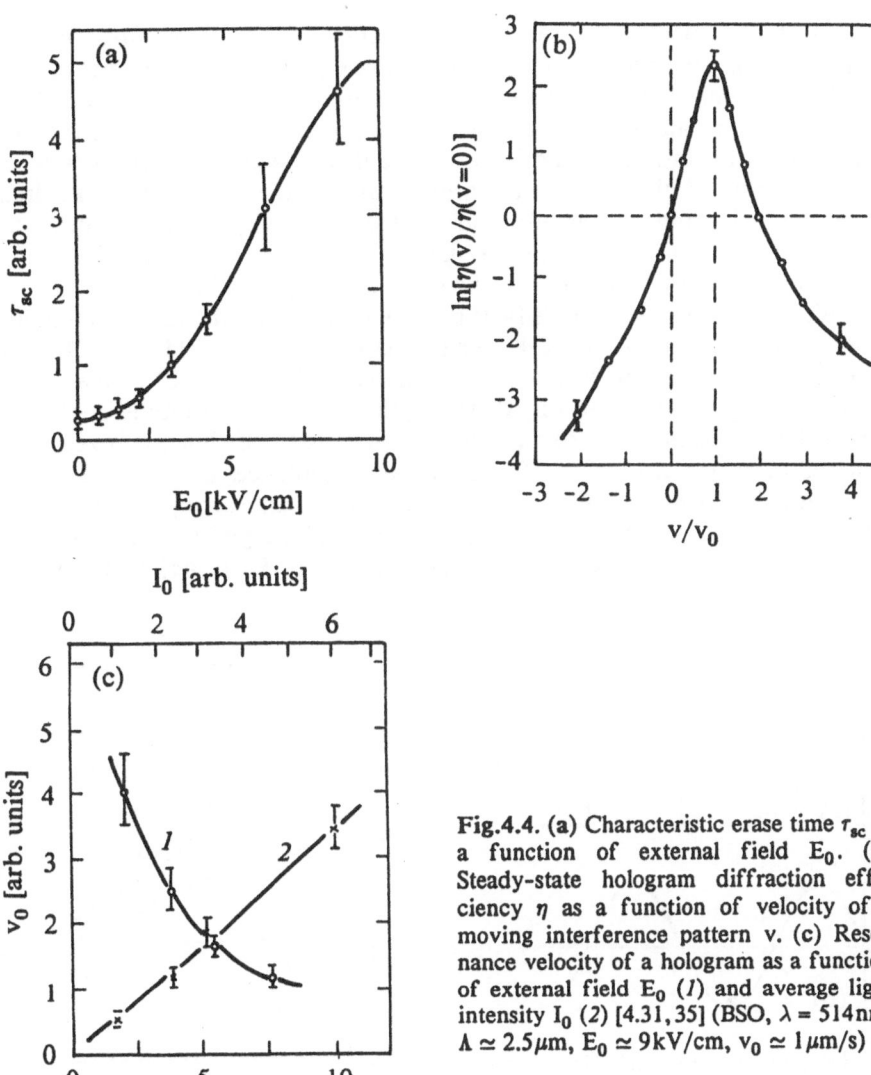

Fig.4.4. (a) Characteristic erase time τ_{sc} as a function of external field E_0. (b) Steady-state hologram diffraction efficiency η as a function of velocity of a moving interference pattern v. (c) Resonance velocity of a hologram as a function of external field E_0 (1) and average light intensity I_0 (2) [4.31, 35] (BSO, $\lambda = 514$nm, $\Lambda \simeq 2.5\,\mu$m, $E_0 \simeq 9$kV/cm, $v_0 \simeq 1\,\mu$m/s)

4.5.3 Recording Under an External Alternating Electric Field

There is an alternative way of synchronizing a non-moving interference pattern with a hologram, namely, the hologram "arrest". It can be achieved by recording in an external alternating field with a temporal period τ_- much shorter than the characteristic time τ_{sc} of the hologram formation [4.35, 41]. The hologram, while moving in opposite directions during two successive half-periods of the field oscillation, remains immobile on the average and is thus synchronized with the stationary interference pattern.

Nonstationary recording in an alternating field can be regarded, in a certain sense, as an analog or further development of the diffusion recording mechanism. Indeed, in both cases an entirely symmetric "spreading" of the photoinduced electrons occurs around the interference pattern maxima. In the diffusion model, spreading of photoelectrons is, however, caused by usual thermal diffusion, while in the nonstationary mechanism it results from a far more efficient drift in the external electric field.

Without going into details [4.35], we note here that the maximum amplitude of the hologram and K^{opt} for a square-wave AC field (Fig.4.5a) are also described by (4.36-39). A sinusoidal AC field (Fig.4.5b) proves to be less efficient, and even for long drift lengths (4.29) the amplitude of the recorded hologram does not exceed that of the usual drift hologram recorded in a DC electric field (mE_0). Typical field dependences of the steady-state amplitude of the hologram recorded in BTO in an alternating fields are depicted in Fig.4.6. Efficient recording of shifted phase gratings in alternating electric fields was performed in BSO [4.35], BTO [4.35, 42, 43], and in GaAs [4.44].

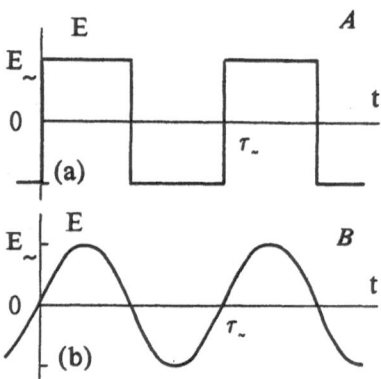

Fig.4.5. Alternating electric fields used for nonstationary holographic recording (A: square-wave field, B: sinusoidal field)

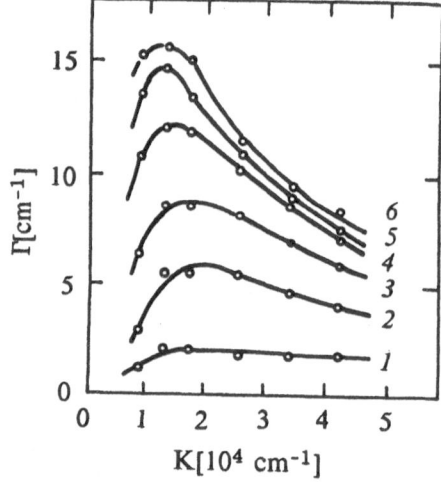

Fig.4.6. Experimental curve for gain coefficient Γ ($\Gamma \propto Im\{E_{sc}\}$) vs. spatial frequency K for different amplitudes of a square-wave alternating electric field E [4.35, 42] (BTO, $\lambda = 633$nm, $\tau_\sim \simeq 25$ms, E being 1: 2.5, 2: 5, 3: 7.5, 4: 10, 5: 12.5, 6: 15 [kV/cm])

Although both nonstationary recording mechanisms discussed can provide the same maximum efficiency, the practical implementation of recording in an AC field is simpler than recording of a moving pattern. The latter requires constant mean levels of the light intensity I_0, electric field E_0, and spatial frequency K throughout the sample volume. Recording in an AC field does not need these. Like the conventional diffusion recording, it is characterized, however, by a higher noise level due to the efficient two-wave enhancement of light scattered from the inhomogeneities of the crystal [4.35].

4.6 Effect of Contact-Induced Phenomena on the Dynamics of Charge and Field Formation

Equations (4.4-7), which describe the photoinduced charge formation in PRCs, must be supplemented by boundary conditions. In Sect.4.1 they were solved for cyclic boundary conditions, which require that the desired solutions be periodic functions with the period equal to the recorded grating period. It was assumed that the crystal is infinite along the x axis, i.e., along the external field direction. The effects of the processes associated with the finite sizes of the crystal along this direction were ignored. Such an approximation is not, however, always applicable, and the need for a solution of the problem with other boundary conditions arises.

Among such situation is the use of a PRC in a SLM where the photoinduced charge formation is strongly affected by the crystal boundaries. The recording light in the SLM enters the crystal along the electric field direction through the transparent electrodes, i.e., referring to the orientation of the axes, taken in Sect.4.1, along the x axis. The incident light intensity is modulated in the yz plane, the light intensity changes along x, being determined only by the light absorption in the crystal. Let us consider the problem when the electrodes on the crystal surface are blocking; i.e., the photoelectrons leaving the region adjacent to the contact are not compensated through injection.

The other example is related to the holographic recording through the drift mechanism in a DC electric field. The electrodes deposited on the PRC surface for this purpose can also be blocking. As a consequence, immediately after the voltage is applied and the illumination is switched on, the field redistribution can occur in the crystal. That is why conditions for the current of a necessary magnitude should be created to avoid charge accumulation near the crystal boundaries.

To demonstrate the major specific features of the charge formation associated with the contact-induced phenomena, we consider the case when the crystal has blocking electrodes on the surface and assume that electrons are not injected into the crystal from the electrode, but can leave the crystal by the electrode. This leads to the boundary condition n = 0 (n being the free-electron density) on the crystal surface with the negative electrode (x =

0). In addition, we assume, for simplicity, that the crystal is uniformly illuminated by weakly absorbed recording light and neglect the absorption ($\alpha \rightarrow 0$). As will be shown in Sect.7.1, the basic results obtained here for the uniform illumination, are applicable to the process of image recording in SLMs.

The theoretical and experimental analysis [4.45-57] allowing for conditions in the regions adjacent to the crystal surfaces with electrodes reveals that there are three stages of evolution of internal fields. The first stage corresponds to low exposures and can be regarded as linear. At this stage, the charge field can be considered small, as compared with the external field, and the internal field is equal to the external one. If we neglect diffusion and photovoltaic fields, (4.5-7) reduce to

$$\frac{\partial \rho(x)}{\partial t} = - e\mu E_0 \frac{\partial n(x)}{\partial x} ; \quad \frac{n(x)}{\tau} - g = - \mu E_0 \frac{\partial n(x)}{\partial x} . \tag{4.41}$$

The solution for the initial conditions $\rho = 0$ at $t = 0$ is

$$\rho(x,t) = egt \int_0^x \exp\left(- \frac{x-x'}{L_0}\right) dx' , \tag{4.42}$$

where $L_0 = \mu\tau|E_0|$. Then by solving Poisson's equation, we obtain

$$E(x,t) = E_0 - \epsilon\epsilon_0 egt L_0[1 - \exp(-x/L_0)] . \tag{4.43}$$

As seen from (4.43), the field $E(x,t)$ near the negative electrode ($x = 0$) equals the external field E_0 and decreases with increasing separation from the electrode because of screening of the external field by the positive charge. The positive charge is accumulated in the region adjacent to the electrode, since the photoelectrons leave it and there is no compensation from the electrode. The positively-charged-layer thickness is equal to the drift length L_0 according to (4.43).

As exposure grows, the external field is screened to such an extent that $E(x,t) \simeq 0$ in a certain region of the crystal. The so-called "bottleneck" (the weak-field region) arises. The photoelectrons drift from the region adjacent to the electrode to the bottleneck, slow down, and recombine intensively. As a result, the positive charge is compensated in the bottleneck, thereby decreasing the positively-charged-layer thickness and leading to a displacement of the bottleneck toward the negative electrode.

Simultaneously, the positive-charge density near the electrode grows. The process behaves nonlinearly and can be regarded as the second stage of the internal-field variation in PRCs. The nonlinear dynamics of charge formation has mathematically been treated in [4.50]. For convenience, we assume that all photoelectrons leave the positive-charge region, since the bottleneck coordinate coinciding with the boundary of the positively-charged region is $x_0(t) < L_0$. Then

$$\rho(x,t) = \begin{cases} \rho_0(t) = egt , & 0 < x < x_0(t) \\ 0 , & x_0(t) < x < d . \end{cases} \tag{4.44}$$

From (4.44), the net charge in the crystal $Q = \int_0^d \rho(x,t)dx = egtx_0(t)$. By solving Poisson's equation and taking into account (4.44), we obtain

$$
E(x,t) = \begin{cases} E_0 - \dfrac{egt}{\epsilon\epsilon_0}\left[x_0(t) - x - \dfrac{x_0^2(t)}{2d}\right], & 0 < x < x_0(t) \\[3mm] E_0 + \dfrac{egt}{\epsilon\epsilon_0}\dfrac{x_0^2(t)}{2d}, & x_0(t) < x < d . \end{cases} \tag{4.45}
$$

Using (4.45) we derive the equation for the bottleneck coordinate

$$
\frac{\partial(tx_0(t))}{\partial t} = \mu\tau E_0\left[1 + \frac{egt}{\epsilon\epsilon_0 E_0}\frac{x_0^2(t)}{2d}\right]. \tag{4.46}
$$

This Rikkarti equation reduces to the linear second-order differential equation for Bessel functions. If we ignore $\partial x_0(t)/\partial t$ in (4.46) (the so-called quasi-classical approximation), the approximate solution of (4.46) can be obtained:

$$
x_0(t) = \frac{\epsilon\epsilon_0}{egtL_0}\left[\sqrt{1 + \frac{2L_0 egt}{\epsilon\epsilon_0 U}} - 1\right], \tag{4.47}
$$

where U is the voltage drop across the crystal. It is evident from (4.47) that for time intervals $t < U/L_0^2 eg = t_k$ corresponding to the first stage of the internal field formation, $x_0 \simeq L_0$ and is time independent. Note that the parameter t_k defines the applicability of the linear approximation used for the first stage. At $t > t_k$, according to (4.43), $E(x,t) > 0$; i.e., the field in the crystal proves to be directed opposite to the external one; this is physically meaningless.

According to (4.47), for $t > t_k$ (the second stage)

$$
x_0(t) \simeq \sqrt{\frac{2\epsilon\epsilon_0 U}{egt}} . \tag{4.48}
$$

From (4.45, 48), the field on the negative electrode is

$$
|E(0,t)| \simeq \sqrt{\frac{2Uegt}{\epsilon\epsilon_0}} ; \tag{4.49}
$$

i.e., it grows as $\sqrt{t}$.

For the second stage, the electric field is strong only in the positively-charged layer whose thickness $(x_0(t) < L_0)$ decreases with exposure. Equations (4.48, 49) do not include such parameters as μ and E_0 just for this reason.

The basic conclusions of the above analysis were verified in experiments using BSO crystals with Pt or In_2O_3 electrodes [4.48, 49]. Figure 4.7 shows the curves for the internal electric field versus x for different ex-

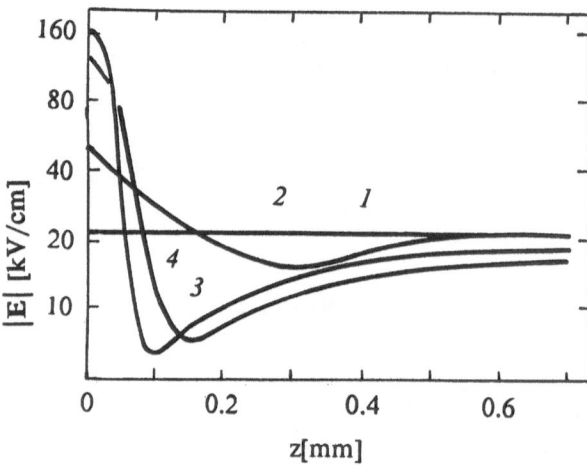

Fig.4.7. Distribution of the longitudinal field across the BSO crystal thickness for different exposures W with light at 442 nm (E_0 = 23 kV/cm, W being *1*: 0, *2*: 1, *3*: 5, *4*: 10 [μJ/cm²])

posures and recording light intensities, such that the internal field evolution occurs during several seconds. Analysis of the data confirms that injection from the electrode can be ignored and the near-electrode region of the positive charge and the bottleneck are formed in the crystal. The bottleneck moves toward the negative electrode with increasing exposure and $x_0(t) \propto \sqrt{t}$.

As (4.49) predicts, if the injection is absent, the field near the negative electrode increases infinitely with increasing exposure, which is certainly impossible in practical situations. As shown in [4.52,53], the field strength of several hundred kV/cm near the electrode in BSO crystals causes an efficient injection of electrons into the crystal. This influences, to a great extent, further evolution of the field and allows the third stage of charge formation to be separated out. At this stage, the electron injection from the electrode slows down and then stops the growth of the positive charge density in the region adjacent to the electrode. This stabilizes the field near the contact at the level when the injection current is equal to the photocurrent in the crystal. The negative charge injected from the electrode partly compensates for the positive charge in the region adjacent to the electrode. The field in the bottleneck grows and a fraction of the injected electrons moves behind the bottleneck, to form there a negative charge layer. As a consequence, a double charge layer is formed in the region adjacent to the electrode; i.e., the positive charge is located in the immediate vicinity of the electrode and the negative charge layer follows it. The field behind this double layer is not screened. Thus, the internal field far from the electrode can again increase to a magnitude close to E_0 at the third stage. Then the injection current through the electrode equals the photocurrent in the sample volume.

The processes of field redistribution discussed here play a dominating role in image recording in photorefractive SLMs (Chaps. 7, 8). In addition, these processes are essential for establishing the fields and currents in the crystals during holographic recording under an external DC field. In particular, this can reduce the effective electric field in the volume of the sample.

Note in conclusion that the nonlinear processes associated with the contact-induced effect - for instance, of the bottleneck type - leads to a number of interesting phenomena. For instance, it has been shown that internal field formation can be accompanied by oscillatory time dependences of the photocurrent [4.51]. The bottleneck in the crystal is equivalent, in a certain sense, to a virtual blocking contact. Therefore, multiple formation of the bottleneck is possible, giving rise to alternating layers of positive and negative charges, i.e., to the so-called charge stratification [4.54-57].

4.7 Additional Remarks

We have discussed the basic mechanisms and most important, specific features of holographic recording in PRCs. Some related problems beyond the scope of this discussion should be mentioned, too.

1) Pulsed holographic recording in PRCs was theoretically analyzed in [4.58].
2) Holographic recording in PRCs with a complicated structure of the impurity centers was studied in [4.17, 59, 60].
3) Recording in crystals with ionic conductivity and also thermal fixing were investigated in [4.61, 62].
4) Alternative mechanisms of the photorefractive effect in ferroelectric crystals not directly related to the diffusion, drift, and photovoltaic mechanisms were considered in [4.9, 63-67].
5) Holographic currents through photorefractive samples for the steady-state recording regime were studied in [4.68-70], and for the nonstationary regime in [4.71].
6) Holographic recording through the circular photovoltaic effect in ferroelectric photorefractive crystals of the $LiNbO_3$ type was investigated in [4.73-76].

5. Light Diffraction
from Anisotropic Volume Phase Gratings

The main characteristic feature of a photorefractive phase grating is an anisotropy. It arises from an anisotropy of the linear electrooptic effect through which the space-charge field E_{sc} is converted to phase relief. So the phase grating in PRC is a spatially periodic distribution of the optical anisotropy of the crystal, and its amplitude is described by a tensor.

In addition, PRCs exhibit linear birefringence or/and natural optical activity (circular birefring-ence) in the initial state (with no hologram recorded). Therefore, light propagates through PRCs in the form of orthogonally polarized eigenwaves with different refractive indices.

In this chapter we analyze light diffraction from anisotropic gratings recorded in anisotropic media. In particular, tensor amplitudes of photorefractive gratings are calculated, and amplitude and polarization properties of the diffracted light waves, and also selective properties of the holograms in PRCs are considered. In contrast to the next chapter, we neglect here the influence of light waves on the grating. In other words, we confine the analysis to the grating with a constant amplitude.

5.1 Principles of Light Diffraction
from Isotropic Volume Phase Gratings

Let us briefly discuss the major approaches used to describe light diffraction from a volume isotropic phase grating in an optically isotropic transparent medium [5.1-4]. The spatial distribution of the dielectric permeability $\epsilon^\omega(\mathbf{r})$, which is a scalar in this case, is given by

$$\epsilon^\omega(\mathbf{r}) = \epsilon^\omega + \Delta\epsilon^\omega \cos(\mathbf{K}\cdot\mathbf{r}) , \tag{5.1}$$

where ϵ^ω is the average dielectric permeability of a homogeneous unperturbed medium, and $\Delta\epsilon^\omega$ is the amplitude of the phase grating. Vector $\mathbf{K}$ is usually referred to as the wave vector (or simply vector) of a volume grating. $\mathbf{K}$ is normal to the grating layers with equal indices of refraction, and its modulus $|\mathbf{K}|$ is given by $K = 2\pi/\Lambda$, Λ being the grating spacing (Fig. 5.1).

Efficient diffraction of a plane wave from such a grating is observed only when the Bragg condition is satisfied:

$$2\Lambda\sin\theta_B = \frac{\lambda}{n} . \tag{5.2}$$

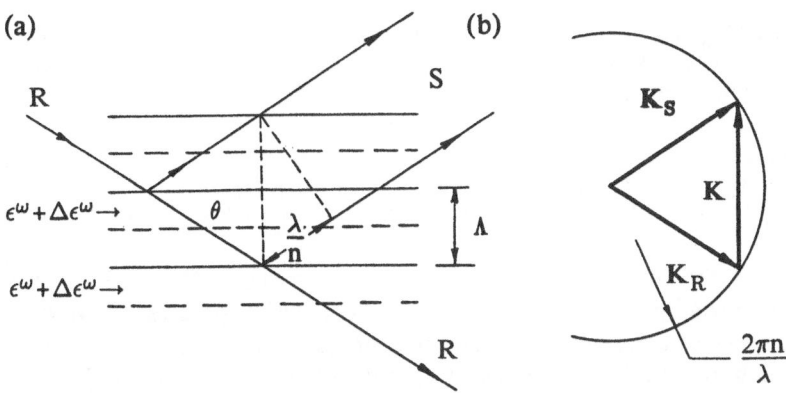

Fig.5.1. (a) Geometrical illustration of Bragg condition. (b) Vector diagram illustrating Bragg condition in an optically isotropic medium

here λ/n is the wavelength of the light within the medium, and θ_B is the Bragg incidence angle. In a more general vector form, this condition is given by

$$\mathbf{K_S} = \mathbf{K_R} \pm \mathbf{K} , \qquad (5.3)$$

where $\mathbf{K_R}$ and $\mathbf{K_S}$ are the wave vectors of plane readout and diffracted (signal) waves, respectively. This means that efficient Bragg diffraction is observed if $\mathbf{K}$ connects two points of the wave vector surface of the original, spatially homogeneous medium (Fig.5.1b) [5.5].

The main parameter to be determined is the maximum intensity of the diffracted beam or the diffraction efficiency of the grating η. Another major question which needs to be answered is how the deviation of the readout-beam parameters (incidence angle θ and wavelength λ) from their Bragg values (5.2) affects the diffraction intensity. This implies an analysis of selective properties (angular and wavelength) of a volume hologram. For discussing these and related problems, two approaches are extensively used at present, namely, the kinematic and the dynamic approximations.

5.1.1 Kinematic Approximation

The kinematic approximation [5.6-8] is applicable only to the case of a low diffraction efficiency ($\eta \ll 1$), when changes in the amplitude of the readout beam, as it propagates through the grating, can be neglected. Of primary concern here is the diffracted-wave amplitude at the grating output rather than the analysis of the light-propagation process within its volume. The approximation provides a simple calculation procedure and an obvious geometrical interpretation. It proves to be an efficient tool for analyzing a wide variety of diffraction phenomena.

The key point of the approach is that we pass from a consideration of the grating (5.1) to an analysis of its three-dimensional spectrum in wave vector space (K_x', K_y', K_z') (Fig.5.2a). In particular, for a homogeneous grating with the vector $\mathbf{K}$ confined to a rectangular box of sizes $L_x \times L_y \times L_z$, the distribution of the spatial Fourier components is described by [5.2,8,9]

$$\Delta\tilde{\epsilon}^\omega(K_x',K_y',K_z')$$
$$\propto \mathrm{sinc}\left[\frac{L_x}{2}(K_x' - K_x)\right]\mathrm{sinc}\left[\frac{L_y}{2}(K_y' - K_y)\right]\mathrm{sinc}\left[\frac{L_z}{2}(K_z' - K_z)\right]. \quad (5.4)$$

Diffraction of the plane light wave with the wave vector $\mathbf{K_R}$ from such a grating is regarded as a sum of independent diffraction processes involving the Bragg spectral components of distribution (5.4), i.e., those components $\mathbf{K_i}$ for which the Bragg condition is rigorously satisfied: $\mathbf{K_{S_i}} = \mathbf{K_R} \pm \mathbf{K_i}$. Their complex amplitudes are proportional to the amplitudes of the respective Fourier components (Fig.5.2b).

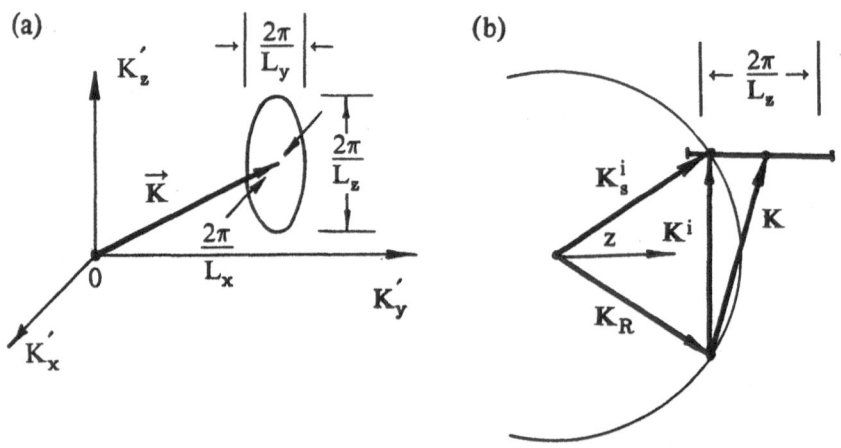

Fig.5.2. (a) Uncertainty of the wave vector of a sinusoidal grating recorded within a limited volume $L_x \times L_y \times L_z$. (b) Determination of Bragg components in the spectrum of spatial frequencies of a volume grating limited along the z axis (L_x, $L_y \rightarrow \infty$)

5.1.2 Dynamic Approximation

Using the dynamic approximation [5.1,3,10-12] we can describe thick gratings with a diffraction efficiency approaching unity, which can ensure a total conversion of the readout beam into the reconstructed one.

The analysis in the dynamic approximation can be carried out in many ways, among which the coupled-wave theory has found most extensive use. This approach consists of analysing a set of coupled linear differential equations describing the amplitudes of the readout and the diffracted waves. As a rule, only two Bragg waves are included in the consideration, and in most cases the results agree strikingly well with the experimental data.

Let us derive a set of coupled-wave equations for the diffraction of a plane light wave from an elementary sinusoidal transmission phase grating (Fig.5.1a). The analysis will be carried out for the Bragg incidence angle when the vector equality (5.3) is strictly satisfied. A general solution for a wave within the grating volume will be sought in the form of a sum of two plane waves, namely a readout and a diffracted wave,

$$A(r) = R\exp(-iK_R \cdot r) + S\exp(-iK_S \cdot r) . \tag{5.5}$$

Their complex amplitudes R and S are considered here to vary across the grating thickness (i.e., they are functions of the z coordinate).

The light field $A(r)$ should evidently satisfy the wave equation [5.5, 13] for a medium with a sinusoidal phase grating (5.1):

$$\{\Delta + (2\pi/\lambda)^2 \ [\epsilon^\omega + \Delta\epsilon^\omega\cos(K \cdot r)]\}A(r) = 0 . \tag{5.6}$$

Independent consideration of the terms with the exponential factors $\exp(-iK_R \cdot r)$ and $\exp(-iK_S \cdot r)$ allows (5.6) to be transformed into a set including two equations:

$$\left[\Delta + \left(\frac{2\pi}{\lambda}\right)^2 \epsilon^\omega\right]R(z)\exp(-iK_R \cdot r) + \left(\frac{2\pi}{\lambda}\right)^2 \frac{\Delta\epsilon^\omega}{2}S(z)\exp(-iK_R \cdot r) = 0 ,$$
$$\tag{5.7}$$
$$\left[\Delta + \left(\frac{2\pi}{\lambda}\right)^2 \epsilon^\omega\right]S(z)\exp(-iK_S \cdot r) + \left(\frac{2\pi}{\lambda}\right)^2 \frac{\Delta\epsilon^\omega}{2}R(z)\exp(-iK_S \cdot r) = 0 .$$

A further simplification can be done for a low grating amplitude $\Delta\epsilon^\omega \ll \epsilon^\omega$, when the complex amplitudes R and S vary rather slowly through the grating thickness. In this case we can neglect the second derivatives $\partial^2 R(z)/\partial z^2$ and $\partial^2 S(z)/\partial z^2$ in comparison with $(2\pi/\lambda)\partial R(z)/\partial z$ and $(2\pi/\lambda) \times \partial S(z)/\partial z$. As a result, for a transmission grating with layers normal to the sample's front face (5.7) is transformed into a fairly simple set of equations widely known as Kogelnik's equations [5.10]:

$$\frac{\partial R(z)}{\partial z} = - i\chi_i S(z) , \tag{5.8a}$$

$$\frac{\partial S(z)}{\partial z} = - i\chi_i R(z) .$$

Here

$$\chi_i = (2\pi/\lambda)^2 \Delta\epsilon^\omega /4K_{Rz} , \tag{5.8b}$$

with $K_{Rz} = (2\pi n/\lambda)\cos\theta_B$.

A direct solution of (5.8) for the conventional boundary conditions $S(0) = 0$, $R(0) = R_0$ yields the well-known expression for the diffraction efficiency

$$\eta_E = \left| \frac{S(d)}{R_0} \right|^2 = \sin^2(\chi_i d) = \sin^2 \left(\frac{\pi \Delta n d}{\lambda \cos\theta_B} \right), \qquad (5.9)$$

where the refractive-index grating amplitude is $\Delta n = \Delta \epsilon^\omega / 2n$. Strictly speaking, (5.9) is valid only for diffraction of E-polarized light waves. As it was shown in [5.10], in an optically isotropic medium the H-polarized and E-polarized waves diffract from the grating (5.1) independently, and

$$\eta_H = \sin^2(\chi_i d \cos 2\theta_B) = \sin^2 \left(\frac{\pi \Delta n d \cos 2\theta_B}{\lambda \cos\theta_B} \right). \qquad (5.10)$$

To obtain η_H, (5.8a) should apparently be replaced by a more general set of vector equations:

$$\mathbf{e}_R \frac{\partial R(z)}{\partial z} = - i\chi_i \mathbf{e}_S S(z) ,$$

$$\mathbf{e}_S \frac{\partial S(z)}{\partial z} = - i\chi_i \mathbf{e}_R R(z) , \qquad (5.11)$$

where $\mathbf{e}_R$ and $\mathbf{e}_S$ are normalized polarization vectors of the corresponding light waves.

5.2 Basic Types of Light Diffraction

Unlike isotropic holographic media, the wave vector surface in the PRC is usually split into two shells inserted one into the other [5.5]. In birefringent (in particular, in ferroelectric) PRCs these shells correspond to the ordinary and extraordinary linearly polarized light waves (linear birefringence $\Delta n_l = n_e - n_o \neq 0$). Such a splitting can also arise in cubic PRCs from both the natural optical activity (circular birefringence $\Delta n_c = n_r - n_l \neq 0$) and linear birefringence induced by an external electric field. In the former case, the split shells correspond to the right and left circularly polarized light waves.

Let us assume that a phase grating ($\mathbf{K}$) is formed in a birefringent (or optically active) medium. As shown in the preceding section, the maximum diffraction efficiency is reached when the Bragg condition (5.3) is satisfied. In contrast to the optically isotropic medium (Fig.5.1b), condition (5.3) can, however, be fulfilled here for four different combinations of wave vectors of the readout ($\mathbf{K}_R$) and diffracted ($\mathbf{K}_S$) plane waves (Fig.5.3), similar to the Bragg diffraction from acoustic waves in birefringent crystals [5.14,15].

As a rule, splitting between the shells of the wave vector surface in PRC ($\simeq 2\pi \Delta n_{l,c}/\lambda$) exceeds appreciably the uncertainty of the grating wave vector resulting from the finite linear sizes of the hologram ($\simeq 2\pi/d$). Thus,

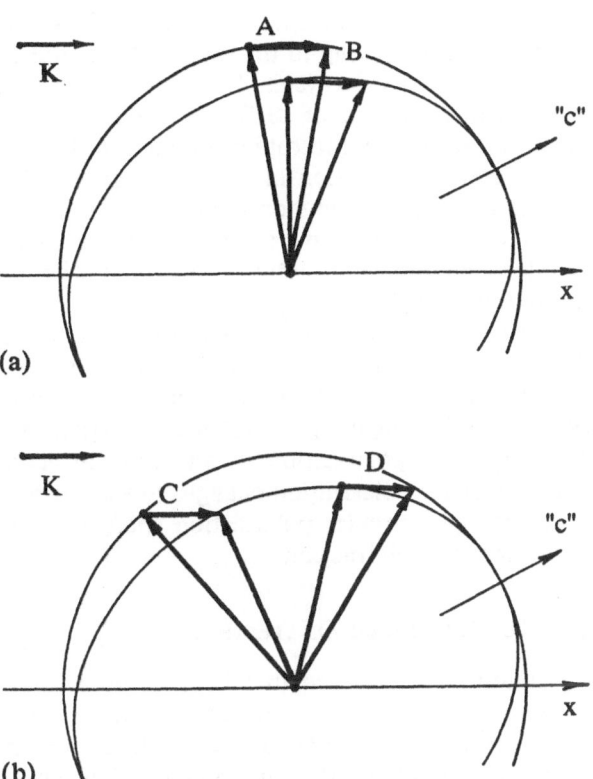

Fig.5.3. Diffraction eigenprocesses in a birefringent crystal (A, B: intramode (isotropic) diffraction; C, D: intermode (anistroptic) diffraction)

all four diffraction processes shown in Fig.5.3 can be regarded as proceeding independently. Each of these "eigenprocesses" of diffraction is unambiguously characterized by the Bragg incidence angle of the readout plane wave and the polarization states of the readout and the diffracted beams.

A particular case of diffraction is the diffraction of linearly H- and E-polarized light waves in an isotropic medium (Fig.5.1). Owing to the absence of the wave-vector surface splitting, the number of diffraction eigenprocesses is here reduced to two. Both of them are observed at the same incidence angle of the readout wave. Along with this, the H and E components diffract independently and thus can be attributed to different diffraction eigenprocesses. As is shown in Sect.5.5, these polarization eigenstates of diffraction in optically isotropic PRCs are not the only ones possible. They are determined, in a general case, not only by the orientation of the incidence plane, but also by anisotropic properties of the photorefractive phase grating itself.

For two of the four diffraction eigenprocesses in a birefringent or optically active photorefractive crystal (A and A in Fig.5.3), the grating wave vector **K** connects the points belonging to the same shell of the wave vector surface. This is the so-called intramode (or isotropic) diffraction, inasmuch as both the readout and diffracted waves belong to the same polarization eigenstate (or mode) of the light waves in the crystal. Two other diffraction processes where the vector **K** connects points of different shells (*C* and *D* in Fig.5.3) are referred to as intermode (or anisotropic) diffraction, because they involve light waves of different polarization modes of the crystal. For a long time, beginning with the first experiments on holographic recording in PRCs, only intramode diffraction processes were investigated and used. The possibility of efficient intermode light diffraction in PRCs and its unique features were first reported in [5.16-20].

One can say that the polarization type of the light wave is preserved in the intramode diffraction and changes in the intermode one. For instance, the intramode diffraction occurs with a fixed type of linear polarization (ordinary or extraordinary) in a birefringent photorefractive crystal and with a fixed direction of the circular polarization in an optically active crystal. On the other hand, the ordinary polarization changes to the extraordinary one (or vice versa) or the right circular polarization changes to the left one (or vice versa) in the intermode diffraction.

5.2.1 The Bragg Conditions for Intermode Diffraction

To illustrate how the Bragg condition is fulfilled in PRCs, let us consider the simplest model example when the wave-vector surface is split into two concentric spheres. Let us also assume that in order to record an elementary sinusoidal transmission grating light waves of the same polarization with a higher refractive index n_1 are used in a symmetric geometry (Fig.5.4).

The intramode diffraction of light waves with the same polarization will obviously be observed at the incidence angle θ' of the recording beams. The second intramode diffraction process (for orthogonally polarized light beams with the refractive index n_2) is also observed at the same angle of incidence θ' outside the crystal (Fig.5.4a).

The intermode diffraction processes from the grating must be observed at other incidence angles. A simple geometrical consideration [5.16] gives the following Bragg incidence angles of the readout waves with higher (θ_1') and lower (θ_2') refractive indices (Fig.5.4b):

$$\theta_1' = \pm \arcsin\left[\sin\theta' + \frac{(n_1 - n_2)(n_1 + n_2)}{4\sin\theta'} \right] ,$$

$$\theta_2' = \pm \arcsin\left[\sin\theta' - \frac{(n_1 - n_2)(n_1 + n_2)}{4\sin\theta'} \right] .$$

(5.12)

Note that at

$$\theta' = \arcsin\sqrt{(n_1 - n_2)(n_1 + n_2)/8} ,$$

(5.13)

θ' proves to be equal to θ_2'. This means [5.16] that during recording of a grating by waves with a low refractive index n_2 under these angles of inci-

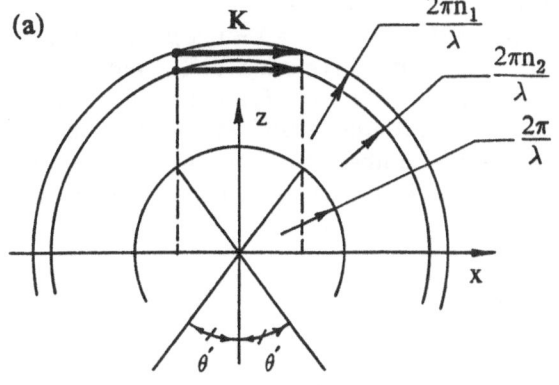

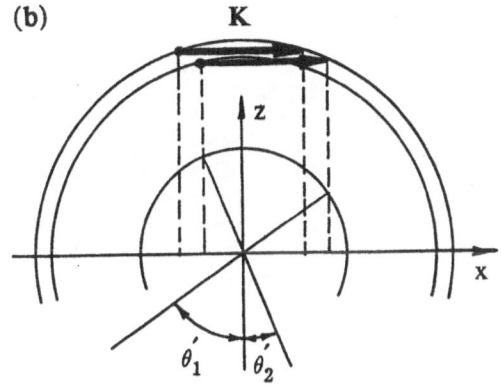

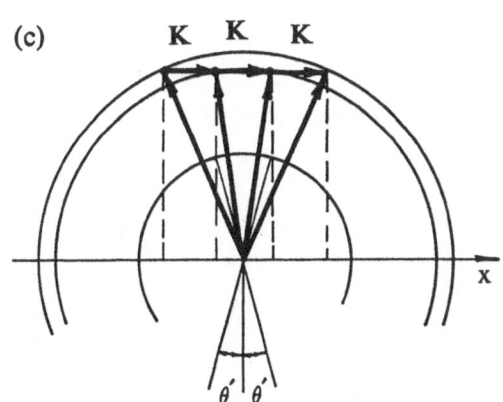

Fig.5.4. (a) Equal Bragg incidence angles θ' for different intramode diffraction processes in a symmetrical experimental geometry. (b) Different Bragg incidence angles for intramode (θ') and intermode ($\theta'_{1,2}$) diffraction processes. (c) Simultaneous observation of Bragg intramode and intermode diffraction for each of the light beams recording the hologram (anisotropic self-diffraction)

dence, two intermode diffraction processes (Fig.5.4c) are also observed simultaneously. In current literature this process is typically called *anisotropic self-diffraction*. Experimentally, it was observed in ferroelectric $LiNbO_3$ [5.16, 22], $BaTiO_3$ [5.23], and $KNbO_3$ [5.24].

In birefringent crystals $\Delta n_1 \sim 10^{-1}$-10^{-2}, and the Bragg incidence angles for intermode and intramode diffraction differ appreciably [5.16,

22-24] (Fig.5.5). Splitting of the wave-vector surface in cubic, optically active crystals is rather small ($\Delta n_c \simeq 10^{-4}$ in BSO at $\lambda = 0.5~\mu m$). Therefore, the Bragg incidence angles for intermode diffraction differ only slightly from the incidence angles of the light beams used for hologram recording or, what is the same, from those for intramode diffraction. This leads to splitting of the Bragg maximum into a number of closely spaced peaks with different polarizations (Fig.5.6) observed, in particular, in experiments with BSO [5.20, 21].

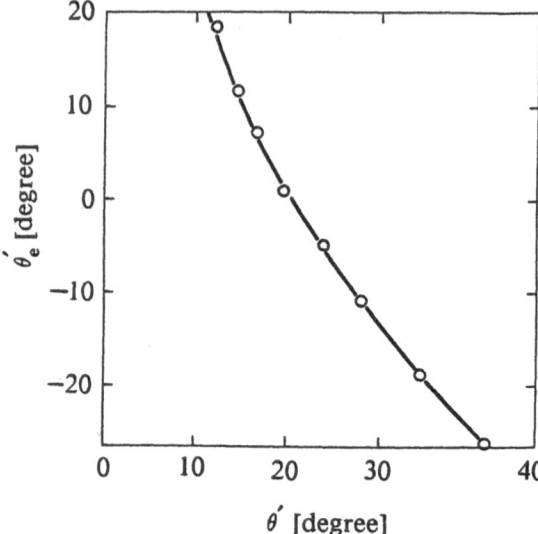

Fig.5.5. Bragg incidence angle θ_e' as a function of incidence angles of recording light beams ($\pm \theta'$) for intermode diffraction in LiNbO$_3$: Fe [5.4] (symmetrical recording geometry, c-axis is normal to the incidence plane, $\lambda = 442$ nm)

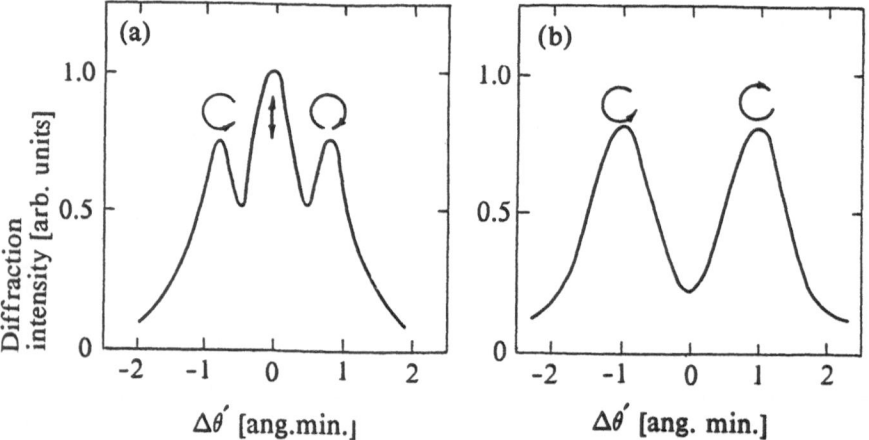

Fig.5.6. Fine structure of the Bragg maximum observed for diffraction of a linearly polarized light beam from an elementary sinusoidal grating in a cubic optically active (110)-cut BSO crystal [5.20]: (a) $\mathbf{K} \parallel [001]$, (b) $\mathbf{K} \parallel [1\bar{1}0]$ ($\lambda = 633$ nm, $d \simeq 7$ mm, arrows show light polarization in the respective diffraction maximum)

5.3 Anisotropic Phase Holograms in PRCs

The second distinguishing feature of PRCs as holographic media is an essentially anisotropic nature of recorded phase holograms. This is a direct consequence of the anistropy of the linear electro-optic effect [5.15,25], which is responsible for transformation of the space-charge electric field $E_{sc}(r)$ into a phase relief. Formally, this means that the amplitude of a photorefractive grating is described by the tensor $\Delta\hat{\epsilon}^\omega$. Such an anisotropic phase grating - in contrast to the refractive index grating (5.1) - is, in fact, nothing else but periodic variations of the local optical anisotropy of the crystal.

As shown above, when an elementary sinusoidal grating is recorded, local photoionization, spatial redistribution, and subsequent trapping of electrons in a PRC result in the formation of the "frozen-in" electric charge grating

$$\rho(r) = \rho\cos(Kr) , \tag{5.14}$$

giving rise to a spatially periodic distribution of the hologram electric field:

$$E_{sc}(r) = E_{sc}\sin(K{\cdot}r) = \rho(K/|K|^2)(\hat{\epsilon}\epsilon_0)^{-1}\sin(K{\cdot}r) . \tag{5.15}$$

Here $\hat{\epsilon}$ is the static dielectric tensor of the crystal.

In turn, the optical properties of the medium are described by the dielectric tensor $\hat{\epsilon}^\omega(r)$ at optical frequencies [5.5,13]. For a photorefractive crystal exhibiting the linear electro-optic effect and for a given spatial distribution of the electric field (5.15) $\hat{\epsilon}^\omega(r)$ can be written as

$$\hat{\epsilon}^\omega(r) = \hat{\epsilon}^\omega + \Delta\hat{\epsilon}^\omega\sin(K{\cdot}r) . \tag{5.16}$$

Here $\hat{\epsilon}^\omega$ is the initial dielectric tensor of a homogeneous crystal without grating, and the phase grating amplitude $\Delta\hat{\epsilon}^\omega$ proves also to be a tensor [5.4,16,19,26] , see also (1.15),

$$\Delta\hat{\epsilon}^\omega = -\hat{\epsilon}^\omega[\hat{r}E_{sc}]\hat{\epsilon}^\omega . \tag{5.17}$$

Note that in this expression $\hat{r}$ is the linear electro-optic tensor of the crystal involved; and the conventional contracted notation is implied:

$$\Delta\epsilon_{ik}^\omega = - \sum_{l=1}^{3} \sum_{n=1}^{3} \sum_{m=1}^{3} \epsilon_{il}^\omega r_{lnm}(E_{sc})_m \epsilon_{nk}^\omega . \tag{5.18}$$

For convenience, we refer to phase gratings of the type (5.16) as anisotropic gratings to distinguish them from isotropic phase gratings (refractive index gratings) (5.1), whose amplitude is a scalar.

5.3.1 Space–Charge Gratings in LiNbO₃

For the sake of illustration, we sketch here the calculation of the grating tensor amplitude $\Delta \hat{\epsilon}^\omega$ for a LiNbO₃ crystal (point group 3m) in a typical holographic orientation with the optical c–axis lying in the sample plane. The grating vector **K** will also be assumed lying in this plane at an arbitrary angle β to the c–axis (Fig.5.7). It is convenient to use the crystallographic

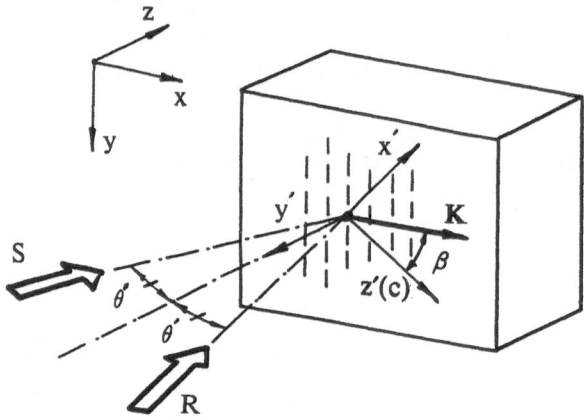

Fig.5.7. Orientation of coordinate and crystallographic axes in the holographic experiment with ferroelectric uniaxial photorefractive crystal of the LiNbO₃ type

coordinate system (x', y', z') where the matrices of tensor $\hat{\epsilon}$, $\hat{\epsilon}^\omega$ and $\hat{r}$ have the simplest form (Table A.2). Under an additional simplifying assumption that the crystallographic y' axis coincides with the normal to the sample surface (Fig.5.7), $\mathbf{E}_{sc}$ is described by the column vector

$$\mathbf{E}_{sc} : \frac{\rho}{K\epsilon_0} \begin{vmatrix} \epsilon_a^{-1}\sin\beta \\ 0 \\ \epsilon_c^{-1}\cos\beta \end{vmatrix} . \tag{5.19}$$

Here ϵ_c and ϵ_a are the static dielectric permeabilities of the crystal along and perpendicular to the c axis. As a result, (5.17) yields the tensor of the anisotropic phase grating amplitude:

$$\Delta\hat{\epsilon}^\omega : \frac{\rho}{K\epsilon_0} \begin{vmatrix} \dfrac{r_{13}n_o^4}{\epsilon_c}\cos\beta & -\dfrac{2r_{22}n_o^4}{\epsilon_c}\sin\beta & \dfrac{r_{51}n_o^2 n_e^2}{\epsilon_a}\sin\beta \\[2mm] -\dfrac{2r_{22}n_o^4}{\epsilon_c}\sin\beta & \dfrac{r_{13}n_o^4}{\epsilon_c}\cos\beta & 0 \\[2mm] \dfrac{r_{51}n_o^2 n_e^2}{\epsilon_a}\sin\beta & 0 & \dfrac{r_{33}n_e^4}{\epsilon_c}\cos\beta \end{vmatrix} . \tag{5.20}$$

74

An analogous expression for $\Delta\hat{\varepsilon}^\omega$ in a similar orientation of the birefringent photorefractive BaTiO$_3$ (point group 4mm) can immediately be obtained from (5.20) by substituting $r_{22} = 0$.

As shown in Sect.4.2, gratings with a fixed amplitude of the charge distribution $\rho(r)$ are formed in PRCs when we can ignore the influence of the hologram electric field on the spatial redistribution of the photoinduced charge carriers. Among these cases are, for instance, the initial stage of hologram recording where the amplitude E_{sc} grows linearly with exposure, (4.15), or a steady-state regime of recording with a complete filling of vacant trapping centers (4.24) (violation of quasi-neutrality).

5.3.2 Electric-Field Gratings in LiNbO$_3$

Of major practical interest, however, is the steady-state regime of recording without quasi-neutrality violation. It takes place when the normal component (along the grating vector $\mathbf{K}$) of $E_{sc}(r)$ is high enough for compensation of spatially nonuniform currents of mobile carriers excited by the interference pattern $I(r)$. Because of the drift currents perpendicular to $\mathbf{K}$, the tangential component of $E_{sc}(r)$ vanishes and a spatial distribution of the hologram field almost parallel to $\mathbf{K}$ is formed. Its steady-state value is independent of the crystallographic orientation of the sample and is determined by the grating spacing Λ (for diffusion recording) or the magnitude of the normal component (parallel to $\mathbf{K}$) of the field E_0 or E_G alone.

It is quite natural that for the grating with the field amplitude E_{sc} directed along $\mathbf{K}$ the tensor $\Delta\hat{\varepsilon}^\omega$ differs markedly from that calculated in Sect.5.3.1. Specifically, for the orientation discussed above, (5.20) transforms for the LiNbO$_3$ crystal into

$$
\Delta\hat{\varepsilon}^\omega : E_{sc} \begin{vmatrix} r_{13}n_o^4\cos\beta & -2r_{22}n_o^4\sin\beta & r_{51}n_o^2n_e^2\sin\beta \\ -2r_{22}n_o^4\sin\beta & r_{13}n_o^4\cos\beta & 0 \\ r_{51}n_o^2n_e^2\sin\beta & 0 & r_{33}n_e^4\cos\beta \end{vmatrix} . \tag{5.21}
$$

The difference between (5.20 and 21) obtained for a specified amplitude of charge ρ and field E_{sc} is caused by the anisotropy of the static dielectric permeability of the crystal. Therefore, it is typical only of birefringent crystals and is absent in originally isotropic cubic PRCs.

Along with this, the cases of charge and field gratings in birefringent crystals discussed are the limiting approximations that are most easily analyzed theoretically. Generally, formation of anisotropic phase gratings occupying an intermediate position between (5.20 and 21) is possible in such crystals. Moreover, all photorefractive crystals are piezoelectrics, and as a result $E_{sc}(r)$ induces spatially nonuniform mechanical stresses in the volume of a sample. That is why for an accurate calculation of $\Delta\hat{\varepsilon}^\omega$ the contribution of the photoelastic effect should also be taken into account [5.27].

5.4 Diffraction Efficiency of Anisotropic Phase Holograms in Birefringent PRCs

As shown in the preceding section, an original anisotropy of a birefringent crystal results in four independent processes of diffraction from the same sinusoidal grating (Fig.5.3). In the weak grating approximation when $\Delta\epsilon^\omega/2n \ll \Delta n_1$, the diffraction grating can be regarded as a perturbation of the next order of smallness, as compared with crystal's natural anisotropy. Thus the linearly polarized (ordinary or extraordinary) eigenwaves of the uniform crystal can be used in the diffraction analysis as the basic waves. As a consequence, polarizations of the readout (e_R) and diffracted (e_S) light beams are predetermined and rigidly specified for each of the four possible eigenprocesses of diffraction.

Substitution of $\Delta\hat{\epsilon}^\omega$ for $\Delta\epsilon^\omega$ in (5.8b, 11) and obvious simplifying manipulations yield the coupled-wave equations

$$\frac{\partial R(z)}{\partial z} = -i\left[\frac{\pi(e_R^* \Delta\hat{\epsilon}^\omega e_S)}{2n\lambda\cos\theta_R}\right]S(z) ,$$

$$\frac{\partial S(z)}{\partial z} = -i\left[\frac{\pi(e_S^* \Delta\hat{\epsilon}^\omega e_R)}{2n\lambda\cos\theta_S}\right]R(z) . \tag{5.22}$$

By solving them, we obtain the general expression for the efficiency of the diffraction eigenprocess [5.4, 16, 19]:

$$\eta = \sin^2[|\chi|d] = \sin^2\left[\frac{\pi|e_R^* \Delta\hat{\epsilon}^\omega e_S|d}{2n\lambda}\right] . \tag{5.23}$$

Here we used the symmetry of tensor $\Delta\hat{\epsilon}^\omega$ $[(e_R^* \Delta\hat{\epsilon}^\omega e_S) = (e_S^* \Delta\hat{\epsilon}^\omega e_R)^*]$ and ignored the difference between $\cos\theta_{R,S}$ and 1.

5.4.1 Orientation Dependences of Amplitudes of Diffraction Eigenprocesses in LiNbO₃

Calculation of η for a given process of diffraction, in an arbitrary orientation, is fairly complicated [5.28]. We therefore present here only the plots of the coefficients χ for different eigenprocesses of diffraction observed in the most common photorefractive birefringent crystal LiNbO₃. For simplicity, the geometry of recording is assumed to be symmetric ($K\|x$), and the spatial frequency of the grating is assumed to be low ($\theta \ll 1$).

The holographic recording using LiNbO₃ is usually performed in X or Y cuts, where the c axis is in the sample plane (Fig.5.7). Both the mechanism of hologram recording via the linear photovoltaic effect ($E_G\|c$-axis) and the electro-optic coefficient r_{33}, which is the largest in LiNbO₃, are efficiently utilized in this geometry.

Let us analyze the behavior of amplitudes of all diffraction processes – intramode diffraction with extraordinary (χ_e) and ordinary (χ_o) light polarizations, and also intermode diffraction (χ_a) – on rotation of the LiNbO$_3$ sample about the normal to its front face. It is known that the steady–state regime is not usually reached in holographic recording in LiNbO$_3$ crystals. Owing to a highly efficient (though not too "fast") photovoltaic recording mechanism and fairly good electro-optic properties, the diffraction gratings with high efficiencies are here produced at the initial stage of recording. Thus, the grating with a specified amplitude of charge distribution (5.20) appears to be most appropriate for LiNbO$_3$ crystals. As a result, direct substitution of the matrix $\Delta\hat{\epsilon}^\omega$ (5.20) into (5.23) yields the following amplitudes of the diffraction eigentypes (Fig.5.8a) [5.19, 28]:

$$\chi_o \simeq \frac{\pi\rho}{2K\epsilon_c\epsilon_0}\frac{r_{13}n^3}{\lambda}\cos\beta \,,$$

$$\chi_e \simeq \frac{\pi\rho}{2K\epsilon_c\epsilon_0}\frac{r_{33}n^3}{\lambda}\cos\beta \,, \tag{5.24}$$

$$\chi_a \simeq \frac{\pi\rho}{2K\epsilon_a\epsilon_0}\frac{r_{51}n^3}{\lambda}\sin\beta \,.$$

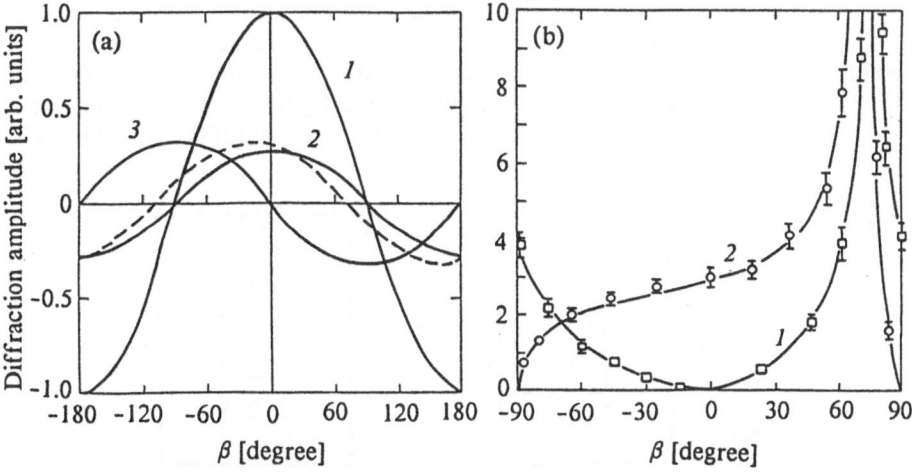

Fig.5.8. (a) Theoretical angular dependences of amplitudes of different diffraction processes in LiNbO$_3$ for a specified charge grating amplitude in the orientation shown in Fig.5.7. *1* intramode diffraction (extraordinary polarization); *2* intramode diffraction (ordinary polarization); *3* intermode diffraction. When x′ ∥ z, curve *2* behaves, as shown by a dashed line [5.28]. (b) Experimental relative efficiencies of different diffraction processes in LiNbO$_3$:Fe [5.19, 28] (λ = 633nm, d = 2mm, K = 5.5 10^{-4}cm^{-1}). *1*: $\sqrt{\eta_a}/\sqrt{\eta_o}$; *2*: $\sqrt{\eta_e}/\sqrt{\eta_o}$

Here we took n_o, $n_e \simeq n$, and also took into account the fact that for a small angle θ the extraordinary wave is nearly polarized along the c-axis (z'), and the ordinary wave along the x' axis (Fig.5.7).

Thus intramode diffraction from a charge grating of a specified amplitude ρ in LiNbO₃ is most efficient when the optical c-axis lies in the incidence plane ($\beta = 0$). On the other hand, the efficiency of the intermode diffraction is the highest in the orthogonal orientation of the sample ($\beta = \pm 90°$). Note that the experimental data [5.19, 28] on the orientation dependences of χ_o, χ_e and χ_a are in fairly good agreement with the result of the analysis (Fig.5.8b).

5.4.2 Efficient Intramode Diffraction in BaTiO₃

In conclusion, we consider the question of primary importance for applications of BaTiO₃ crystals. When PRCs are used as dynamic holographic media, of major interest are the intramode diffraction processes with preserved polarization of the diffracted light. For the orientation discussed above (Fig.5.7), an extremely high electro-optic coefficient r_{51} of BaTiO₃ proves to be useless from the point of view of intramode diffraction processes. It can, however, be used in an essentially asymmetric recording geometry, as it was shown in [5.29].

To illustrate this technique, let us discuss a simple symmetric recording geometry with the grating vector **K** parallel to the x axis. The c axis is also assumed to be in the incidence plane, yet at an arbitrary angle δ to the sample's surface (Fig.5.9). When dealing with the problem of diffraction from a grating with a specified field amplitude $\mathbf{E}_{sc} \| \mathbf{K}$, which is most typical of the diffusion recording in BaTiO₃, the tensor amplitude of the phase

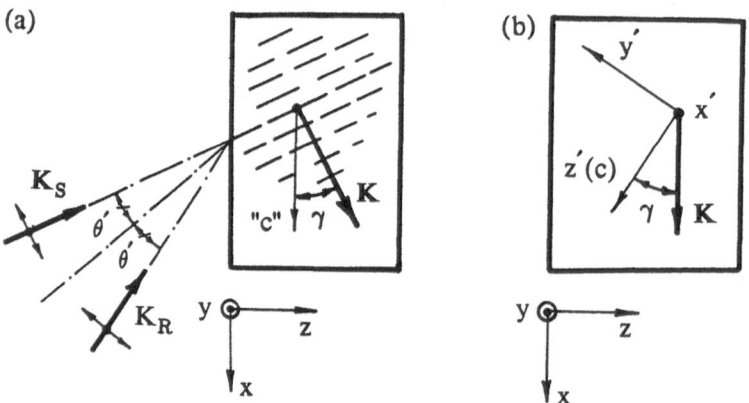

Fig.5.9. (a) Asymmetrical geometry of holographic recording in photorefractive BaTiO₃ [5.29]. (b) Orientation of coordinate and crystallographic axes for theoretical consideration of the intramode diffraction efficiency for extraordinary light beams in BaTiO₃

grating, $\Delta\hat{\varepsilon}^\omega$, in the crystallographic coordinate system is

$$\Delta\hat{\varepsilon}^\omega: E_{sc} \begin{vmatrix} r_{13}n_o^4\cos\delta & 0 & 0 \\ 0 & r_{13}n_o^4\cos\delta & r_{51}n_o^2n_e^2\sin\delta \\ 0 & r_{51}n_o^2n_e^2\sin\delta & r_{33}n_e^4\cos\delta \end{vmatrix}. \qquad (5.25)$$

A direct substitution of (5.25) and the polarization vector components of the extraordinary waves $e_{R,S}$ $(0, \sin\delta, \cos\delta)$ into (5.25) yields the amplitude of the intramode diffraction process for the light waves of extraordinary polarization:

$$\chi_e \simeq \frac{\pi}{2}E_{sc}n^3(r_{13}\cos\delta\sin^2\delta + 2r_{51}\sin^2\delta\cos\delta + r_{33}\cos^3\delta). \qquad (5.26)$$

Since the electro-optic coefficient r_{51} for BaTiO$_3$ is well above r_{13} and r_{33}, the maximum value

$$\chi_e^{max} \simeq \frac{\pi}{2}E_{sc}n^3r_{51}\frac{4}{3\sqrt{3}} \qquad (5.27)$$

is reached at $\delta = \pm\arcsin\sqrt{2/3} \simeq \pm55°$, maximizing the product $\sin^2\delta\cos\delta$.

5.5 Diffraction Efficiency of Anisotropic Phase Holograms in Cubic PRCs

In contrast to birefringent crystals, different approaches for the description of diffraction in cubic PRCs are to be used. For instance, though cubic crystals do not exhibit linear birefringence, a natural optical activity (i.e., circular birefringence Δn_c) can be observed in crystals of point group 23. In widely used cubic BSO and BGO crystals, Δn_c can be as high as 10^{-4} (at $\lambda \simeq 0.5$ μm). The amplitude of the phase grating does not here exceed $(n^3r_{41}/2)E_0 \simeq 2\cdot10^{-5}$, where $E_0 \sim 10$ kV cm^{-1} is a typical value of the electric field applied to the sample. That is why the diffraction processes in these cubic optically active crystals can be treated using the approach employed in Sect.5.4 for birefringent PRCs.

On the other hand, wave-vector surface splitting is absent in cubic, optically inactive crystals of point group $\bar{4}$3m (where photorefractive semiconductors GaAs, InP, and CdTe belong) in the absence of external field. The only difference from the case discussed in Sect.5.1 is evidently that the phase grating in the PRC is anisotropic (5.16).

We shall demonstrate later how to calculate the diffraction efficiency of anisotropic phase holograms in cubic PRCs for these two limiting approximations. Along with these, the third intermediate case when the phase-grating amplitude and birefringence of the original crystal are of the same order of smallness ($\Delta n_c \sim \Delta\epsilon^\omega/2n$) is also fairly typical of cubic PRCs. Such a situation can arise when, for instance, an external electric field is applied to a crystal to make it optically anisotropic, as well as in the case of an insufficiently high optical activity. Calculations of the amplitude-polarization characteristics of the diffracted light in BSO presented in [5.21] can be given as an example of the analysis in this approximation. A more thorough analysis of diffraction processes in optically active cubic PRCs with an externally induced linear birefringence, was given in recent papers [5.30].

5.5.1 Cubic Crystals with High Optical Activity

Let us consider now a cubic optically active BSO crystal in the most typical holographic orientation (Fig. 5.10) when the sample is cut in the (110) crystallographic plane [5.31]. Allowing for the fact that in cubic crystals $\mathbf{E_{sc}} \parallel \mathbf{K}$, and the tensor $\hat{r}$ has the form given in Table A.6, the grating amplitude $\Delta\hat{\epsilon}^\omega$ in the crystallographic coordinate system ($[001]$, $[1\bar{1}0]$, $[110]$) is

$$\Delta\hat{\epsilon}^\omega : E_{sc}\, n^4 r \begin{vmatrix} 0 & -\sin\beta & 0 \\ -\sin\beta & -\cos\beta & 0 \\ 0 & 0 & \cos\beta \end{vmatrix}. \tag{5.28}$$

Here β is the angle between the vector $\mathbf{K}$ that lies in the incidence plane and the crystallographic [001] axis.

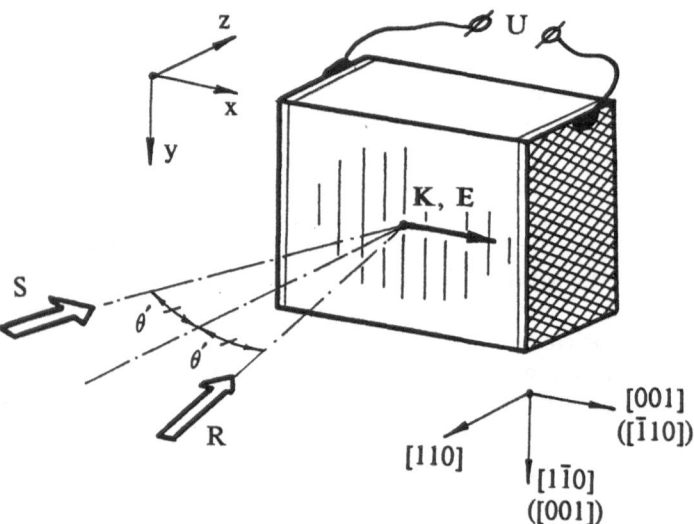

Fig.5.10. Conventional holographic orientations of cubic photorefractive crystals (BSO, BTO, GaAs, etc.) [5.31]

For modest incidence angles ($\theta \ll 1$), the normalized polarization vectors for the left and right circularly polarized waves in optically active cubic crystal are:

$$\mathbf{e}_l : \frac{1}{\sqrt{2}} \begin{vmatrix} 1 \\ i \\ 0 \end{vmatrix} ; \quad \mathbf{e}_r : \frac{1}{\sqrt{2}} \begin{vmatrix} 1 \\ -i \\ 0 \end{vmatrix} . \tag{5.29}$$

Direct substitution of (5.28,29) into (5.23) yields the expressions [5.21] for the amplitudes of intramode (for right-polarized (χ_r) and left-polarized (χ_l) light) and intermode diffraction process (χ_a):

$$\chi_r = \chi_l = -\frac{\pi}{2} E_{sc} \left(\frac{r_{41} n^3}{2\lambda} \right) \cos \beta ,$$

$$\chi_a = \frac{\pi}{2} E_{sc} \left(\frac{r_{41} n^3}{2\lambda} \right) (\cos \beta \mp 2i \sin \beta) . \tag{5.30}$$

Thus, all four diffraction processes prove to be equally efficient at $\beta = 0$ (K $\parallel$ [001]). For the orthogonal grating orientation (K $\parallel$ [1$\bar{1}$0], $\beta = 90°$), the amplitudes of intramode processes equal zero; but the intermode processes become twice as efficient as before. The imaginary intermode diffraction amplitude in (5.30) means an additional quarter-wave phase shift in the reconstructed wave with respect to the readout one. Note that the experimental evidence on the fine structure of the Bragg maximum experimentally observed in BSO (Fig.5.5) [5.20] is consistent with the conclusions of this theoretical analysis.

5.5.2 Cubic Crystals with a Nonsplit Wave–Vector Surface

Let us consider the same holographic orientation of a cubic PRC with a nonsplit wave-vector surface. As discussed in Sect.5.1, the diffraction from an isotropic grating (5.1) may be analyzed in the dynamic approximation. Two diffraction eigenprocesses, namely, H-type (both waves $\mathbf{K}_R$ and $\mathbf{K}_S$ are linearly polarized in the incidence plane) and E-type (the waves are polarized perpendicular to the incidence plane) can be considered separately. It is natural to suppose that a similar pair of orthogonally polarized diffraction eigenprocesses exists in the case under discussion as well. But the relevant waves are polarized in a different manner, owing to the anisotropy of the photorefractive grating (5.16).

Detailed calculations of their polarizations and the efficiencies of the diffraction processes for a general case were given elsewhere [5.4,19]. For the case of small incidence angles considered here, the readout and diffracted waves are, however, polarized nearly in the crystal plane (x, y). As a result, the diffraction eigenprocesses prove to be polarized along the principal axes of the matrix (5.28)

$$\begin{vmatrix} 0 & -\sin\beta \\ -\sin\beta & -\cos\beta \end{vmatrix}.$$

The theoretical dependences of their polarizations, respective diffraction amplitudes, and also data on their experimental verification in BTO are given in Fig.5.11 [5.32].

Two typical orientations of as-cut cubic PRC samples, i.e., $\beta = 0$ or $\beta = 90°$ [5.21, 31, 33] are usually used in practice. In the former case ($K \parallel [001]$), the diffraction eigenprocesses are H-polarized ($\chi_H = 0$) and E-polarized ($\chi_E = -(\pi/2)E_{sc}r_{41}n^3/\lambda$). With the orthogonal grating orientation ($K \parallel [1\bar{1}0]$), the polarization eigenstates of the light waves are at $\pm 45°$ to the incidence plane, the amplitudes of both diffraction processes being equal in magnitude and opposite in sign ($\chi_{+45°} = -\chi_{-45°} = -(\pi/2)E_{sc}r_{41}n^3/\lambda$).

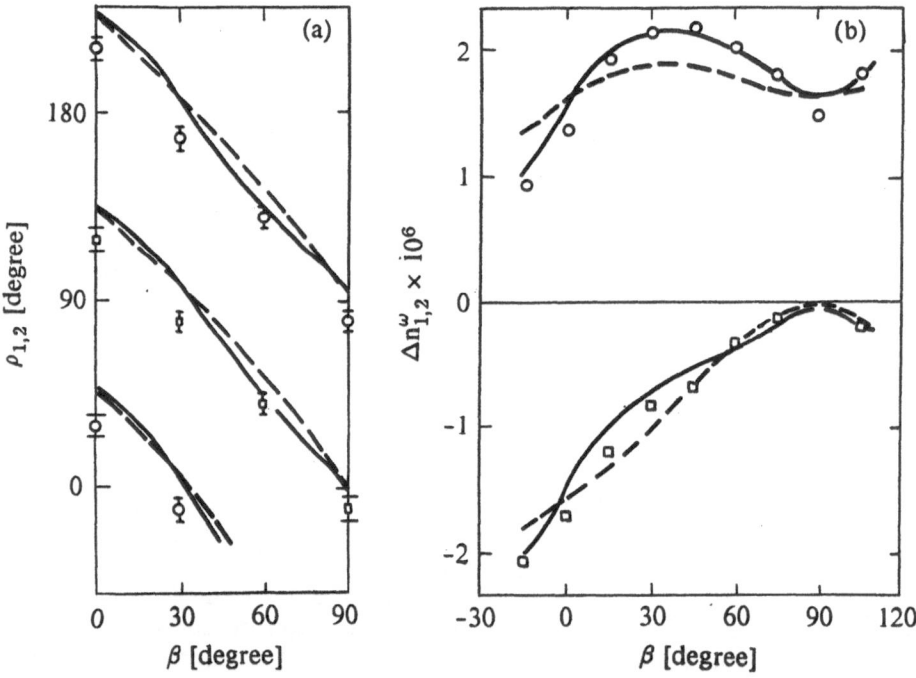

Fig.5.11. Experimental polarizations (a) and amplitudes (b) of diffraction eigenprocesses vs. angle of the sample rotation for an anisotropic grating in BTO [5.32] (β is the angle between the $[1\bar{1}0]$ axis and the incidence plane, polarization angles ψ are also counted off the incidence plane, $\lambda = 633$ nm, $E_0 = 0$, $\Lambda = 3.9$ μm, $E_D \simeq 0.4$ kV/cm, $\rho d/2 \simeq 10°$). Solid and dashed lines show theoretical curves calculated for $E_{sc} = 0.4$ kV/cm with and without taking into account contribution from photoelasticity

Upon reconstruction of this anisotropic grating with a linearly polarized (H or E) light beam, the diffracted +45° and -45° components prove to be out-of phase. It results obviously in a 90° rotation of initial polarization of the reconstructing beam [5.21,34]. Note that in this case, rotation of polarization in a diffracted beam is not connected with an intermode diffraction, but is a result of two simultaneous intermode diffraction processes.

During recording of a dynamic hologram in a cubic PRC oriented in such a manner, this effect enables observation of a "polarization transfer" [5.35] in the recording waves. For a PRC with diffusion mechanism this means simultaneous rotation of linear output polarizations of the recording beams from their initial state (H or E) [5.35.36]. For a drift hologram recorded in an external DC electric field E_0, this effect results in an efficient reduction of ellipticity in their output polarizations [5.35].

Note that the maximum efficiency of the diffraction process in cubic PRCs is achieved in this cut at $\mathbf{K} \parallel [1\bar{1}1]$ for H-polarized waves ($\chi_H^{max} = -(\pi/\sqrt{3})E_{sc}r_{41}n^3/\lambda$) [5.32].

5.6 Selective Properties of Volume Phase Holograms

Conventional volume holograms are characterized by a single diffraction order that propagates, as a rule, in the direction of the signal beam participating in recording of the hologram. The intense diffraction is observed only in a neighborhood of the Bragg condition (5.3), and therefore the volume diffraction structure selects some typical ("Bragg") combinations from all possible wavelengths and incidence angles of the readout beam. In particular, if the incidence angle of the readout light is kept constant, the volume grating selects a corresponding Bragg wavelength, and we can speak about the wavelength (spectral) selectivity of the grating. On the other hand, if the readout wavelength is specified, the volume grating demonstrates its angular selective properties.

The major quantitative parameters related to the properties under discussion are the ranges of allowable deviations of wavelengths or incidence angles from their Bragg values. We mean actually the spectral and angular widths of the Bragg maximum which were given in Sect.2.2 for the isotropic phase grating in an isotropic medium (2.15,16). In most cases, these relations can also be used for the volume holograms in PRCs.

Some comment should be made on the angular selectivity of holograms recorded in PRCs. Indeed, (2.15) was obtained for the volume grating that is infinite in the xy plane, while the linear sizes of the hologram volume in PRCs are, as a rule, finite and nearly the same along all three orthogonal directions ($L_x \simeq L_y \simeq L_z = d$). It proves nevertheless to be applicable because of a small angular divergence ($\simeq \lambda/d$) of the wave diffracted from the sample aperture (d x d) as compared with (2.15).

5.6.1 Anomalous Selective Properties of Intermode Diffraction

There are, however, some specific geometries of intermode (anisotropic) diffraction with selective properties differing dramatically from those of isotropic diffraction [5.20, 24, 37]. One of the most important examples is given in Fig. 5.12a. While the angular selectivity of this anisotropic diffraction process remains the same, its wavelength selectivity upon readout by the wave S (with a lower refractive index n_2) is twice as high as that predicted by (2.16). On the other hand, the wavelength selectivity on readout by the wave R (with a higher refractive index n_1) proves to be much poorer as compared with (2.16). For the first time, this phenomenon was experimentally verified using the birefringent $LiNbO_3$:Fe (Fig. 5.13).

A general theoretical analysis of selective properties of anisotropic diffraction [5.20] gives the following half-widths of the Bragg maximum:

$$\Delta\theta = \frac{\lambda}{2nd\sin\theta_1}, \quad \frac{\Delta\lambda}{\lambda} \simeq \frac{\lambda}{2d\sin\theta_2}. \tag{5.31}$$

Here the characteristic angles θ_1 and θ_2 are shown in Fig. 5.12b. Note that the consideration was performed in the kinematic approximation for the grating of equal linear sizes along all directions ($L_x = L_y = L_z = d$) and for

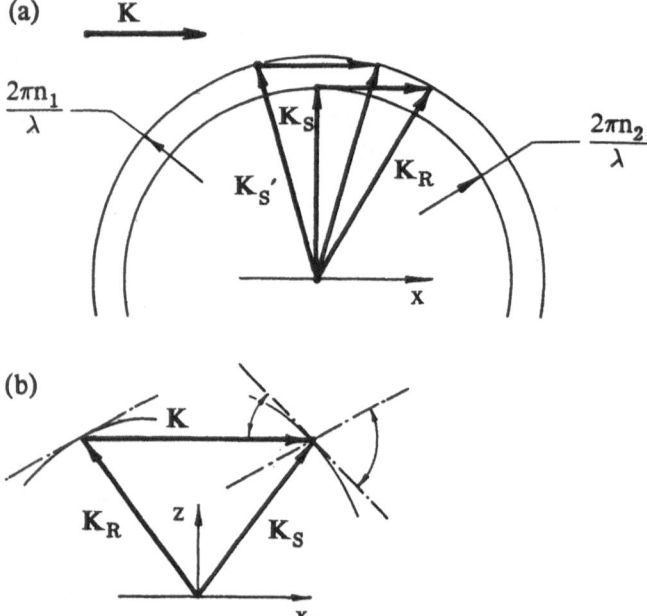

Fig. 5.12. (a) Bragg intermode diffraction with a reduced wavelength selectivity. (b) Vector diagram illustrating the theoretical analysis of angular and wavelength selectivity of diffraction processes in a PRC

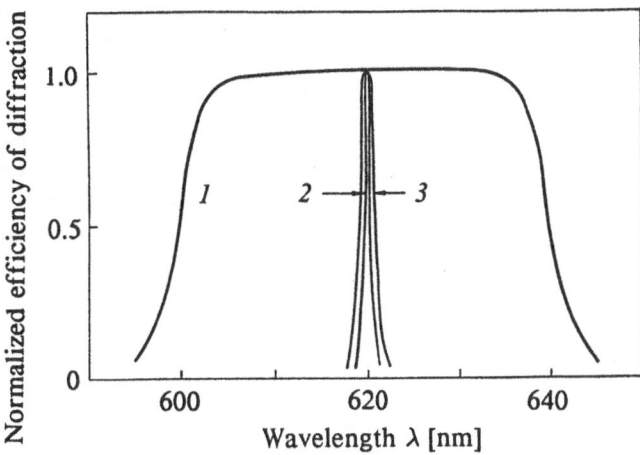

Fig.5.13. Wavelength dependences of the light intensity diffracted from an elementary sinusoidal grating in the geometry shown in Fig.5.12a [5.20]. *1*: intermode diffraction, readout by the wave R; *2*: intermode diffraction, readout by the wave S; *3*: intramode diffraction readout by the wave S′ ($LiNbO_3$:Fe, $K \simeq 6.8 \cdot 10^4 cm^{-1}$, $d \simeq 2mm$)

not too small angles θ_1 and θ_2, when the curvature of segments l_1 and l_2 in Fig.5.12b can be neglected.

Specifically, for the geometry given in Fig.5.12b upon reconstruction by the wave R, $\theta_2 = 0$ and $\Delta\lambda$ should go to infinity in accordance with (5.31). On the other hand, upon reconstruction by the wave S, $\theta_2 \simeq 2\theta_B$ and the wavelength selectivity turns out to be twice as high as that for the usual isotropic diffraction when $\theta_2 = \theta_B$. Note that in all cases $\theta_1 \simeq 2\theta_B$ and the angular selectivity remains unchanged.

5.6.2 Wideband Reconstruction of Volume Holograms at Different Wavelengths

In conclusion, we discuss one of the major problems of volume holography, namely, reconstruction of a hologram of a complicated wavefront at a different wavelength. Let us consider the simplest case of an elementary sinusoidal grating **K**. A change for another wavelength λ_2 requires that the incidence angle of the readout plane wave be markedly altered to satisfy the new Bragg condition (Fig.5.14a). The needed change in the incidence angle depends, however, not only on λ_1 and λ_2, but also on the spatial frequency of the grating.

The volume hologram of a complicated image consists of a whole set of elementary sinusoidal gratings with different magnitudes and orientations of the wave vectors. Therefore, it is impossible to reconstruct it by merely changing the incidence angle. The angular half-width of the object beam reconstructed at λ_2 in such a way is determined by an angle $\Delta\theta$ within which the misalignment between the crossing surfaces of the wave vec-

(a)

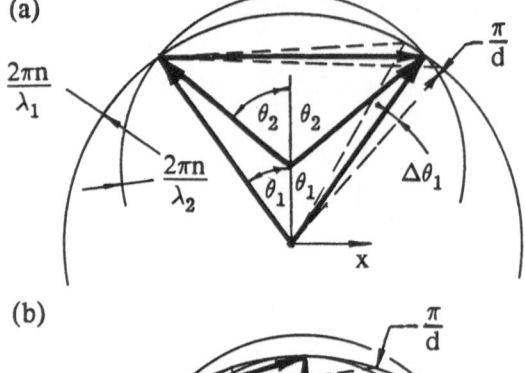

$\dfrac{2\pi n}{\lambda_1}$

$\dfrac{2\pi n}{\lambda_2}$

$\theta_2 \; \theta_2$

$\theta_1 \, \theta_1$

$\Delta\theta_1$

$\dfrac{\pi}{d}$

x

Fig.5.14. (a) Vector diagram illustrating calculation of spatial frequency band of a volume hologram for readout at a different wavelength. (b) Similar diagram for a wideband geometry (readout via intermode diffraction)

(b)

$\dfrac{2\pi \Delta n}{\lambda_2}$

$\dfrac{\pi}{d}$

$\Delta\theta_1$

x

tors corresponding to λ_2 and λ_1 does not reach $\simeq \pi/d$ (Fig.5.14a). For small Bragg angles ($\theta_1, \theta_2 \ll 1$) when $\theta_2 \simeq \theta_1 \lambda_2/\lambda_1$, the angular half-width of the reconstructed light beam does not exceed [5.17, 18]

$$\Delta\theta \simeq \frac{\pi}{d} \bigg/ \frac{2\pi n}{\lambda_2}\,(\theta_2 - \theta_1) \simeq \frac{\lambda_1 \lambda_2}{2nd\theta_1(\lambda_2 - \lambda_1)}. \tag{5.32}$$

For typical values $\theta_1 \simeq 0.05$, $\lambda_1 = 0.5\ \mu m$, $\lambda_2 = 1.0\ \mu m$, n = 2, and d = 2mm, $\Delta\theta$ is as small as $\simeq 2\cdot10^{-3}$. This corresponds to a resolution of about 10 line pairs/mm in the reconstructed image.

As (5.32) predicts, the maximum resolution in the reconstructed image is reached at $\theta_2 - \theta_1 \to 0$. For an isotropic medium this condition means the coaxial Gabor geometry [5.1]. Note that to estimate $\Delta\theta$ in this limiting case, we must take into account the difference in the curvatures of wave surfaces at the point of their contact, this yields

$$\Delta\theta \simeq 2\sqrt{\frac{\lambda_1^2}{nd(\lambda_2 - \lambda_1)}}. \tag{5.33}$$

For the values of n, d, λ_1, and λ_2 used in the previous estimate, the resolution in the reconstructed image reaches $\simeq 100$ line pairs/mm, which is quite sufficient for most practical situations. A strong zero diffraction order in the center of the reconstructed image restricts, however, the applicability of this geometry.

As was shown in [5.17, 18], by using anisotropic diffraction in birefringent PRCs, the highest possible bandwidth of a volume hologram (5.33) read out at a different wavelength can also be obtained in an essentially different geometry (Fig.5.14b). Both signal waves (recorded at λ_1 and reconstructed at λ_2) are collinear in this configuration, as in the conventional Gabor scheme. In contrast to the latter geometry the signal and reference waves (at λ_2) are, however, noncollinear, and the zero diffraction order and the reconstructed image are separated. Note that an additional SNR improvement can be obtained here using an output polarizer adjusted to suppress the zero diffraction order, which is polarized orthogonally to the reconstructed signal wave.

The angles of incidence of the recording ($\lambda_1 = 442$nm) and reconstructing ($\lambda_2 = 633$nm) waves, and orientation of the birefringent photorefractive $LiNbO_3$: Fe in the configuration under discussion have been given in [5.18]. The experimental evidence for the improvement in the quality of the reconstructed image of the resolution chart is presented in Fig.5.15.

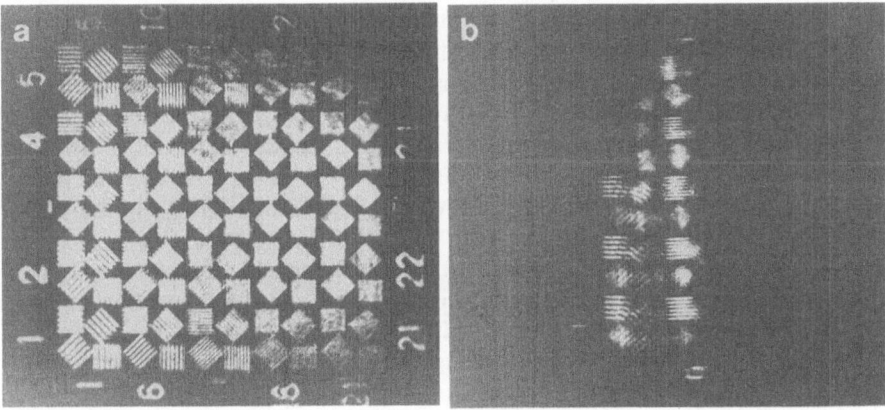

Fig.5.15. Results of reconstruction of a complicated image from the volume hologram recorded in $LiNbO_3$:Fe at $\lambda_1 = 442$ nm in the geometry of Fig.5.14b [5.17, 18]. (a) Readout at $\lambda_2 = 633$ nm using intermode diffraction, and (b) the same, using intramode diffraction

6. Fundamentals of Dynamic Holography

The photorefractive crystals discussed in this book belong to dynamic holographic media. This means that the recorded holograms (the so-called dynamic holograms) do not require development and can be reconstructed using an additional auxiliary beam immediately in the process of recording. This somewhat simplified approach was characteristic of the first studies [6.1-3], where dynamic holograms formed in PRCs were described using the parameters introduced earlier for conventional stationary holograms, such as diffraction efficiency η and sensitivity S.

The turning point in the development of dynamic holography was connected with understanding of the fact that the recording light waves themselves diffract from the recorded dynamic hologram. This changes the process of hologram formation noticeably because, by affecting the recording beams, the hologram changes the course of its further recording. Owing to the presence of the above effects, the PRC may be regarded as a specific case of a nonlinear optical medium where an effect of the "light by light scattering" [6.4] is observed. As will be shown later, a more adequate approach requires that new parameters be used instead of traditional holographic parameters. In further analysis we use a complex coupling constant γ as a universal parameter which, with due account for its dependence on frequency detuning between recording beams ($\Delta\omega$), allows description of a wide range of phenomena in dynamic holography.

Staebler and *Amodei* [6.5] were the first to demonstrate using $LiNbO_3$ the fundamental phenomenon of dynamic holography in PRCs, i.e., the intensity transfer between the light beams recording a diffusion phase hologram (Fig.6.1a). Another example is the effect of coherent erasure or enhancement of a diffusion hologram during the reconstruction processes (Fig.6.1b), which was initially observed in $LiNbO_3$ [6.6], too.

This chapter discusses the most important effects of dynamic holography. To introduce the reader to this field, we shall briefly consider diffraction of an interference pattern from a fixed phase hologram [6.5, 7].

6.1 Diffraction of an Interference Pattern of Two Plane Waves from a Fixed Matched Phase Grating

Let us assume that two plane light waves S and R with the wave vectors $\mathbf{K_S}$ and $\mathbf{K_R}$ satisfy the Bragg condition for diffraction from an elementary si-

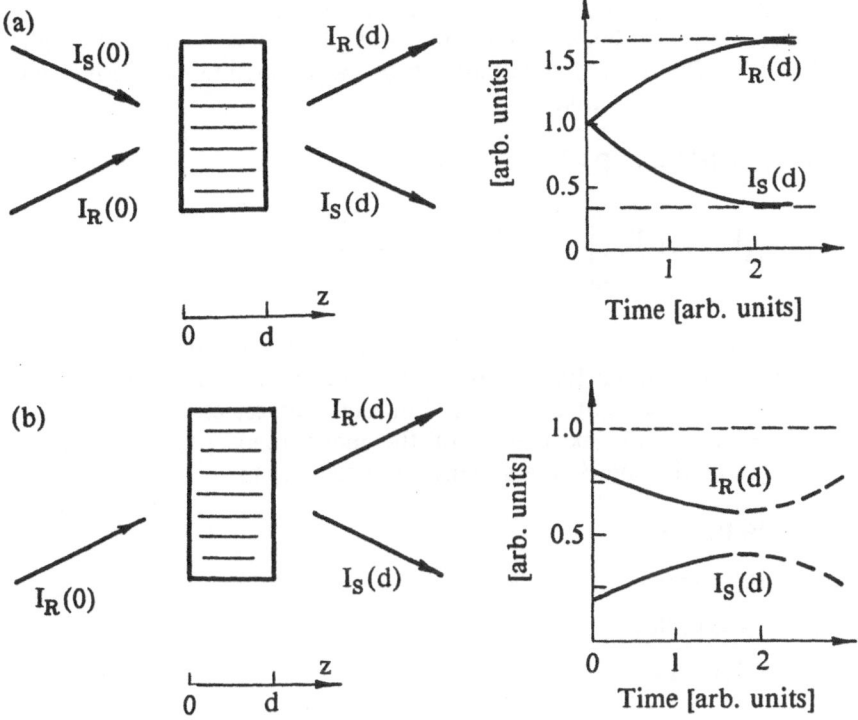

Fig.6.1. (a) Energy transfer during recording of a shifted dynamic phase grating. (b) Self-enhancement of a shifted dynamic phase grating during its readout

nusoidal phase grating

$$\Delta\epsilon^{\omega}(x) = |\Delta\epsilon^{\omega}|\cos(Kx + \psi) = \frac{\Delta\epsilon^{\omega}}{2}\exp(iKx) + \frac{(\Delta\epsilon^{\omega})^*}{2}\exp(-iKx) . \quad (6.1)$$

Here $\Delta\epsilon^{\omega} = |\Delta\epsilon^{\omega}|\exp(i\psi)$ is regarded as the complex amplitude of the phase grating under consideration. Let us see how intensities and phases of the waves of interest change after passing through a thin layer of such a grating with thickness Δz.

6.1.1 Finite Increments of Intensities and Phases of the Light Waves Behind a Thin Grating Layer

The set of coupled-wave equations (5.8a) written as finite increments of the complex amplitudes of respective light waves behind a thin layer of grating with thickness Δz yields

$$\Delta R = - i\chi S\Delta z ,$$

$$\Delta S = - i\chi^* R\Delta z , \quad (6.2)$$

where χ (5.8b) is regarded, like $\Delta\epsilon^\omega$, as a complex value. Let us rewrite (6.2) as increments of intensities (I_S, I_R) and phases (ψ_S, ψ_R) of the light beams S and R. The desired equations can be easily obtained by taking into account the following relations:

$$\Delta I_R = \Delta(RR^*) = (\Delta R R^* + R \Delta R^*) ,$$

$$\Delta \psi_R = \Delta\left[\frac{-i}{2} \ln\left(\frac{R}{R^*}\right)\right] = -\frac{i}{2}\left(\frac{\Delta R}{R} - \frac{\Delta R^*}{R^*}\right), \qquad (6.3)$$

and a similar pair of relations for ΔI_S and $\Delta\psi_S$. Note that the numerical factor (Sect.2.1) is omitted for simplicity in (6.3a), as well as in the formula given below including light intensities. For the simplest case $R = R^*$, $S = S^*$, i.e. when one of the extremums of the interference pattern coincides with the origin of the coordinare system, (6.3) reduces to

$$\Delta I_R = 2R\,\mathrm{Re}\{\Delta R\} ,$$

$$\Delta \psi_R = R^{-1} \mathrm{Im}\{\Delta R\} ,$$

$$\Delta I_S = 2S\,\mathrm{Re}\{\Delta S\} , \qquad (6.3')$$

$$\Delta \psi_S = S^{-1} \mathrm{Im}\{\Delta S\} .$$

Direct substitution of (6.3') into (6.2) yields the desired equations

$$\Delta I_R = -\Delta I_S = 2RS\,\mathrm{Im}\{\chi\}\Delta z ,$$

$$\Delta \psi_R = -\frac{S}{R}\mathrm{Re}\{\chi\}\Delta z , \qquad (6.4)$$

$$\Delta \psi_S = -\frac{R}{S}\mathrm{Re}\{\chi\}\Delta z .$$

6.1.2 Intensity Transfer via Shifted Phase Grating

Let us inspect the first of the derived equations (6.4). The equality $\Delta I_R = -\Delta I_S$ expresses the energy conservation law for the light diffraction off a phase grating in a nonabsorbing medium. An increase in the intensity of one of the beams can result only from an equivalent decrease in the other beam. In the specific case when $\mathrm{Im}\{\chi\} = 0$, no changes in intensities I_R, I_S after light propagation through the grating layer are, however, observed. "Intensity transfer" or "energy coupling" between beams R and S is usually said to be absent here. This situation evidently occurs when there is no phase shift between the grating and the interference pattern produced by the plane waves involved. As far as the position of the diffraction grating relative to the interference pattern $I(x)$ is concerned, this is the case of the so-called unshifted grating ($\psi = 0; \pm \pi$, Fig.6.2a).

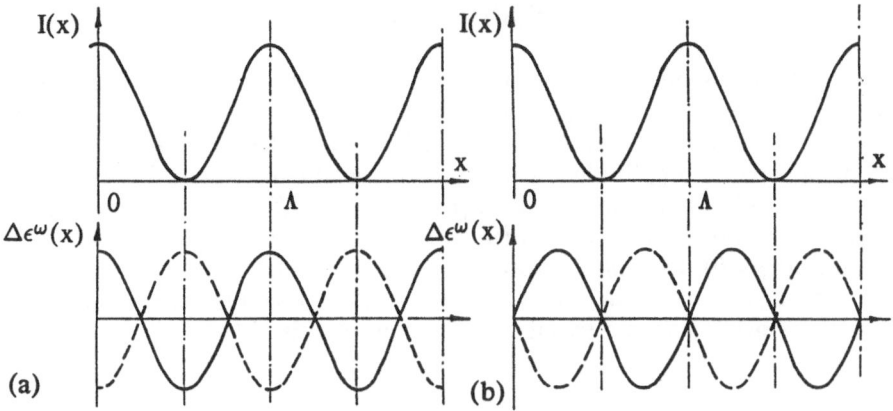

Fig.6.2. Unshifted (a) and shifted (b) phase gratings recorded within a PRC volume by an interference pattern I(x)

For all other shifts between the grating and interference pattern ψ, the energy coupling does take place and peaks at $\psi = \pm\pi/2$, when $\chi = \pm i|\chi|$. In this limiting case, the grating is shifted by $\frac{1}{4}$ of the grating spacing Λ from the interference pattern and is therefore called a "shifted" grating (Fig. 6.2b). Irrespective of the beam intensity ratio, the phase grating maxima are shifted from the interference pattern maxima in the direction of the beam that experiences gain. The direct consequence of this fact is that the energy coupling via a shifted grating changes the sign when the light propagation directions are reversed (Fig.6.3).

Note that in the simplest case of equal beam intensities ($I_S = I_R = I$) a relative change $\Delta I_{R,S}/I$ is:

$$\Delta I_{R,S}/I = \pm\, 2|\chi|\Delta z = \pm\, \frac{\pi|\Delta\epsilon^\omega|\Delta z}{n\lambda\cos\theta_B}.$$ (6.5)

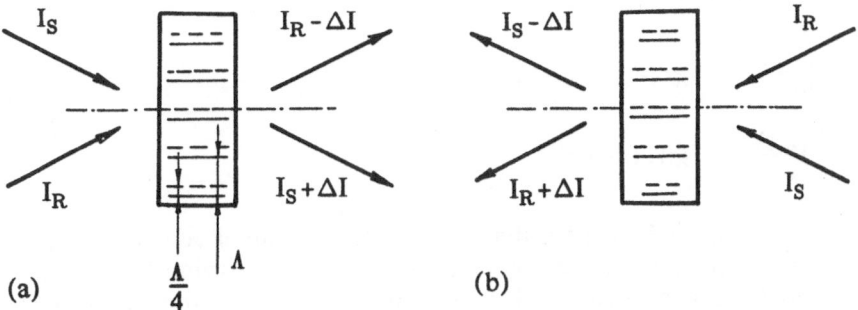

Fig.6.3. Sign inversion of energy transfer via shifted phase grating on reversal of light beams' propagation directions (dashed lines show the interference pattern maxima, solid lines indicate the phase grating maxima)

6.1.3 Phase Transfer via Unshifted Phase Grating

The phase shifts due to "phase transfer" in the light waves involved can be analyzed in a similar fashion. As (6.4) predicts, they prove to be of the same sign for both waves, the larger phase delay being impressed on the beam of smaller amplitude. In contrast to the intensity transfer, the phase transfer is zero for a purely shifted phase grating ($\psi = \pm \pi/2$) when $\mathrm{Re}\{\chi\} = 0$; and vice versa, for an unshifted phase grating (that is, at $\psi = 0, \pm \pi$, when $\mathrm{Im}\{\chi\} = 0$), the phase transfer is the maximum.

For beams of equal intensities, the highest possible phase shift for a thin phase grating with a specified amplitude $\Delta\epsilon^\omega$ is given by

$$\Delta\psi_S = \Delta\psi_R = \pm \left|\chi\right|\Delta z = \pm \frac{\pi\left|\Delta\epsilon^\omega\right|\Delta z}{2n\lambda\cos\theta_B} . \tag{6.6}$$

6.2 Self-Diffraction from a Dynamic Volume Phase Hologram

In contrast to diffraction from a fixed grating considered in Sect.6.1, the grating in PRCs is a result of recording a pattern of interference between two waves under consideration. As a result, the holographic recording in PRCs is dramatically affected by the recorded grating. Such a complicated process of diffraction of recording waves from the recorded volume holo-gram in a dynamic holographic medium is called self-diffraction or two-wave mixing.

An adequate description should involve a self-consistent consideration of the phase grating formation process and diffraction of light beams from the grating [6.8-13]. The problem is greatly simplified if we take into ac-count that the characteristic time for the hologram recording in PRCs is typically well above the time needed for the light waves to pass through the crystal thickness ($\simeq$ dn/c). Thus we can regard the light diffraction proces-ses as occurring in the quasi-stationary regime and employ the set of cou-pled-wave equations

$$\frac{\partial S(z,t)}{\partial z} = - i\chi^*(z,t)R(z,t) ,$$

$$\tag{6.7}$$

$$\frac{\partial R(z,t)}{\partial z} = - i\chi(z,t)S(z,t) ,$$

identical to (5.8a), where both the complex grating amplitude and complex amplitudes of light waves are assumed to be slowly varying functions of time t and the z coordinate. This is evidently valid only for a one-dimen-sional case when the initial photorefractive medium with the input and output surfaces in the xy plane is spatially homogeneous and the light beams R and S are plane waves.

6.2.1 Steady-State Two-Wave Mixing in PRCs

The problem proves to be simpler if we consider the steady-state regime of recording when both the light field involved (which is in fact the pattern of interference between R and S) and the phase grating reach steady states matching each other. This means that the time dependences of R, S, and χ can be eliminated in (6.7). We restrict our attention in this section to this important case.

To solve (6.7) for conventional boundary conditions at the input of the medium (S(0) = S_0, R(0) = R_0), it should be supplemented with the material equation describing the relation between the complex steady-state grating amplitude $\chi(z)$ and complex amplitudes of the light waves R(z) and S(z). This equation has the simplest form for the linear regime of holographic recording [6.14] when the grating amplitude is proportional to the modulation depth m(z) of the interference pattern being recorded:

$$\chi^*(z) = \frac{i}{2}\gamma m(z) = i\gamma \frac{R^*(z)S(z)}{|R(z)|^2 + |S(z)|^2} , \tag{6.8}$$

where the coupling constant γ = const(m).

For a simultaneous solution of (6.7, 8) one has to substitute (6.8) into (6.7). Thereafter it is necessary to multiply the first equation of (6.7) by $R^*(z)$, and the second by $S^*(z)$. By substracting the second conjugated equation from the first one we obtain

$$R^*(z)\frac{\partial S(z)}{\partial z} - S(z)\frac{\partial R^*(z)}{\partial z} = \gamma S(z)R^*(z) ,$$

and further

$$\frac{\partial}{\partial z}\left[\frac{S(z)}{R^*(z)}\right] = \gamma \frac{S(z)}{R^*(z)} .$$

This yields the final expression

$$\frac{S(z)}{R^*(z)} = \frac{S_0}{R_0^*}\exp(\gamma z) . \tag{6.9}$$

For the low-intensity signal beam ($I_S \ll I_R$), the reference beam amplitude R can be assumed to be constant across the hologram thickness (an undepleted pump approximation). As a consequence, self-diffraction from a dynamic hologram reduces to an exponential variation of the signal beam amplitude

$$S(z) = S_0\exp(\gamma z) . \tag{6.10}$$

Of major practical interest are the intensity variations of the light beams as they propagate through the crystal thickness. The expression pertaining here is readily obtainable from (6.9) by multiplying its right- and left-hand sides by their complex conjugates:

$$\frac{I_S(z)}{I_R(z)} = \frac{|S_0|^2}{|R_0|^2} \exp(\Gamma z) , \tag{6.11}$$

where

$$\Gamma = 2\text{Re}\{\gamma\} . \tag{6.12}$$

The expression obtained predicts the well-known fact [6.5, 9, 12] that a shifted component of the phase grating should be necessarily present to observe energy coupling between the recording beams, as with a fixed phase grating discussed in Sect. 6.1. Indeed, if the recorded grating is of a purely unshifted type, γ (6.8) is imaginary, $\Gamma = 0$, and the intensity ratio I_S/I_R proves to be constant throughout the sample thickness.

6.2.2 Gain Factor of PRCs

Note that Γ (6.12), which is measured in cm^{-1}, is referred to in the literature as a steady-state gain factor or simply a gain factor of PRC. This term is evidently attributable to the fact that at $\Gamma > 0$ the photorefractive sample illuminated by a powerful pump beam can be used for enhancement of a weak signal beam. So it can be considered an analog of an active gain medium with population inversion. Such photorefractive amplifiers, however, require a narrow-band pumping and operate effciently only if the frequency detuning between the signal and pump beams $\Delta\omega$ does not exceed τ_{sc}^{-1}, where τ_{sc} is the characteristic time for hologram recording under given conditions.

Note also that it is possible to observe enhancement using the PRC with unshifted phase holograms as well. Enhancement is observed for signal beams detuned by $\Delta\omega \simeq \tau_{sc}^{-1}$ [6.12, 15, 16] from the pump beam and also in the transient regime of holographic recording (the so-called transient energy coupling) [6.17-20].

The coefficient γ (coupling constant) which is of primary importance for PRC applications, can be obtained from (5.8b, 17) and (6.8):

$$\gamma \simeq - \frac{i\pi E_{sc}^{st}}{m} \frac{rn^3}{\lambda} \frac{1}{\cos\theta_B} . \tag{6.13}$$

Here r is the relevant electro-optic coefficient and E_{sc}^{st} is determined from (4.14, 24, 27, 36), depending on the dominating mechanism and regime of holographic recording in a particular PRC. Specifically, γ is real for diffusion and nonstationary recording or for the drift mechanism under the condition of quasi-neutrality violation, and γ is imaginary for drift recording if the quasi-neutrality condition is fulfilled.

As follows from the data presented in the Appendix, the basic mechanisms of holographic recording in presently known PRCs give the gain factors Γ of up to 10-100 cm^{-1} that is clearly superior to the respective gain factors obtained in conventional laser media such as gases, metal vapors, solids and dyes, and is inferior only to those of semiconductors [6.21].

6.3 Four-Wave Mixing via an Isotropic Phase Grating

Another important geometry of dynamic holography is the four-wave mixing configuration where a second plane pump wave R_2 traveling opposite to R_1 is introduced (Fig.6.4).

The interest in this geometry is connected with the fact that the fourth wave S_2 arising within the volume of the medium is a phase-conjugate replica of the original signal wave S_1 [6.22,23]. In this section we do not dwell upon phase conjugation itself, which was convincingly demonstrated in a similar geometry in early holographic experiments [6.24-26]. Here we are mainly interested in the interaction between the four light waves within the volume of a dynamic holographic medium. Even for an elementary sinusoidal transmission hologram (S_1 like R_1 is a plane wave) this process proves to be fairly complicated.

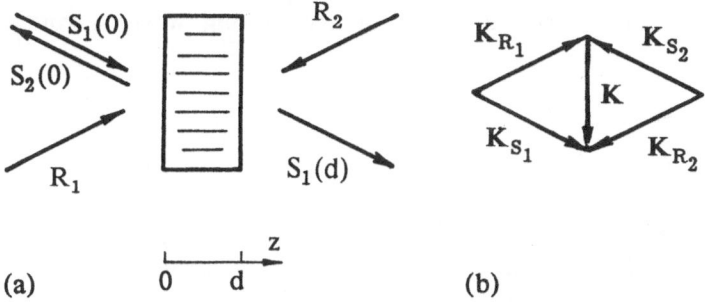

Fig.6.4. (a) Basic four-wave mixing geometry with two counterpropagating collinear pump beams R_1, R_2. (b) Bragg conditions for selfdiffraction of two pairs of light waves (R_1,S_1 and R_2,S_2) from a common transmission grating K

6.3.1 Steady-State Four-Wave Mixing via an Isotropic Transmission Phase Grating in PRCs

Indeed, the waves S_1 and R_1 responsible for the initial recording of the transmission hologram change their amplitudes and phases because of self-diffraction from the recorded hologram. In turn, generation of a conjugate wave S_2 and its interference with the reference wave R_2 that has created it give rise to an additional (secondary) hologram recording process. Thus we are in fact dealing with two processes of two-wave mixing of waves S_1, R_1 and S_2, R_2. Each of them involves both the hologram recording and diffraction from it. Even without being directly related (we ignore here possible recording of reflection holograms), they are nevertheless involved in a complicated four-wave mixing process, owing to the presence of a common hologram.

The coupled-wave equations have here the form

$$\frac{\partial S_1(z)}{\partial z} = - i\chi^*(z)R_1(z) ,$$

$$\frac{\partial R_1(z)}{\partial z} = - i\chi(z)S_1(z) ,$$

$$\frac{\partial S_2(z)}{\partial z} = i\chi(z)R_2(z) ,$$

$$\frac{\partial R_2(z)}{\partial z} = i\chi^*(z)S_2(z) .$$

(6.14)

As in the preceding section, we consider here a steady-state regime when none of the values involved - R_1, R_2, S_1, S_2, χ - varies with time. Attention should be paid to the inversion of sign of the right-hand sides of the second pair of equations (6.14) caused by direction reversal of waves S_2 and R_2 relative to the z axis. An additional complex conjugation of the grating amplitude $\chi(z)$ is due to the difference in the Bragg conditions for counter-propagating waves for a fixed direction of the grating wave vector **K** (Fig.6.4b)

$$\mathbf{K}_{S_1} = \mathbf{K}_{R_1} + \mathbf{K} ,$$

$$\mathbf{K}_{S_2} = \mathbf{K}_{R_2} - \mathbf{K} .$$

The steady-state grating amplitude $\chi(z)$ in the linear regime of holographic recording during four-wave mixing is obviously equal to

$$\chi^*(z) = \frac{i}{2}\gamma m(z) = i\gamma \frac{R_1^*(z)S_1(z) + R_2(z)S_2^*(z)}{|R_1(z)|^2 + |S_1(z)|^2 + |R_2(z)|^2 + |S_2(z)|^2} . \quad (6.15)$$

A general analysis of (6.14,15) for arbitrary boundary values of the amplitudes $S_1(0)$, $R_1(0)$ and $R_2(d)$ ($S_2(d) = 0$) is fairly complicated [6.27,28]. So we treat here the simplest version of the problem in the approximation of given amplitudes of pump waves (undepleted pumps approximation) with equal intensities ($R_1(z) = R_2(z) = R \gg |S_1(z)|$; $|S_2(z)|$). The complex grating amplitude (6.15) is then given by

$$\chi^*(z) = i\gamma \frac{S_1(z) + S_2^*(z)}{2R} \quad (6.16)$$

and the coupled wave equations (6.14) reduce to

$$\frac{\partial S_1(z)}{\partial z} = \frac{\gamma}{2}[S_1(z) + S_2^*(z)] \,,$$

$$\frac{\partial S_2^*(z)}{\partial z} = \frac{\gamma}{2}[S_1(z) + S_2^*(z)] \,. \qquad (6.17)$$

Solving this set of equations by conventional methods, we obtain for conventional boundary conditions $S_1(0) = S_0$, $S_2(d) = 0$

$$S_1(z) = S_0 \frac{\exp(\gamma z) + \exp(\gamma d)}{1 + \exp(\gamma d)} \,,$$

$$S_2^*(z) = S_0 \frac{\exp(\gamma z) - \exp(\gamma d)}{1 + \exp(\gamma d)} \,. \qquad (6.18)$$

6.3.2 Reflectivity of a Four-Wave Mixing Geometry

Among the salient parameters characterizing any four-wave mixing geometry are transmittivity (T) and reflectivity (R), which determine the intensity of the corresponding signal beam at the output of the sample

$$T = \left| \frac{S_1(d)}{S_0} \right|^2 \,,$$

$$R = \left| \frac{S_2(0)}{S_0} \right|^2 \,. \qquad (6.19)$$

As follows from (6.18) for the case under discussion these values are [6.29]

$$T = \left| \frac{\exp(\gamma d/2)}{ch(\gamma d/2)} \right|^2 \,,$$

$$R = |th(\gamma d/2)|^2 \,. \qquad (6.20)$$

These expressions reveal that reflection with gain (R > 1) is obtainable in the steady-state only for unshifted phase gratings. The reflectivity for shifted phase gratings, that can produce a noticeable enhancement of a weak signal beam (6.11) in the two-wave geometry, is given by

$$R = th^2\left(\frac{\Gamma d}{4}\right) \,; \qquad (6.21)$$

so it cannot exceed unity for any product of Γd. This contradiction can be explained by the fact that the counterpropagating pump wave R_2 performs a coherent erasure of the hologram (Fig.6.1b). In other words this means that the secondary hologram recorded within the crystal by interfering S_2 and R_2 is exactly opposite in phase to the initial hologram recorded by S_1 and R_1, thereby leading to its efficient suppression. Thus we can speak about a negative feedback arising in the crystal volume; i.e., formation of the primary hologram by beams R_1 and S_1 gives rise to a wave S_2, whose interaction with R_2 leads to suppression of the initial hologram.

This type of feedback is unavoidable in the conventional four-wave mixing configuration. It follows directly from the symmetry properties of two-wave mixing process via a shifted phase grating. Indeed, as shown in Sect.6.1.2 (Fig.6.3), the process is bound to occur in the opposite direction with reversed signal and reference waves. In other words, if a weak signal beam is enhanced in one process, it is suppressed in the other and vice versa. This means actually that one of the pump beams corresponding to the two-wave mixing process with the signal beam decay is sure to erase the hologram coherently.

The problem can be solved at least in part by introducing an asymmetric pump when one of the pump beams performing coherent hologram erasure is inferior in intensity to the other beam ($r = |R_2|^2/|R_1|^2 \lesssim 1$). As shown in [6.29], the highest reflectivity is attained here for optimum $r = \exp(-\Gamma d/2)$ and equals

$$R = \text{sh}^2\left[\frac{\Gamma d}{4}\right].$$ \hfill (6.22)

If we have symmetric pumping ($r = 1$), reflection with gain via shifted phase gratings is possible as well. It is observed for weak signal waves shifted in frequency by $\Delta\omega \simeq \tau_{sc}^{-1}$ relative to the pump beams [6.30], or in case of slightly noncollinear pumping beams R_1 and R_2 [6.31]. The problem of increasing the efficiency of four-wave mixing by shifted gratings can, however, be most adequately solved by using anisotropic properties of phase gratings in PRCs (Sect.6.4).

Quite a different situation is found in the four-wave mixing process when a transmission phase grating of an unshifted type is involved (γ is imaginary). Without going into details, we note only that self-diffraction of a counterpropagating pump beam R_2 results in an additional tilt of the grating. As a consequence, the component of the grating shifted with respect to the forward two-wave mixing process between S_1 and R_1 arises. It leads to enhancement of S_1, the hologram amplitude, and, hence, of the reflected wave. Note that as $\gamma d \rightarrow \mp i\pi$, reflectivity and transmittivity of this geometry, see (6.20), tend to infinity, which means that the oscillation regime is reached.

6.4 Four-Wave Mixing via Anisotropic Phase Gratings

A major specific feature of anisotropic phase gratings (5.16) that distinguishes them from isotropic ones (5.1) is the presence of two orthogonally
polarized diffraction eigenprocesses with remarkably different diffraction
efficiencies (Chap.5). Indeed, the effective amplitudes $\chi_{1,2}(z)$ of one and
the same anisotropic grating can differ both in amplitude and even in sign
for different diffraction processes, since they are usually determined by entirely different electro-optic coefficients.

If polarizations of all light waves S_1, R_1, R_2, and S_2 in four-wave
mixing via an anisotropic phase grating are the same and correspond to the
polarization eigenstate of the grating, the results of the analysis given in the
preceding section are entirely applicable. Four-wave mixing can proceed
quite differently if polarization of one pair of beams (R_1, S_1) corresponds
to one diffraction eigentype and that of the second pair (R_2, S_2) corresponds to the other one. This means that the first two-wave mixing process
(between S_1 and R_1) has one coupling constant (γ_1) and the second process
(between S_2 and R_2) has the other one (γ_2).

For undepleted pump waves R_1 and R_2 of equal intensities $(r = 1)$, the
set of equations describing the four-wave mixing differs from (6.17) only
by the coupling constants

$$\frac{\partial S_1(z)}{\partial z} = \frac{\gamma_1}{2}[S_1(z) + S_2^*(z)] \, ,$$

$$\frac{\partial S_2^*(z)}{\partial z} = \frac{\gamma_2}{2}[S_1(z) + S_2^*(z)] \, . \tag{6.23}$$

We confine the analysis to two typical cases $\gamma_1 = -\gamma_2 = \gamma$ and $\gamma_2 = \gamma$, $\gamma_1 = 0$. The third limiting case $(\gamma_1 = \gamma_2 = \gamma)$ was discussed in detail in the
preceding section. The case $\gamma_2 = 0$ is of no particular interest because of the
absence of a conjugate wave S_2.

6.4.1 Four-Wave Mixing via Anisotropic Phase Grating
with Positive Feedback

Solution of (6.23) for $\gamma_1 = -\gamma_2 = \gamma$ and conventional boundary conditions
$S_1(0) = S_0$, $S_2(d) = 0$ yields

$$S_1 = S_0 \frac{2 + \gamma(d - z)}{2 - \gamma d} \, ,$$

$$S_2^* = S_0 \frac{\gamma(d - z)}{2 - \gamma d} \, . \tag{6.24}$$

Transmittivity and reflectivity are here

$$T = \left| \frac{2}{2 - \gamma d} \right|^2 ,$$

$$R = \left| \frac{\gamma d}{2 - \gamma d} \right|^2 .$$

(6.25)

For a shifted phase grating when $\gamma = \Gamma/2$ [6.32, 33],

$$T = \left(\frac{4}{4 - \Gamma d} \right)^2 ,$$

$$R = \left(\frac{\Gamma d}{4 - \Gamma d} \right)^2 .$$

(6.26)

More general equations for T and R, taking into account optical activity of cubic PRCs, nonequality of pumping beams intensities and their possible depletion were obtained in recent papers [6.34].

As shown in Fig.6.5a, if Γd is fixed, the obtained values (6.26) are always superior to T and R, (6.20, 21), observed in four-wave mixing experiments via an isotropic grating. Moreover, for $\Gamma d \gtrsim 3$, T, (6.26), exceeds the two-wave transmittivity (T = exp(Γd), (6.10)) achieved under similar conditions. For $\Gamma d \to 4$, T and R tend to infinity, thus indicating that the self-oscillation regime is reached.

A drastic increase of transmittivity and reflectivity observed for $\Gamma > 0$ in this geometry has a fairly obvious qualitative interpretation. Indeed, the opposite signs of coupling constants γ_1 and γ_2 mean that one and the same shifted phase grating has opposite signs for the light waves traveling through the crystal in opposite directions. As shown in the preceding section, this causes both two-wave mixing processes to occur now in one direction. That is, at $\Gamma > 0$ they lead to a simultaneous enhancement of weak signal beams S_1 and S_2 and hence to an increase of the grating amplitude throughout the entire sample thickness. This process can be termed four-wave mixing with positive feedback, since the introduction of a counter-propagating pump beam R_2 leads to an additional enhancement of the hologram.

Figure 6.5b shows typical experimental transmittivity and reflectivity curves obtained in a similar geometry of four-wave mixing via shifted phase gratings recorded in an external alternating field in cubic BTO [6.35]. Analogious geometry resulting in an amplified phase-conjugate wave was also used in a cubic GaAs [6.36].

Note that attaining the maximum possible reflectivities in this geometry leads to more severe requirements on the alignment of the pump beams

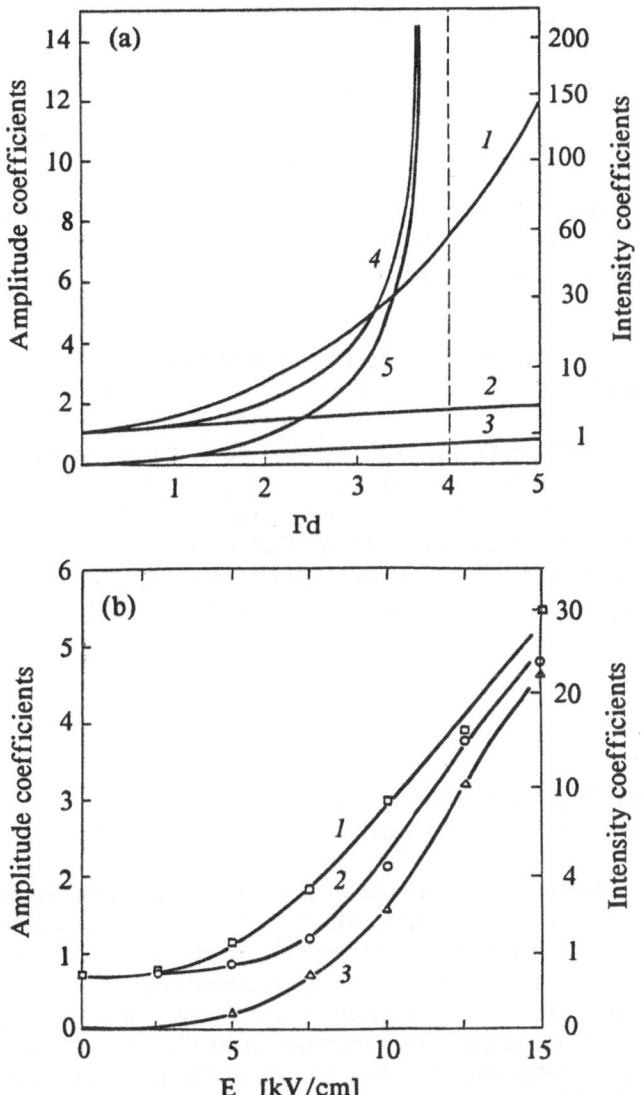

Fig.6.5. (a) Transmittivity (T) and reflectivity (R) for two- and four-wave mixing via a shifted phase grating (Two-wave mixing *1*: T; four-wave mixing with negative feedback *2*: T, *3*: R; four-wave mixing with positive feedback *4*: T, *5*: R). **(b)** Experimental curves for transmittivity (*1*) in two-wave mixing, and transmitttivity (*2*) and reflectivity (*3*) in four-wave mixing with positive feedback versus amplitude of an alternating electric field E~ [6.35] (BTO, $\lambda = 633$ nm, $d \simeq 4$ mm, $\Lambda \simeq 5\,\mu$m, a square-wave alternating field, $\tau \simeq 25$ ms, original intensity ratio between the signal and the reference waves $\simeq 2\cdot10^{-5}$)

101

and phase homogeneity of the sample. As shown in [6.33], in the most interesting range $2 < \Gamma d < 4$,

$$\frac{\Delta\theta_1}{\Delta\theta} \sqrt{R} \simeq 2 . \tag{6.27}$$

Here $\Delta\theta_1$ is the allowable angular misalignment between the propagation directions of R_1 and R_2 in the incidence plane and $\Delta\theta$ is the half-width of the Bragg maximum for a volume hologram with given K and d (5.31).

6.4.2 Four-Wave Mixing via Anisotropic Phase Gratings Without Feedback

The third of the cases listed above ($\gamma_1 = 0$, $\gamma_2 = \gamma$) obviously occupies an intermediate position between the cases of negative ($\gamma_2 = \gamma_1$) and positive ($\gamma_2 = -\gamma_1$) feedback discussed so far. The zero coupling constant for the waves traveling in the forward direction means that they do not diffract from the hologram, in whose recording they nevertheless do participate. Solution of (6.23) for conventional boundary conditions yields [6.33]

$$\begin{aligned} T &= 1, \\ R &= |\exp(-\gamma d/2) - 1|^2 . \end{aligned} \tag{6.28}$$

This means that fairly high reflectivities can also be obtained here for purely shifted gratings at $\gamma < 0$ when wave S_2 is enhanced through two-wave mixing with R_2.

This case can be observed for example in a conventional holographic recording geometry in $BaTiO_3$ (Sect.5.4.2), where diffraction efficiency of the hologram under reconstruction by an ordinary (E) light wave is dramatically inferior to that under reconstruction by an extraordinary (H) wave. To increase the efficiency of a four-wave mixing via shifted phase grating in $BaTiO_3$, this geometry was used experimentally in [6.37]. Note, that as far as we know, it was used for the first time in one of the first papers on holographic recording in this PRC [6.38], where phase-conjugate reflectivity $R \sim 10$ was experimentally observed.

6.5 Optical Oscillators and Self-Pumped Phase Conjugators

Like any other light amplifier, the photorefractive crystal pumped by an external laser can be transformed into an optical oscillator. To this end, the PRC should be placed into an optical resonator providing an optical feedback for the signal beam.

6.5.1 Ring Oscillator Using Two-Wave Mixing in PRCs

In particular, if two-wave mixing in a PRC is involved, such a resonator can be an ordinary ring cavity (Fig.6.6). Oscillation is established here, as in an ordinary laser, if decay of a given cavity eigenmode is compensated for by the energy coupling with the pump beam in the PRC volume:

$$M \exp(\Gamma d) > 1. \tag{6.29}$$

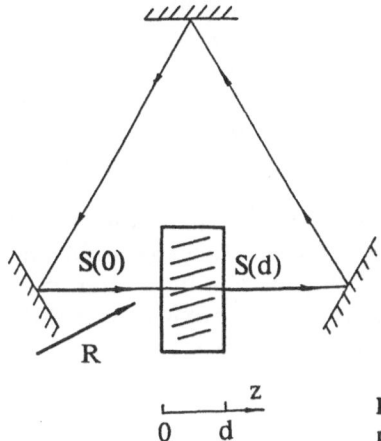

Fig.6.6. Ring oscillator operating through two-wave mixing in PRC

Here transmittivity M characterizes the net ring cavity round-trip light losses at mirrors, within the sample volume, due to Fresnel reflections from sample faces, at beam splitters, etc.

Oscillations through two-wave mixing in a ring cavity were experimentally achieved in $BaTiO_3$ pumped by a He-Ne laser [6.39] and an argon laser [6.40, 41], and also in BSO at the krypton laser wavelength (λ = 568 nm) [6.42]. The efficiency of pump-to-oscillating wave power conversion in the experiments mentioned above was 20-60%.

One of the most interesting specific features of photorefractive optical oscillators is frequency detuning ($\Delta\omega$) between the laser pump beam and the light wave excited in the cavity. A detuning of the order of the reciprocal characteristic time for hologram formation in a PRC ($\sim\tau_{sc}^{-1}$) was first experimentally observed in the ring resonator under discussion [6.41, 42]. The first explanation of this effect as arising from frequency detuning between the reference and signal beams needed for the most efficient energy exchange in two-wave mixing by an unshifted grating was given in [6.42]. The presence of a similar effect in $BaTiO_3$ [6.41], where a shifted hologram is formed through the diffusion mechanism and the most efficient two-wave energy exchange is observed at zero frequency detuning, indicates, however, that there must be a more general explanation. One was suggested later in [6.43]. It states that not only the energy balance condition (6.29) but also a phase condition must be satisfied in the geometry shown in Fig.6.6, as well as in any other oscillator with a cavity. The latter implies that the round-trip phase shift in the oscillation beam is to be equal to integer 2π.

The gain linewidth in conventional laser media is typically much greater than the frequency offset between two neighboring resonator modes (= $2\pi c/L$, where c is the velocity of light and L is the ring cavity length). Thus the phase condition is met here for the laser modes, which are almost identical to the resonator modes. An insignificant effect of mode "pulling" toward the center of the gain line [6.44] is usually ignored.

The situation is strikingly different if a PRC is used as an active gain medium. The effective gain linewidth centered near the pump frequency ω_0 is determined by the characteristic time for hologram formation in PRCs ($1/\tau_{sc} \sim 1\text{-}10^3$ Hz for reasonable pump levels) and proves to be much smaller than the intermode shift $2\pi c/L \gtrsim 10^9$ Hz. The energy balance condition (6.29) evidently allows generation of a light wave with the frequency ω near the pump frequency ω_0, which can markedly differ from the nearest eigenfrequency of the resonator ω_i ($|\Delta\omega_i| = |\omega_i - \omega_0| \gg 1/\tau_{sc}$).

The phase condition can be satisfied here only if the extra phase shift $\Delta\omega_i L/c$ will be compensated for by the offset $\Delta\omega$ between the oscillation frequency ω and the pump frequency ω_0. Indeed, the amplitude transmittivity for signal wave S_1 in nearly degenerate two-wave mixing via a shifted phase grating in the undepleted pump approximation is given [6.43] by

$$\frac{S_1(d)}{S_0} = \exp\left[\frac{\Gamma d}{2(1 + i\Delta\omega\tau_{sc})}\right]$$

$$= \exp\left[\frac{\Gamma d}{2(1 + \Delta\omega^2\tau_{sc}^2)}\right]\exp\left[\frac{-i\Gamma d\Delta\omega\tau_{sc}}{2(1 + \Delta\omega^2\tau_{sc}^2)}\right]. \tag{6.30}$$

This implies that the nonzero frequency detuning $\Delta\omega$ results in a decrease of the effective gain factor of the PRC and in an extra phase shift due to phase transfer effect. The phase condition required for building up oscillations in the geometry shown in Fig.6.6 is given in this case by

$$\frac{\Gamma d\Delta\omega\tau_{sc}}{2(1 + \Delta\omega^2\tau_{sc}^2)} = \frac{\Delta\omega_i L}{c}. \tag{6.31}$$

Frequency detuning $\Delta\omega$ between a generated and a pumping wave as a function of the ring cavity length was experimentally studied [6.43] in a BaTiO$_3$ crystal pumped by an argon laser ($\lambda = 514$ nm). The zero frequency detuning was observed for the maximum generated power, which was obviously achieved when the frequency of one of the resonator modes ω_i was equal to the pump frequency ω_0. In the neighborhood of this point $\Delta\omega$ was linearly related to the cavity length change.

6.5.2 Linear Geometry of Passive Phase Conjugators

A similar photorefractive oscillator can also be built using a two-mirror ("linear") Fabry-Perrot resonator (Fig.6.7a). Unlike the ring resonator (Fig. 6.6), a counterpropagating signal wave S_2, which is a phase-conjugate replica of the forward wave S_1, also passes through the PRC sample in this geometry. Its diffraction from the hologram produces a fourth wave R_2, which in turn begins to participate in formation of the hologram together with S_2. A thorough quantitative analysis of such an optical oscillator should obviously include inspection of a set of nonlinear equations describing the four-wave mixing process [6.45-47].

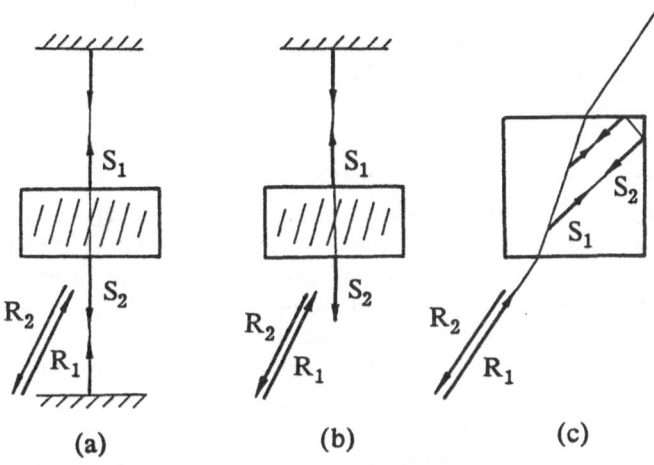

Fig.6.7. Optical oscillators using four-wave mixing in a photorefractive crystal (a) with a "linear" cavity, (b) with a "semilinear" cavity, and (c) oscillator using internal reflection from a corner of the crystal)

An essential feature of the "linear" oscillator is that the reflected wave R_2 is also a phase-conjugate of the original wave R_1. Thus, this geometry of the optical oscillator may also be regarded as a passive (or self-pumped) phase-conjugating mirror. When pumped by a signal external beam R_1 of a complicated shape, it produces a conjugate wave $R_2 \propto R_1^*$. The required auxiliary phase-conjugate waves S_1 and S_2 are generated in the geometry automatically, because of the buildup of oscillation in the resonator.

Oscillation in the linear resonator was experimentally studied using BaTiO$_3$ ($\lambda = 633$nm) [6.39] and LiNbO$_3$:Fe ($\lambda = 442$ nm) [6.48]. The phase-conjugating properties of the geometry under discussion in the experiments using BaTiO$_3$ [6.49] and the presence of frequency detuning between the generated and pump beams [6.50] were also demonstrated.

6.5.3 Single-Mirror and Mirrorless Geometries of Passive Phase Conjugators

Because of extremely high gain factors Γ, photorefractive media also can provide oscillation buildup when used in single-mirror (or "semilinear") resonators (Fig.6.7b). This geometry, like the two-mirror Fabry-Perrot resonator, can also be regarded as a self-pumped (or passive) phase-conjugating mirror. Oscillation in this geometry was studied in experiments with $BaTiO_3$ pumped by either one or two beams [6.38, 39, 49].

The self-pumped phase-conjugating mirror using internal reflection near the edge of a PRC (Fig.6.7c) [6.51] can also be looked upon as a modification of the semilinear geometry (Fig.6.7b). The role of the mirror in this case is played by an internal corner cube reflector formed by two polished adjacent crystal faces. The geometry automatically chooses the optimum angles for auxiliary waves S_1 and S_2, which ensure the maximum efficiency of four-wave mixing.

6.5.4 Ring Geometry of Passive Phase Conjugator

Another geometry of a self-pumped (passive) phase-conjugating mirror involving buildup of two auxiliary light waves in a ring geometry (Fig.68a) was proposed in [6.52-54]. In contrast to passive phase conjugation in PRCs discussed above, in this geometry a powerful pump beam passes twice (in opposite directions) through the interaction region. The conditions for amplification of weak signal waves S_1 and S_2 by a shifted transmission hologram are simultaneously fulfilled for both elementary two-wave mixing processes, keeping the oscillation threshold at the minimum ($\Gamma d^{th} \simeq 2$). Theoretical analysis [6.28, 53] predicts also that at $\Gamma d \gtrsim 2$ reflectivity of such a phase-conjugating mirror almost approaches the optical transmittivity M of the entire system for a single round-trip of the pump beam.

Two major specific features of the geometry should be noted. First, phase conjugation is observed here only for pump beams with a complicated structure [6.53]. Therefore, an auxilliary aberrator (Fig.6.8a) - for instance, a length of a multimode optical fiber - is needed [6.55]. Second, since beams R_2 and S_2, which interact within the medium, travel in opposite directions along the same optical path inside the ring cavity, the requirements on the coherence length of the pump source and mechanical stability of the resonator are least severe [6.56]. The use of $BaTiO_3$ [6.53, 56], SBN [6.53.57], and BTO [6.58] in this geometry of the optical oscillator were reported.

6.5.5 Double Phase-Conjugate Mirror

Efficient interaction of mutually noncoherent laser beams is also used in a geometry of a double phase-conjugate mirror (Fig.6.8b) proposed in [6.59, 60]. Photorefractive crystal is illuminated here by two noncollinear light beams R_1 and R_2 from independent lasers. Oscillations resulting in two signal waves S_1 and S_2 are also possible here when some threshold value of Γd is reached.

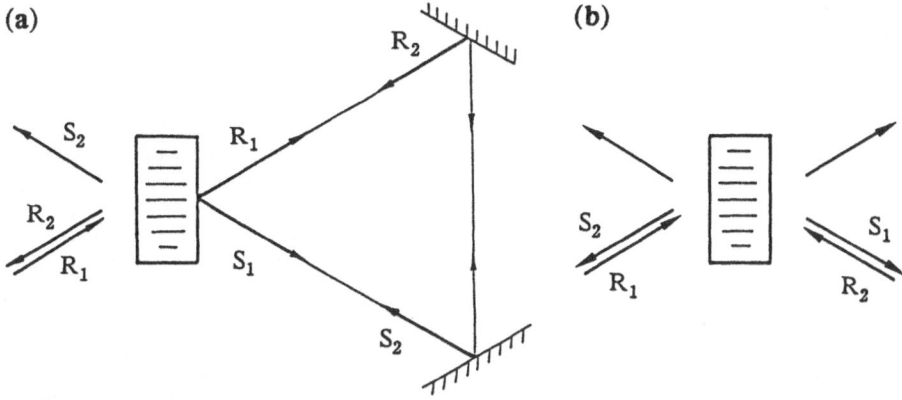

(a)   (b)

Fig.6.8. (a) Simplified configuration of a passive ring phase-conjugate mirror using PRC. (b) Double phase-conjugate mirror geometry

In a case of complicated wavefronts of R_1 and R_2, generated signal waves prove to be phase-conjugate replicas of corresponding counterpropagating pump beams ($S_1 \propto R_2^*$ and $S_2 \propto R_1^*$). This is not true, however, for simple plane pump waves when S_1 and S_2 are complicated noise-like waves propagating along conical surfaces on which wave vectors K_{R_1} and K_{R_2} are lying [6.61].

For equal intensities of pump beams ($r = I_{R_1}/I_{R_2} = 1$) threshold value $\Gamma d^{th} = 4$. It is clearly equal to that of a four-wave mixing geometry with positive feedback (Sect.6.4.1). For nonsymmetric pumping, a threshold grows as [6.35]:

$$\Gamma d^{th} = 2\frac{r+1}{r-1}\ln r \qquad\qquad (6.32)$$

This geometry was experimentally investigated mainly with $BaTiO_3$ [6.59,60] as a photorefractive medium and, in particular, using semiconductor GaAlAs lasers [6.62,63]. Efficient interaction in this arrangement was also observed in a cubic photorefractive BTO crystal at $\lambda = 633\,nm$ for recording in external alternating electric field [6.64]. Fig.6.9 represents intensities of the output, $I_{R_1}(d)$, and phase-conjugate, $I_{S_1}(d)$, light beams versus alternating field amplitude $E_\sim$, and also a threshold value of Γth versus pump beam ratio r, observed experimentally in this PRC.

6.5.6 Concluding Remarks

In conclusion we would like to note that investigations of different types of photorefractive oscillators and passive phase-conjugate geometries are in progress in different laboratories at present. In addition to the arrangements considered above, one can mention oscillators with two photorefractive crystals [6.60,65,66], oscillators using vector interactions via photogalvanic

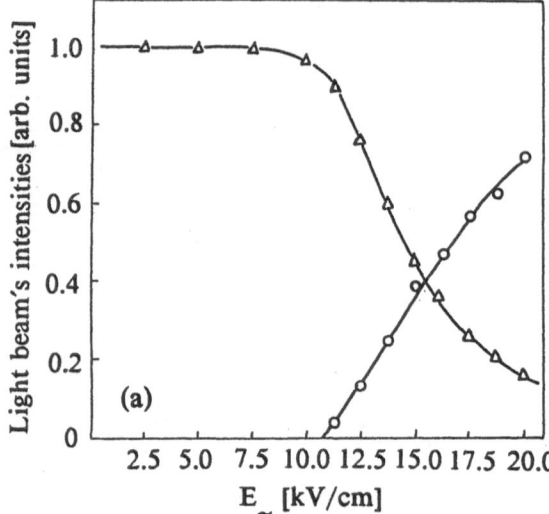

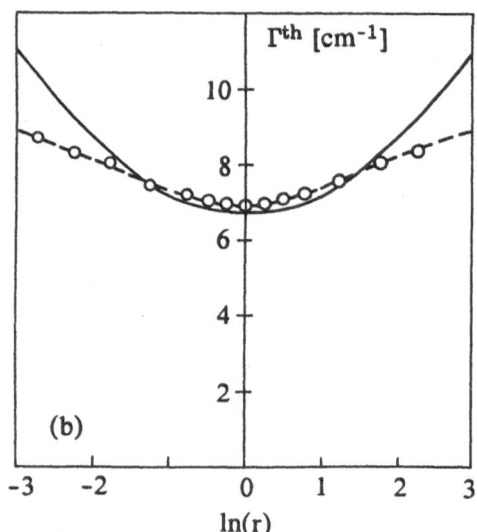

Fig.6.9. (a) Energy transfer between pump beam R_1 ($\triangle$) and diffracted beam S_1 (O), observed in a double phase-conjugate mirror arrangement at the output of a BTO sample versus amplitude of the external alternating electric field $E_\sim$ [6.64]. (b) Threshold gain factor Γ^{th} experimentally observed in a double phase-conjugate BTO mirror as a function of the pump-beam intensity ratio r [6.64]; the solid line represents a theoretical value of Γth ($\lambda = 633\,nm$, $\Lambda = 5\,\mu m$, $d = 6\,mm$)

mechanisms of hologram formation [6.67-69], and passive phase-conjugate geometries [6.70, 71] similar to those used in Brillouin phase-conjugate mirrors [6.72]. Serious attempts were also made to understand complicated spatial-time evolution of the light beams generated by photorefractive oscillators [6.73-76].

For more details of theory and experimental studies of photorefractive oscillators, we can recommend review papers by different authors published recently in [6.77] and also a book [6.78] by M.S. Soskin with co-workers, devoted especially to this topic.

7. The Physics of Electro-Optic Spatial Light Modulators

In this chapter we discuss the mechanisms of recording and readout of images from photorefractive Spatial Light Modulators (SLMs) and consider the factors determing their parameters. Many of the conclusions here are applicable to SLMs using electro-optic crystals that do not exhibit the photorefractive effect. Such modulators include the Titus, Phototitus, and the microchannel SLM.

7.1 Dynamics of Charge Formation in Photorefractive SLMs Under Nonuniform Illumination

In Sect. 4.6 we analyzed the volume charge formation within the photorefractive crystal under uniform illumination with recording light. We assumed that the charge injection from the electrode into the crystal is limited because the contacts are not ohmic. This is a significant feature of the recording mechanism in PRCs when they are used in SLMs. In this section we shall inspect how the dynamics of charge formation in SLMs affects the readout light modulation amplitude during image recording, i.e., when the crystal is illuminated nonuniformly with recording light.

The rigorous calculation of the charge distribution for nonuniform illumination, which requires solution of the set of equations (4.5-7) for the three-dimensional case, is a fairly involved task. Therefore, some researchers [7.1-3] have employed an approximate method using the results obtained for uniform illumination, which involves the solution of a one-dimensional problem; the charge density varies along the z coordinate only.[1]

Every recording light exposure creates a corresponding charge distribution $\rho(z)$. It was supposed that a similar dependence on z takes place with nonuniform illumination [7.1-3]. However, now the charge density is a three-dimensional (3D) function such that the $\rho(z)$ dependence at every point in the xy plane corresponds to the local exposure density $W(x,y)$; so the 3D function $\rho(x,y,z)$ can be found. Then $\rho(x,y,z)$ is substituted into Poisson's equation, and the nonuniform field $E(x,y,z)$ arising during image recording is calculated. Such an approach neglects the effect of transverse

---

[1] We assume that z is orthogonal to the crystal surface. The external field is created using a pair of transparent electrodes on the crystal plate surface. Recording and readout light enter the crystal through the transparent electrodes.

components of the internal field, $E_x(x, y, z)$ and $E_y(x, y, z)$, on the charge transport during image recording. However, as the comparison with the experimental evidence indicated, this calculation gives qualitatively correct results.

In Sect.4.6 we showed that the recording light induces a positive charge in the PRC of the BSO type. Both the density and the thickness of the charge layer depend on the exposure W, i.e., a charge with spatially modulated density and thickness arises in the crystal under nonuniform illumination.

In Sect.7.5 we shall give an example of calculation of the readout light modulation for a particular model of charge distribution. Here we illustrate qualitatively how the modulation amplitude varies in the process of recording of a periodic grating in the transverse-effect SLM. We assume that a sinusoidal grating with an average exposure W_0 is recorded, and the photo-induced charge has the density $\rho(x, z) = \rho(z)(1 + \sin 2\pi\nu x)$ with $\rho(z) = \rho_0 =$ const when $0 < z < z_0$ and $\rho(z) = 0$ when $z > z_0$ ($z = 0$ corresponds to the crystal surface). In order to take into account qualitatively the effect of an electrode placed on the crystal surface we use the principle of *mirror reflection*. Accordingly, we calculate the field for the real and virtual charges that have opposite signs and symmetric positions with respect to the electrode plane, which is equivalent to the field of the real charge in the case the electrode is present on the crystal surface. The thickness of the charge layer is, for simplicity, assumed to be $z_0 \ll 1/\nu$. The modulation amplitude A can be obtained by integration of (3.7), so in our case A is proportional to Qz_0, where $Q = \rho_0 z_0$ is the net charge density across the crystal thickness [7.3]. The dependences $Q(W_0)$ and $z_0(W_0)$ were discussed in Sect.4.6. Here we use these results to demonstrate the variation of A with W_0.

At the first stage of recording corresponding to the exposure range $0 < W_0 < W_1$ in Fig.7.1, the charge density increases $\propto W_0$, and thickness of the

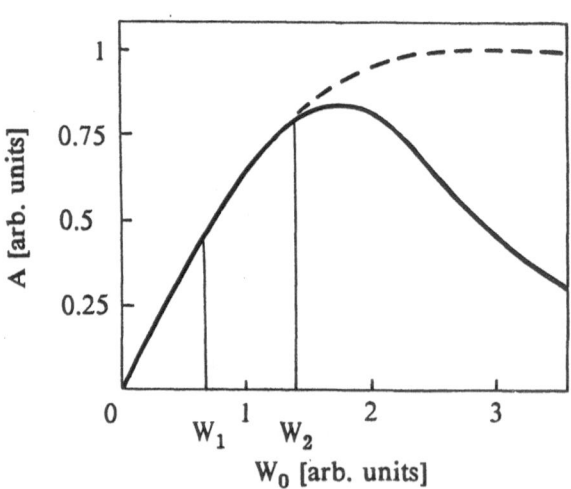

Fig.7.1. Qualitative dependence of the readout-light modulation amplitude A on the recording light exposure W_0. The dashed line shows the curve for the case when electrodes are insulated from the crystal. W_1 and W_2 are the boundaries between recording stages (see the text)

charged layer z_0 in the crystal remains constant. Here $Qz_0 \propto W_0$ and hence $A \propto W_0$; i.e., there is a linear relation between the output and input signals. The thicker the charge layer z_0, the higher A.

In Sect.4.6 we considered the case of low recording light absorption when z_0 equals the drift length of the photoelectrons. It has been shown [7.4] that in the general case

$$z_0 = \alpha^{-1}\chi(\chi - 1)^{-1}\ln\chi , \qquad (7.1)$$

where $\chi = \mu\tau|E_0|\alpha$. Analysis of (7.1) reveals that z_0 is commensurable with the least of the two values, i.e., either with the absorption depth of the recording light α^{-1} or with the electron drift length in the external electric field $L_0 = \mu\tau|E_0|$. Indeed, if $\alpha^{-1} \ll L_0$ the electrons are entirely removed from the layer that absorbs the recording light and the positive charge is accumulated.

In the opposite case, when $L_0 \ll \alpha^{-1}$, the light is absorbed almost uniformly at the electron drift length. The excited electrons leaving are compensated for by the arriving ones from the neighboring regions everywhere, except at the crystal layer of thickness L_0 adjacent to the crystal surface.

At the second stage of recording corresponding to exposures $W_1 < W_0 < W_2$ in Fig.7.1 the net charge grows $\propto W_0^{1/2}$ and the thickness of the layer z_0 decreases $\propto W_0^{-1/2}$. As a consequence, the modulation amplitude is exposure independent and reaches its maximum value.

If the electrode is not insulated from the crystal by the dielectric layer, injection of electrons from the electrode into the crystal can grow with increasing exposure. This compensates in part for the electrons leaving the region adjacent to the surface, thus slowing the growth of the positive charge density. Since its thickness simultaneously decreases, the readout-light modulation amplitude begins to fall. If the electrode is insulated, such a decrease does not occur and the modulation amplitude remains constant (the dashed line in Fig.7.1).

The $A(W_0)$ dependence associated with the dynamics of photoinduced charge formation, which we discussed here qualitatively, is experimentally observed in such photorefractive SLMs as PRIZ and PROM [7.5-7].

7.2 Light Propagation in Anisotropic Crystals

Before discussing spatial light modulation by electro-optic crystals, we analyze in more detail than before, how light travels through the anisotropic medium.

The propagation of light is characterized in crystal optics by the dielectric impermeability tensor $\hat{\alpha}^\omega$ that relates the strength of the electric field of a light wave, $\mathbf{A}$, to its electric induction vector $\mathbf{D}$, $\mathbf{A} = \hat{\alpha}^\omega \mathbf{D}$. Tensor $\hat{\alpha}^\omega$ is the inverse of the dielectric permeability tensor $\hat{\epsilon}^\omega$, $\hat{\alpha}^\omega\hat{\epsilon}^\omega = 1$, and like $\hat{\epsilon}^\omega$, is a Hermitian tensor of the second rank. Let us assume that the light is not absorbed in the crystal. Since there are optically active crystals among the crystals of interest, we shall consider a fairly general case of a birefringent optically active nonabsorbing crystal, for which we may write [7.8].

$$\mathbf{A} = \begin{vmatrix} \alpha_{11}^{\omega} & -i\alpha_{12}^{\omega} & 0 \\ i\alpha_{21}^{\omega} & \alpha_{22}^{\omega} & 0 \\ 0 & 0 & \alpha_{33}^{\omega} \end{vmatrix} \mathbf{D}, \quad \mathbf{B} = \mathbf{H} \tag{7.2}$$

where $\mathbf{B}$ and $\mathbf{H}$ are the vectors of induction and strength of the magnetic field of the light wave and $\alpha_{12}^{\omega} = \alpha_{21}^{\omega}$. For the crystals without optical activity, $\alpha_{12}^{\omega} = \alpha_{21}^{\omega} = 0$.

The field of the light wave in the crystal must obey Maxwell's equations, which can be written for a nonconducting medium as

$$\mathrm{curl}\mathbf{A} = -\frac{1}{c}\frac{\partial \mathbf{H}}{\partial t}, \quad \mathrm{div}\mathbf{H} = 0,$$

$$\mathrm{div}\mathbf{D} = 0, \quad \mathrm{curl}\mathbf{H} = \frac{1}{c}\frac{\partial \mathbf{D}}{\partial t}. \tag{7.3}$$

We assume that the monochromatic light with the circular frequency ω propagates along the z axis. The solution of the set (7.3) consists of the plane waves, which may be given by Maxwell's column vectors

$$\mathbf{D}(z,t) = \begin{vmatrix} D_x \\ D_y \end{vmatrix} e^{i(\omega t - K_L z)}, \quad \mathbf{H}(z,t) = \begin{vmatrix} H_x \\ H_y \end{vmatrix} e^{i(\omega t - K_L z)}, \tag{7.4}$$

where D_x, D_y, and H_x, H_y are the projections of the vectors of electric induction and magnetic field strength on the x and y axes, respectively, and K_L is the wave number of the light wave traveling along the z axis. By combining (7.2-4), we obtain the set of equations for D_x and D_y;

$$\left(\alpha_{11}^{\omega} - \frac{K_L^2 c^2}{\omega^2}\right)D_x - i\alpha_{12}^{\omega}D_y = 0,$$

$$\tag{7.5}$$

$$i\alpha_{12}^{\omega}D_x + \left(\alpha_{22}^{\omega} - \frac{K_L^2 c^2}{\omega^2}\right)D_y = 0.$$

The set (7.5) has a nontrivial solution only if its determinant vanishes. This yields the expression for K_L, i.e.,

$$\left(\alpha_{11}^{\omega} - \frac{K_L^2 c^2}{\omega^2}\right)\left(\alpha_{22}^{\omega} - \frac{K_L^2 c^2}{\omega^2}\right) - {\alpha_{12}^{\omega}}^2 = 0, \tag{7.6}$$

by solving which we find two magnitudes

$$K_{L1,2} = \left\{ \frac{\omega^2}{2c^2} \left[(\alpha_{11}^\omega + \alpha_{22}^\omega) \pm \sqrt{\Delta \alpha^{\omega\,2} + 4\alpha_{12}^{\omega\,2}} \right] \right\}^{1/2} , \qquad (7.7)$$

where $\Delta \alpha^\omega = (\alpha_{11}^\omega - \alpha_{22}^\omega)$. Now (7.4) gives two polarization types (eigenmodes of the light wave), each of which propagates through the crystal with its own phase velocity $v_{1,2} = c/n_{1,2}$, where $1/n_{1,2} = K_{L1,2}c/\omega$ are the refractive indices of the crystal for the eigenmodes

$$\mathbf{D}_1(z,t) = \mathbf{D}' \left| \begin{matrix} 1 \\ ij \end{matrix} \right| e^{i(\omega t + K_{L1} z)} , \quad \mathbf{D}_2(z,t) = \mathbf{D}'' \left| \begin{matrix} 1 \\ -i/j \end{matrix} \right| e^{i(\omega t + K_{L2} z)} , \quad (7.8)$$

where $\mathbf{D}'$ and $\mathbf{D}''$ are the scalar amplitudes, and

$$j = \frac{2\alpha_{12}^\omega}{\Delta \alpha^\omega + (\Delta \alpha^{\omega\,2} + 4\alpha_{12}^{\omega\,2})^{1/2}} . \qquad (7.9)$$

Such waves preserve their polarization as they propagate through the crystal, and the effect of the crystal involves merely a phase change.

According to (7.8,9), the eigenmodes in an optically active ($\alpha_{12}^\omega \neq 0$) birefringent ($\Delta \alpha^\omega \neq 0$) crystal have elliptic polarization. The axes of the ellipse for both modes are oriented along the x and y axes, and the degree of their ellipticity - i.e., the ratio between the lengths of the semimajor and semiminor axes - is the same and is governed by j. The vector $\mathbf{D}(z,t)$ is rotated clockwise in one wave and counterclockwise in the other. If $\Delta \alpha^\omega = 0$, $j = 1$ and the eigenmodes are the right- and left-hand circularly polarized waves with each of the waves having its own refractive index $1/n_{1,2} = (\alpha_{11}^\omega \pm \alpha_{12}^\omega)^{1/2}$. Thus we can say that the optically active crystal is circularly birefringent.

When all the nondiagonal components of the dielectric impermeability tensor are zero ($\alpha_{12}^\omega = \alpha_{21}^\omega = 0$), although $\alpha_{11}^\omega \neq \alpha_{22}^\omega$, the crystal is linearly birefringent. In this case the solution of (7.3) yields linearly polarized waves as eigenmodes. The refractive indices for these waves are $1/n_1 = \sqrt{\alpha_{11}^\omega}$ and $1/n_2 = \sqrt{\alpha_{22}^\omega}$.

If the light with the induction vector $\mathbf{D} = \left| \begin{matrix} D_x \\ D_y \end{matrix} \right|$ is incident on a plate of a linearly birefringent crystal of thickness d and the coordinate axes are chosen, as before, so that the eigenmodes have polarizations $\left| \begin{matrix} D_x \\ 0 \end{matrix} \right|$ and $\left| \begin{matrix} 0 \\ D_y \end{matrix} \right|$, the induction vector of the light immediately behind the crystal plate will be

$$\mathbf{D}' = \left| \begin{matrix} D_x \\ 0 \end{matrix} \right| e^{-i\varphi_1} + \left| \begin{matrix} 0 \\ D_y \end{matrix} \right| e^{-i\varphi_2} = \left| \begin{matrix} D_x e^{-\varphi_1} \\ D_y e^{-i\varphi_2} \end{matrix} \right| , \qquad (7.10)$$

where $\varphi_{1,2} = 2\pi n_{1,2}d/\lambda$. Because $n_1 \neq n_2$ for birefringent crystals, the light behind the crystal will generally be elliptically polarized according to (7.10). This equation may be rewritten as

$$\mathbf{D}' = \hat{\mathrm{T}}\mathbf{D}, \quad \text{with} \quad \hat{\mathrm{T}} = \begin{vmatrix} e^{-\varphi_1} & 0 \\ 0 & e^{-i\varphi_2} \end{vmatrix}. \tag{7.11}$$

The matrix $\hat{\mathrm{T}}$ is called the overall Jones matrix for linearly birefringent crystals. It is diagonal in the coordinate system where the directions of two axes coincide with the polarization directions of the eigenmodes. The dielectric impermeability tensor is also diagonal in this coordinate system. Such a coordinate system is called the principal system.

The Jones matrices can be used not only for birefringent crystals, but also for any other optical element which when traversed by light changes its polarization state. First of all, such elements include polarizers of different types. If the light passes through several optical elements, their resulting effect on the polarization state will be determined by the product of Jones matrices for each separate element. Bear in mind that the Jones matrices do not permute and should be written in a single coordinate system. There are tables for Jones matrices for principal coordinate systems. For an arbitrary çoordinate system the matrices can be derived by using the transformation $\mathrm{T} = \hat{a}^{-1}\mathrm{T}\hat{a}$, where $\hat{a}$ is the rotation matrix making the selected coordinate system coincident with the principal axes. By tradition the Jones matrix representation is used to analyze plane waves passing through an optical system. We shall show later that it is a very useful tool for considering spatial light modulation by electro-optic crystals.

The Jones matrix can be written for crystals, exhibiting optical activity similar, to (7.11). In this case, however, the light wave eigenmodes in the crystal are elliptically polarized and the incident light should therefore be represented as a superposition of two orthogonal elliptical waves given by (7.8,9). By performing transformations we obtain here [7.9]

$$\hat{\mathrm{T}} = \begin{vmatrix} \cos\tfrac{1}{2}\phi - i\cos\chi\sin\tfrac{1}{2}\phi & -\sin\tfrac{1}{2}\phi\sin\chi \\ \sin\chi\sin\tfrac{1}{2}\phi & \cos\tfrac{1}{2}\phi + i\cos\chi\sin\tfrac{1}{2}\phi \end{vmatrix}, \tag{7.12}$$

where $\phi = 2\pi(n_2 - n_1)d/\lambda$, n_1 and n_2 are the refractive indices for the elliptically polarized eigenmodes, $\sin\chi = 2j/(1+j^2)$, and $\cos\chi = (1-j^2)/(1+j^2)$.

In addition to the Jones matrix method, we shall use the notion of the index ellipsoid of the crystal. With it, we can conveniently define the directions of the optical axes, refractive indices, and the polarization directions of the eigenmodes of linearly birefringent crystals. The index ellipsoid is the characteristic surface of the dielectric impermeability tensor α^ω and is given by [7.9]

$$\alpha_{11}^\omega x^2 + \alpha_{22}^\omega y^2 + \alpha_{33}^\omega z^2 + 2\alpha_{12}^\omega xy + 2\alpha_{13}^\omega xz + 2\alpha_{23}^\omega yz = 1. \tag{7.13}$$

If the coordinate axes coincide with the principal ones and the light travels, as before, along the z axis, the intersection of the index ellipsoid by the plane normal to the light propagation direction is given by

$$\alpha_{11}^{\omega} x^2 + \alpha_{22}^{\omega} y^2 = 1 \ . \tag{7.14}$$

This is the expression describing the ellipse with the lengths of semiaxes equal to $1/\sqrt{\alpha_{11}^{\omega}}$ and $1/\sqrt{\alpha_{22}^{\omega}}$. Thus, the lengths of the principal semiaxes are equal to the refractive indices for the eigenmodes, and their directions coincide with the polarization directions of the linearly polarized modes.

If the crystal is placed into an electric field, its dielectric impermeability tensor $\hat{\alpha}^{\omega}$ varies through the electrooptic effect, thus leading to changes in the refractive properties of the crystal. Then, the refractive indices and polarization directions of the eigenmodes can undergo variations that can be treated as deformations of the index ellipsoid. Strictly speaking, the optical indicatrix can be used only in considering linear-birefringent crystals. For optically active crystals, the optical indicatrix can be employed when the circular birefringence is negligibly small compared with the linear one. In practice, this situation is fairly common.

7.3 The Electro-Optic Effect in Cubic Crystals

If an electric field is applied to the crystal, its refractive properties change through the electro-optic effect, i.e., the tensor $\hat{\alpha}^{\omega}$ components alter. According to the expressions given above, the phase velocities and/or polarization states of the eigenmodes in the crystals also undergo variations. By expanding the increments of the tensor $\hat{\alpha}^{\omega}$ caused by the electric field into a power series of E we can write

$$\hat{\alpha}^{\omega}(E) = \hat{\alpha}_0^{\omega} + \hat{r}E + \hat{R}EE + \dots \ , \tag{7.15}$$

where $\hat{\alpha}_0^{\omega}$ is the dielectric impermeability tensor at $E = 0$, $\hat{r}$ is the third-rank tensor of the linear electro-optic coefficients, and $\hat{R}$ is the fourth-rank tensor of quadratic electro-optic coefficients. For centrosymmetric media $\hat{r}$ = 0 [7.10]. Here the major contribution to the change of $\hat{\alpha}^{\omega}$ caused by application of an external field comes from the third term of (7.15).

Among cubic crystals, the noncentrosymmetric crystal classes are $\overline{4}3m$ and 23. As far as the PRCs used in SLMs are concerned, the first class includes such crystals as ZnS and ZnSe, and the second class involves the crystals of the sillenite family ($Bi_{12}SiO_{20}$, $Bi_{12}GeO_{20}$, $Bi_{12}TiO_{20}$, etc.). Crystals of these classes have the following nonzero linear electro-optic coefficients: $r_{41} = r_{52} = r_{63}$.[2] For the initially optically isotropic crystals α_{11}^{ω}

---

[2] For convenience, we employ the contracted indices traditional for electro-optics, here possible because of the permutation symmetry of tensor $\hat{r}$. So we have $r_{11i} = r_{1i}$, $r_{22i} = r_{2i}$, $r_{33i} = r_{3i}$, $r_{23i} = r_{32i} = r_{4i}$, $r_{13i} = r_{31i} = r_{5i}$, $r_{12i} = r_{21i} = r_{6i}$.

$= \alpha_{22}^{\omega} = \alpha_{33}^{\omega}$ and, as is apparent from (7.13), their index ellipsoids are spheres. However, the crystals belonging to the 23 class are optically active; i.e., they exhibit circular birefringence whose magnitude is independent of the light propagation direction.

Let us treat the electro-optic effect in a cubic crystal that does not possess optical activity. The index ellipsoid for a cubic crystal is

$$\alpha_0^{\omega}(x^2 + y^2 + z^2) + 2r(E_x yz + E_y xz + E_z xy) = 1 , \qquad (7.16)$$

where $\alpha_0^{\omega} = \alpha_{11}^{\omega} = \alpha_{22}^{\omega} = \alpha_{33}^{\omega}$, $r = r_{41} = r_{52} = r_{63}$, and E_x, E_y and E_z are the components of the vector $\mathbf{E}$ along the respective axes. Below we discuss three most typical propagation directions with respect to the crystal axes.

In our first example, the light travels along the [001] axis, which corresponds to z for (7.16). By setting z = 0, we find the intersection of the index ellipsoid with the plane normal to the propagation direction from (7.16)

$$\alpha_0^{\omega}(x^2 + y^2) + 2rE_z xy = 1 . \qquad (7.17)$$

This equation immediately indicates that the refractive index changes for the light passing along [001] may be caused by the electric field component E_z directed along this axis alone, and do not depend on E_x and E_y. In electro-optics such a component (E_z) is traditionally said to be longitudinal. Since other electric field components do not give rise to refractive-index variations, it is generally said that only the longitudinal effect is possible with this light propagation direction.

To find the refractive indices n_1 and n_2, the equation for the index ellipsoid should be rewritten for the principal axes. This can be done by rotating the coordinate system by 45° about the z axis. The equation for the index-ellipsoid intersection in the new coordinate system is

$$(\alpha_0^{\omega} - rE_z)x^2 + (\alpha_0^{\omega} + rE_z)y^2 = 1 . \qquad (7.18)$$

Hence, $n_{1,2} = (\alpha_0^{\omega} \mp rE_z)^{-1/2}$. Since the initial refractive index of the crystal is $n_0 = 1/\sqrt{\alpha_0^{\omega}}$ and in real situations $|rE_z| \ll \alpha_0^{\omega}$, the method of approximations yields the new refractive indices

$$n_{1,2} = n_0 \pm \tfrac{1}{2} n_0{}^3 rE_z . \qquad (7.19)$$

Let a light wave polarized along the [010] axis be incident on the crystal. The electric induction vector of this wave at the crystal input $\mathbf{D} = \dfrac{D}{\sqrt{2}} \begin{vmatrix} 1 \\ 1 \end{vmatrix}$ can be represented as a sum of two eigenmodes, where D is the wave amplitude. By substituting (7.19) into (7.10) we obtain for the emerging wave

$$\mathbf{D'} = \frac{D}{\sqrt{2}} e^{-[i\pi(n_1 + n_2)d/\lambda]} \begin{vmatrix} e^{-\frac{1}{2} i\Delta\varphi} \\ e^{\frac{1}{2} i\Delta\varphi} \end{vmatrix} , \qquad (7.20)$$

116

where $\Delta\varphi = \varphi_1 - \varphi_2 = 2\pi n_0{}^3 rU/\lambda = \pi U/U_{\lambda/2}$, $U = E_z d$ is the potential difference between the input and output crystal surfaces, and $U_{\lambda/2}$ is the half-wave voltage of the crystal introduced in Chap.2. The phase difference at $U = U_{\lambda/2}$ is π. Here the wave at the crystal output has a linear polarization orthogonal to the initial one.

In our second example the light wave travels along the [110] axis. To find the intersection of the index ellipsoid by the plane normal to the propagation direction, the coordinate axis x is directed along the [1$\bar{1}$0] crystallographic axis, y is along [00$\bar{1}$], and z is along [110]. By using the coordinate transformation and taking z = 0, we obtain from (7.16) the equation for the index-ellipsoid intersection:

$$(\alpha_0^\omega + rE_y)x^2 + \alpha_0^\omega y^2 + 2rE_x xy = 1 \ . \tag{7.21}$$

This equation shows that the shape of the index-ellipsoid cross section is affected only by the components E_x and E_y of the electric field induced in the crystal. These components are orthogonal to the propagation direction and are typically termed transverse components. Since the longitudinal field component does not affect the magnitudes of refractive indices and the polarization states of eigenmodes, we say that only the transverse electrooptic effect is observed when the light travels along the [110] axis of a cubic crystal.

We introduce the projection E_t of vector E on the surface parallel to the plane of the wave front ($E_t = |E_t| = (E_x{}^2 + E_y{}^2)^{1/2}$). Next we transform the coordinate system to the principal axes through rotation by angle

$$\psi = \frac{1}{2}\arctan(2\cot\gamma) \ , \tag{7.22}$$

where γ is the angle between the direction of vector E_t and the [1$\bar{1}$0] crystal axis (Fig.7.2), and ψ is the angle of the coordinate system rotation measured off the [110] axis. As a result we obtain[3]

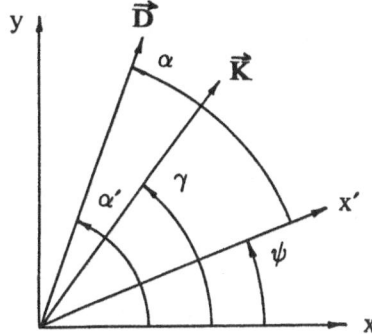

y

$\vec{D}$

α

$\vec{K}$

α'

γ

x'

ψ

x

Fig.7.2. Mutual orientations of crystal axes, one of the principal axes of the index ellipsoid cross-section x', wave vector of sinusoidal grating K_x, and readout-light polarization direction D. (When the crystal plate has the (111) orientation, the [110] axis coincides with x, for the (110) crystal orientation the [001] axis is along x)

---

[3] An expression similar to (7.23) has been given for the left coordinate system [7.11,12].

$$[\alpha_0^\omega + rE_t(\sin\gamma\cos^2\psi + \cos\gamma\sin2\psi)]x^2 +$$

$$[\alpha_0^\omega + rE_t(\sin\gamma\sin^2\psi - \cos\gamma\sin2\psi)]y^2 = 1 \ ,$$

$$n_1 = n_0 - \tfrac{1}{2}n_0{}^3 rE_t(\sin\gamma\cos^2\varphi + \sin2\varphi\cos\gamma) \ , \tag{7.23}$$

$$n_2 = n_0 - \tfrac{1}{2}n_0{}^3 rE_t(\sin\gamma\sin\varphi - \sin2\varphi\cos\gamma) \ .$$

Thus the refractive indices and polarization directions of the eigenmodes depend on the direction of the projection E_t of the electric field strength vector.

In addition, (7.23) predicts one more difference from the longitudinal effect. The electric field changes the phase difference between the eigenmodes $\Delta\varphi$, thereby changing the polarization state of the light wave at the output, while the wave phase defined by the average refractive index $(n_1 + n_2)/2$ remains unaltered. However, when the light travels along the [1$\bar{1}$0] axis, the electric field changes the average refractive index $(n_1+n_2)/2$ in accordance with (7.23). As a consequence, not only the polarization state, but also the phase of the light wave at the crystal output varies (7.20).

In our third example, the light travels along the [111] crystal axis. Let us write the equation for the index ellipsoid cross section in the coordinate system with x along the [1$\bar{1}$0] axis, y along [11$\bar{2}$], and z along [111]. As usual, by setting z = 0, we obtain [7.13]

$$\left[\alpha_0^\omega - \frac{r}{\sqrt{3}}(E_z - \sqrt{2}E_y)\right]x^2 + \left[\alpha_0^\omega - \frac{r}{\sqrt{3}}(E_z + \sqrt{2}E_y)\right]y^2 - 2\sqrt{2/3}\,rE_x xy = 1 \ . \tag{7.24}$$

The principal index ellipsoid cross section here can be obtained by rotating the coordinate axes by the angle

$$\psi = \tfrac{1}{2}(\tfrac{1}{2}\pi - \gamma) \ . \tag{7.25}$$

So we have

$$\left[\alpha_0^\omega - \frac{r}{\sqrt{3}}(E_z - \sqrt{2}E_t)\right]x^2 + \left[\alpha_0^\omega - \frac{r}{\sqrt{3}}(E_z + \sqrt{2}E_t)\right]y^2 = 1 \ , \tag{7.26}$$

which gives the refractive indices

$$n_1 = n_0 - \frac{n_0{}^3 r}{2}\sqrt{\frac{2}{3}}\,E_t + \frac{rn_0{}^3 E_z}{2\sqrt{3}} \ ,$$

$$\tag{7.27}$$

$$n_2 = n_0 + \frac{n_0{}^3 r}{2}\sqrt{\frac{2}{3}}\,E_t + \frac{rn_0{}^3 E_z}{2\sqrt{3}} \ .$$

Therefore, when light travels along the [111] axis, the refractive indices are affected by both the transverse (E_t) and longitudinal (E_z) components of the electric field. However, they influence the light passing through the crystal in different manners. The longitudinal component changes only the average refractive index and, hence, causes variation of the transmitted light phase, but leaves its polarization state unaltered. The transverse component E_t changes the polarization state, but does not affect the phase. In addition, the direction of the vector of the transverse component E_t determines, according to (7.25), the orientation of the principal axes of the index-ellipsoid cross section and, hence, the polarization directions of the light wave eigenmodes. If the component E_t is rotated, but its magnitude is unchanged, the ellipsoid cross section is rotated one half as fast without altering the shape. For instance, if E_t reverses its direction, the semiminor and semimajor axes of the ellipsoid cross section exchange positions. This means that variations of the refractive indices become of opposite sign, as should indeed be in the linear electro-optic effect.

Note that here we considered the examples where light travels strictly along one of the three axes of a cubic crystal. If the light deviates from these axes, the conclusions may not be valid. For instance, in some cases misalignment between propagation direction and the [111] axis by 1° leads to the observation of the longitudinal effect together with the tranverse one.

7.4 Light Diffraction in a Thin Cubic Electro-Optic Crystal

The electro-optic effect was discussed in Chap.3 and Sect.7.3 for uniform electric fields. Here we shall consider the diffraction of light passing through the electro-optic crystal where a sinusoidal electric field grating is recorded (Fig.7.3):

$$E(x, z) = E(z)\sin Kx . \tag{7.28}$$

The refractive indices of the crystal for the eigenmodes are given by

$$n_{1,2}(x, z) = n_0 + \Delta n_{1,2}(x, z) , \tag{7.29}$$

where $\Delta n_{1,2}$ are variations of the initial refractive index n_0 caused by the electro-optic effect, the light propagates along the z axis, and x is orthogonal to the grating fringes. The method for calculation of $\Delta n_{1,2}$ and the equations for the three propagation directions were given in the preceding section. If the electric field in the crystal is expressed by (7.28) and we take into account linearity of the electro-optic effect, we can write $\Delta n_{1,2}$ as

$$\Delta n_{1,2}(x, z) = \Delta n_{1,2}(z)\sin Kx , \tag{7.30}$$

where

$$\Delta n_{1,2}(z) = A_{1,2} E_t(z) + B_{1,2} E_z(z) \tag{7.31}$$

is in the form general for all propagation directions discussed above. Here $A_{1,2}$ and $B_{1,2}$ are coefficients that depend on the propagation direction and, in addition, $B_{1,2}$ depends on the direction of the transverse field component $E_t(z)$. According to the preceding section, $A_{1,2} = 0$ when light travels along the [100] axis (the longitudinal effect), and $B_{1,2} = 0$ when the light is along the [110] axis (the transverse effect).

Let us separate the crystal plate into layers by the planes parallel to its surfaces through which the light passes, as shown in Fig.7.3. We choose the number of layers such that the thickness of each layer Δz is sufficiently small and the changes of field $E(x,z)$ along the z axis can be neglected. The effect of the layer labeled by ℓ on the light passing through it can be described by the Jones matrix, the diagonal form of which is given by (7.11). The matrix may be written for our case as

$$\hat{T}(x, z_\ell) = \exp\left[-i\frac{2\pi(n_1+n_2)\Delta z}{2\lambda}\right]\begin{vmatrix} e^{-i\Delta\varphi_1(z_\ell)\sin Kx} & 0 \\ 0 & e^{-i\Delta\varphi_2(z_\ell)\sin Kx} \end{vmatrix}, \quad (7.32)$$

where $\Delta\varphi_{1,2} = 2\pi\Delta n_{1,2}(z)\Delta z/\lambda$, and z_ℓ is the coordinate of the layer labelled ℓ. For simplicity, we shall omit the factor $i2\pi(n_1+n_2)z/2\lambda$ which can be the same for all layers if the light travells along the [001] or [110] axes. If the diffraction effects inside the crystal can be ignored (the case of a thin hologram), the resultant effect of the crystal can be represented by the product of Jones matrices for each separate layer:

$$\hat{T}(x) = \hat{T}(x, z_1)\hat{T}(x, z_2) ... \hat{T}(x, z_L) . \quad (7.33)$$

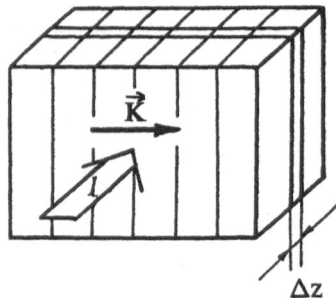

Fig.7.3. Direction of the light propagation (*1*) through an electro-optic crystal and wave vector of a sinusoidal grating **K** recorded in the crystal. In order to calculate light modulation, the crystal is divided into layers of thickness Δz

Let us inspect the case when the transverse field component is parallel to the same line everywhere in the crystal. As follows from the preceding section, this is enough for the eigenmodes to have the same polarization throughout the crystal. Here the matrices $T(x, z_\ell)$ for all layers are diagonal in a single coordinate system. Then multiplication of matrices (7.32) yields summation of the phase shifts acquired by the light wave in each separate layer in the resulting matrix $\hat{T}(x)$. By allowing L to extend to infinity and passing from a summation to an integration, we obtain for the resulting matrix

$$\hat{T}(x) = \begin{vmatrix} e^{-i\varphi_1 \sin Kx} & 0 \\ 0 & e^{-i\varphi_2 \sin Kx} \end{vmatrix}, \quad \varphi_{1,2} = \frac{2\pi}{\lambda} \int_0^d \Delta n_{1,2}(z) dz, \quad (7.34)$$

where d is the crystal-plate thickness. As follows from (7.31,34), for the longitudinal effect

$$\varphi_{1,2} \sin Kx = \frac{2\pi}{\lambda} B_{1,2} \int_0^d E_z(x,z) dz \propto U(x); \quad (7.35)$$

that is, the phase shifts depend only on the potential difference between the points on opposite crystal surfaces through which the light passes. For the transverse effect,

$$\varphi_{1,2} \sin Kx = \frac{2\pi}{\lambda} A_{1,2} \int_0^d E_t(x,z) dz. \quad (7.36)$$

To analyze the diffraction of light transmitted through the crystal, (7.34) is expanded into a Fourier series using the expression

$$e^{i\varphi \sin Kx} = \sum_{m=-\infty}^{\infty} J_m(\varphi) e^{imKx}, \quad (7.37)$$

where J_m is the m-order Bessel function. If the input light wave has the amplitude $D = \begin{vmatrix} D_1 \\ D_2 \end{vmatrix}$, we have at the rear surface of the crystal

$$D(x) = \sum_{m=-\infty}^{\infty} e^{imKx} \begin{vmatrix} J_m(\varphi_1) & 0 \\ 0 & J_m(\varphi_2) \end{vmatrix} \begin{vmatrix} D_1 \\ D_2 \end{vmatrix}. \quad (7.38)$$

Each term of the sum (7.38) describes the plane wave that propagates at a certain angle θ to the z axis, i.e., to the initial propagation direction of the incident light, with $\sin\theta = m\lambda\nu$. At m = 0 this is the zero diffraction order. The terms with m = ±1, ±2 ... describe the first, second, etc. diffraction orders. Equation (7.38) reveals that if the crystal is illuminated by a linearly polarized light, all diffraction orders will have linear polarization. The odd diffraction orders will be polarized orthogonally to the even orders, including the zero one if $\varphi_1 = -\varphi_2$ and $D_1 = D_2$, since the Bessel functions of even orders are even and of odd orders are odd. Here the zero and all even orders can be totally suppressed using an analyzer of polarization, which will pass the first and all other odd diffraction orders unattenuated.

The changes of the polarization state of the light diffracted from the refractive index grating have, in this case, a simple physical interpretation.

Indeed, we can say that the electric field produces two refractive-index gratings in the crystal, one for each orthogonal eigenmode. Each eigenmode diffracts in the crystal independently without changing the polarization states, and then the diffracted waves add. The polarization state of the resultant wave depends on the phase difference between two waves arising from the diffraction from each refractive index grating. If the phase difference is zero, or the crystal is illuminated by only one eigenmode, the polarization of the diffraction order coincides with the initial polarization of the input light. The condition $\varphi_1 = -\varphi_2$ means that the gratings for the ordinary and extraordinary waves are opposite in phase - i.e., they are shifted by π - and, hence, the ordinary and extraordinary waves are also opposite in phase. As a result, the resultant diffracted wave has a linear polarization that differs from the initial one and, if $D_1 = D_2$, is orthogonal to it.

Though we assume that a purely sinusoidal electric field grating is induced in the crystal, (7.38) indicates that not only the first, but also higher diffraction orders must be observed in a general case. This is attributable to the fact that only the electric field and phase shift of the light wave are linearly related in the linear electro-optic effect, while the relation between the complex light wave amplitude and electric-field strength is not linear, as indicated by (7.34). Not only higher diffraction orders result here. If two or more sinusoidal field gratings are produced in the crystal, the diffraction orders with combination frequencies appear, and the amplitudes and polarizations of the fundamental diffraction orders can vary. The expression taking into account the mutual influence of several sinusoidal gratings has been derived in [7.14], and the equation allowing for the effect of the average component of the field grating has been obtained [7.12]. In the latter case it was assumed that the electric field in the crystal was in the form

$$E(x,z) = E_0(z) + E_1(z)\sin Kx , \qquad (7.39)$$

where $E_0(z)$ is the average field component. The phase shifts of the eigenmodes emerging from the crystal, due to the electro-optic effect, are then given by

$$\varphi_{1,2}(x) = \varphi'_{1,2} + \varphi''_{1,2}\sin Kx , \qquad (7.40)$$

where $\varphi'_{1,2}$ and $\varphi''_{1,2}$ depend on $E_0(z)$ and $E_1(z)$, respectively. Similar to the case with (7.38), we can obtain

$$D(x) = \sum_{m=-\infty}^{\infty} e^{iKmx} \begin{vmatrix} J_m(\varphi''_1)e^{i\varphi'_1} & 0 \\ 0 & J_m(\varphi''_2)e^{i\varphi'_2} \end{vmatrix} \begin{vmatrix} D_1 \\ D_2 \end{vmatrix} . \qquad (7.41)$$

Here it follows that, when $\varphi'_1 \neq \varphi'_2 \neq 0$ and the linearly polarized light is incident on the crystal, all diffraction orders will be elliptically polarized.

122

Equation (7.41) yields the expressions for the diffraction efficiency. If there is no analyzer of polarization,

$$\eta = J_1^2(\varphi_1)\cos^2\alpha + J_1^2(\varphi_2)\sin^2\alpha \ , \qquad (7.42)$$

where α is the angle between the direction of initial polarization and the polarization direction of one of the light eigenmodes in the crystal.

If an analyzer is placed behind the crystal, the diffracted light amplitude can be obtained by multiplying (7.41) by the Jones matrix of the analyzer written in the coordinate system whose axes coincide with the polarization directions of the eigenmodes in the crystal. Now the diffraction efficiency can be derived. For instance, for a crossed analyzer

$$\eta = \frac{1}{4}[J_1^2(\varphi_1) + J_1^2(\varphi_2) + J_1(\varphi_1)J_1(\varphi_2)\cos\varphi_0]\sin^2 2\alpha \qquad (7.43)$$

where $\varphi_0 = \varphi_1' - \varphi_2'$. For the transverse effect when $\varphi_1 = -\varphi_2$ and $\varphi_0 = 0$, (7.43) has a simple form

$$\eta = J_1^2(\varphi_1)\sin^2 2\alpha \ . \qquad (7.44)$$

Equations (7.42-44) indicate that the maximum possible diffraction efficiency for a thin birefringent grating, both with the analyzer and without it, is 34% (the maximum value of J_1 is 0.58). This result coincides with the maximum diffraction efficiency of a thin phase hologram.

7.5 The SLM Transfer Function

Equation (3.23) shows that the transfer function of the electro-optic SLMs $\chi(\nu,\xi) \propto \varphi_{1,2}(\nu,\xi)$. Therefore, in this section we shall find the $\varphi_{1,2}(\nu,\xi)$ dependence. The SLM will be regarded as a multilayered structure, as shown in Fig. 7.4. The results obtained for such a structure can be used to interpret the experimental data for most SLMs discussed in this book.

In Chap. 3 the electric field in the electro-optic crystal was considered without taking into account explicitly the boundary conditions. In contrast, here we allow a finite thickness of the crystal, and the presence of electrodes and dielectric layers in the modulator structure. Therefore, boundary conditions must be introduced.

We assume that a volume charge with density $\rho(x, y, z)$ is formed in the electro-optic crystal during image recording. The dielectric permeability of layer 1 is denoted by ϵ_1, of layer 2 by ϵ_2. Here d_1, d_2 are the layer thicknesses, respectively. The coordinate system is chosen so that its origin along the z axis, which is perpendicular to the electrode plane, coincides with the dielectric-layer crystal interface. The electric field potential $\Phi(x, y, z)$ must obey Poisson's equation

$$\text{div grad}\Phi(x, y, z) = - \frac{\rho(x, y, z)}{\epsilon_0 \epsilon_{1,2}} \ . \qquad (7.45)$$

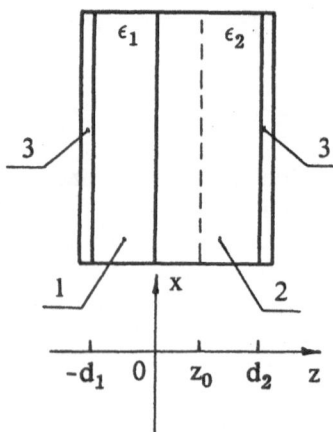

Fig.7.4. Structure of a two-layered SLM. (*1*: insulating layer, the charge $\rho(x,y,z)$ is located in the electro-optic crystal *2*, *3*: electrodes)

To solve the equation, we separate the volume charge into sufficiently thin layers of thickness Δz orthogonal to the z axis. The surface-charge density in the layer with coordinates x,y,z_0 is assumed to be $\rho(x,y,z_0)\Delta z$. Now we derive the potential $\Phi(x,y,z,z_0)$, which produces the charge layer with the coordinate z_0 in the electro-optic crystal, and the overall potential $\Phi(x,y,z)$ obeying (7.45) is obtained by integrating over z_0

$$\Phi(x,y,z) = \int_0^{d_2} \Phi(x,y,z,z_0)dz_0 \ . \tag{7.46}$$

We assume that a one-dimensional volume sinusoidal charge grating with density

$$\rho(x,z) = \rho(z)\sin Kx \tag{7.47}$$

is formed in the SLM [7.11, 12].

The solution of (7.45) for the potential $\Phi(x,z,z_0)$ induced by a selected charge layer z_0 will be derived separately for layer *1* where no charge is present [the potential in this layer is denoted by $\Phi_1(x,z,z_0)$], for layer *2* to the left from z_0 [$\Phi_{21}(x,z,z_0)$], and for layer *2* to the right from z_0 [$\Phi_{22}(x,z,z_0)$]. Equation (7.45) must be supplemented by boundary conditions. They can be given on the assumption that the electrodes are shorted and have a zero potential, and also by writing the relations for the potentials and their derivates at the interface between layers *1* and *2* and in the z_0 plane:

$$\Phi_1(x,-d_1,z_0) = \Phi_{22}(x,d_2,z_0) = 0 \ , \tag{7.48}$$

$$\epsilon_1 \frac{\partial \Phi_1(x,z,z_0)}{\partial z}\bigg|_{z=0} = \epsilon_2 \frac{\partial \Phi_{21}(x,z,z_0)}{\partial z}\bigg|_{z=0} \ , \tag{7.49}$$

$$\frac{\partial \Phi_{21}(x,z,z_0)}{\partial z}\bigg|_{z=z_0} - \frac{\partial \Phi_{22}(x,z,z_0)}{\partial z}\bigg|_{z=z_0} = \frac{\rho(x,z_0)}{\epsilon_0 \epsilon_2} , \qquad (7.50)$$

$$\Phi_1(x,0,z_0) = \Phi_{21}(x,0,z_0) , \qquad (7.51)$$

$$\Phi_{21}(x,z_0,z_0) = \Phi_{22}(x,z_0,z_0) . \qquad (7.52)$$

The solutions will be

$$\Phi_1(x,z,z_0) = C_1 shK(z+d_1)sinKx , \qquad (7.53)$$

$$\Phi_{21}(x,z,z_0) = C_{21} shKz sinKx + C_{22} chKz sinKx , \qquad (7.54)$$

$$\Phi_{22}(x,z,z_0) = C_3 shK(z-d_2)sinKx . \qquad (7.55)$$

The solutions satisfy (7.45) and the boundary conditions (7.48,49) when the electric field is created by the charge $[\rho(z_0)sinKx]dz_0$. The constant factors C_1, C_{21}, C_{22} and C_3 can be inferred from (7.50-52). As a result we find the total field created by a plane sinusoidal charge in the bulk of the two-layer structure. Knowing the field of each layer of the volume charge, we derive the total electric field of the charge - whose density is given by (7.47) - by integrating (7.46).

As discussed in Sect.7.4 where we considered the spatial light modulation by electro-optic crystals, two types of the electro-optic effect exist (the longitudinal and transverse effects). Both phenomena are used in the SLMs. Now we consider these phenomena separately.

7.5.1 The Longitudinal Electro-Optic Effect

The diffraction phenomena arising as the light passes through the SLM crystal are usually ignored. The eigenmodes experience incremental phase shifts φ_1 and φ_2 through the longitudinal effect:

$$\varphi_{1,2} = \frac{2\pi}{\lambda} B_{1,2} \int_0^d E_z(x,y,z)dz = \frac{2\pi}{\lambda} B_{1,2} U(x,y) , \qquad (7.56)$$

where, as before, d is the crystal thickness; E_z is the longitudinal component of the electric field (i.e., the magnitude of projection of the field strength vector on the direction of light propagation - we assume that light travels along z); $U(x,y)$ is the potential difference between the points on the opposite crystal faces with coordinates x,y; and B_1 and B_2 are the electric-field-independent factors. For a cubic crystal, when light travels along the [001] axis, $B_{1,2} = \pm \frac{1}{2} n_0^3 r$ (Sect.7.2).

It follows from (7.56) that the light modulation in the longitudinal SLMs is governed by the potential difference $U(x,y)$. Thus the electro-optic crystal in this SLM structure must be insulated at least from one of the electrodes. If both electrodes are deposited immediately on the plate surfaces, the surfaces prove to be equipotential: $U(x, y) = const.$ Hence, $\varphi_{1,2}$ = const, no matter how complicated the field created in the crystal. Therefore, the crystal not insulated from the electrodes cannot spatially modulate light through the longitudinal effect.

We assumed above that both electrodes have zero potential. The potential differences between the surfaces of layers 1 and 2 have therefore different signs, but equal moduli. As a consequence, the modulation of the phases φ_1 and φ_2 (neglecting the sign) will be the same, regardless of whether the charge is in the crystal or in the dielectric layer. If the electrodes have different potentials what is said above is valid for all spatial frequencies, except the zero one.

For convenience, we assume that the electro-optic crystal is presented by layer 2 (Fig.7.4) and $d = d_2$. Taking into account (7.48) and inserting the expression for C_{22} into (7.54), we obtain the potential difference $U(x, y) = \Phi_{22}(x, 0, z_0)dz_0$ induced by the charge in the z_0 plane. The corresponding phase shifts of the eigenmodes are [7.15]

$$\varphi_{1,2}(x, y) = \varphi_{1,2}(K)\sin Kx , \qquad (7.57)$$

$$\varphi_{1,2}(K) = \frac{2\pi B_{1,2}\sigma(z_0)shKd_1\, shK(d_2 - z_0)}{\lambda\epsilon_0 K(\epsilon_1 chKd_1\, shKd_2 + \epsilon_2\, shKd_1 chKd_2)} , \qquad (7.58)$$

where the amplitude of the surface charge density $\sigma(z_0) = \rho(z_0)dz$.

Equation (7.58), as well as the qualitative treatment of the electric-field gratings in the medium exhibiting the longitudinal electro-optic effect (Chap.3), indicates that the modulation amplitude peaks when the charge is on the crystal surface. In our case this corresponds to $z_0 = 0$, i.e., the interface of the electro-optic crystal-insulating layers, when the function $shK(d_2 - z_0)$ in (7.58) has a maximum.[4] For $z_0 \neq 0$, when the charge plane shifts into the crystal volume or the insulating layer, the light modulation amplitude decreases at all spatial frequencies. Moreover, note that at fairly high spatial frequencies when $\exp[K(d_1 + d_2)] \gg 1$, (7.58) yields

$$\varphi_{1,2}(K) = \frac{2\pi B_{1,2}\sigma(z_0)e^{-Kz_0}}{\lambda\epsilon_0 K(\epsilon_1 + \epsilon_2)} . \qquad (7.59)$$

In other words, $\varphi_{1,2}(K)$ sharply decreases for $z_0 \neq 0$. At $z_0 = 0$, the rate of the modulation amplitude decrease is the lowest and is inversely proportional to the spatial frequency.

Let us treat now the volume-charge distribution. The transfer function is found by integrating (7.58) over z_0. We take the simplest case as an illustration and assume that the charge is located in a layer of thickness d_a

---

[4] Eq.(7.58) for the two-layer structure at $z_0 = 0$ and for the case when the crystal is an anisotropic dielectric (its dielectric permeability is a tensor) was obtained by *Roach* [7.16].

near the interfaces between the insulating layer and the crystal, and that it forms a volume sinusoidal grating. The charge-density amplitude is constant $\rho(z_0) = \rho$ across the charge layer along the z axis. By integrating (7.58), we obtain [7.15]

$$\varphi_{1,2}(K) = \frac{2\pi B_{1,2}\rho[\mathrm{ch}Kd_2 - \mathrm{ch}K(d_2 - d_a)]}{\lambda\epsilon_0 K^2 \mathrm{sh}Kd_2[\epsilon_2 \coth Kd_2 + \epsilon_1 \coth Kd_1]} . \qquad (7.60)$$

We can show that in the limit $d_a \to 0$, (7.60) transforms into (7.58). Equation (7.60) reveals that at fairly high K, when e^{Kd_1}, e^{Kd_2}, and $e^{Kd_a} \gg 1$,

$$\varphi_{1,2}(K) = \frac{2\pi B_{1,2}\rho}{\lambda\epsilon_0 K^2(\epsilon_1 + \epsilon_2)} . \qquad (7.61)$$

That is to say, the modulation decreases at high spatial frequencies inversely proportional to the square of the spatial frequency. This conclusion was inferred here on the assumption that the charge distribution amplitude remains unaltered in the layer of thickness d_a near the crystal surface. Along with this, it has been shown that this conclusion holds for other types of charge distribution across the thickness if the charge layer is near the crystal surface [7.17]. The $\varphi_{1,2}(K)$ dependence of the type given by (7.61) is observed in the PROM device and is an essential feature of the volume-charge distribution [7.6].

Figure 7.5 plots $\varphi_{1,2}$ as a function of K for different magnitudes of d_a. The net charge in the volume of the second layer of the SLM structure was taken to be the same for all curves, irrespective of the charge-layer thickness d_a. For the examples shown in Fig. 7.5, parameters typical of the PROM SLM are chosen (the first layer is parylene, $\epsilon_1 = 3$, $d_1 = 5$ μm; the second layer is the photorefractive BSO crystal, $\epsilon_2 = 56$, $d_2 = 500$ m). Figure 7.5 demonstrates that increasing the charge-layer thickness reduces the modulation amplitude. The higher the spatial frequency, the faster the process. Therefore, resolution decreases with increasing layer thickness.

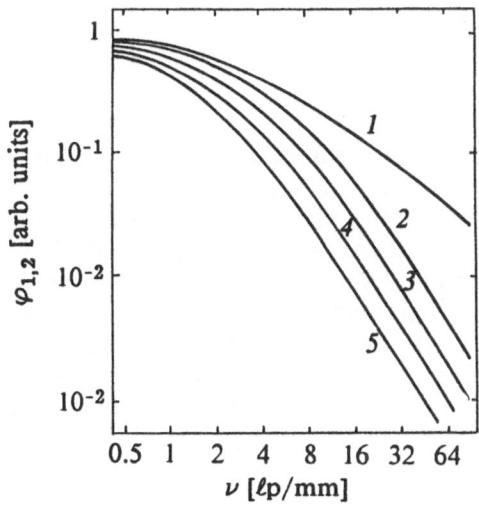

Fig. 7.5. Phase modulation $\varphi_{1,2}$ as a function of spatial frequency ν for the longitudinal electro-optic effect ($d_1 = 5\,\mu$m, $d_2 = 500\,\mu$m, $\epsilon_1 = 3$, $\epsilon_2 = 56$, d_a - 1: 0, 2: 25, 3: 50, 4: 100, 5: 200 [μm])

7.5.2 The Transverse Electro–Optic Effect

In the case of the transverse electo–optic effect, the eigenmodes passing through the crystal experience phase shifts:

$$\varphi_{1,2}(x,y) = \frac{2\pi}{\lambda} A_{1,2} \int_0^d E_t(x,y,z)dz .$$ (7.62)

The coefficients A_1 and A_2 depend on the orientation of the crystal plate planes with respect to the crystallographic axes and on the direction of the transverse component E_t. For the cubic crystal, A_1 and A_2 are determined for the linear electo–optic effect from (7.25-27), and (7.22,23) for the (111) and (110) crystal orientations, respectively.

In contrast to the longitudinal effect, here the modulation is not governed by the potential difference between the points on the opposite crystal faces, as seen from (7.62). Therefore, if a nonuniform electric charge is induced in the crystal, the light can be modulated even when transparent electrodes are directly on the crystal surface. Indeed, let us suppose that a plane sinusoidal charge has the z_0 coordinate in the electro–optic crystal. By differentiating (7.54 and 55), we find the transverse electric field to the left and right from the charge plane (Fig.7.4). Integration of (7.62) yields φ_1 and φ_2. At $d_1 = 0$ (both electrodes are immediately on the crystal surface)

$$\varphi_{1,2}(K) = \frac{2\pi A_{1,2}\,\sigma(z_0)}{\lambda\epsilon_0\epsilon_2 K\,\mathrm{sh}Kd_2}[\mathrm{sh}Kd_2 - \mathrm{sh}Kz_0 - \mathrm{sh}K(d_2-z_0)] .$$ (7.63)

From (7.63) we see that $\varphi_{1,2} = 0$ at $z_0 = 0$. However, the charge grating on the crystal surface can give rise to light modulation if there is no electrode on this surface. This situation can arise, for instance, if the charge is created on one of the crystal surfaces by an electron beam. If there is an electrode on the opposite surface, then

$$\varphi_{1,2}(K) = \frac{2\pi A_{1,2}\,\sigma(0)(1-\mathrm{sh}Kd)}{\lambda K\epsilon_0(\epsilon_1 + \epsilon_2\,\mathrm{cth}Kd)\mathrm{sh}Kd} ,$$ (7.64)

where ϵ_1 is the dielectric permeability of the medium ($\epsilon_1 = 1$ for vacuum).

From (7.63) $\varphi_{1,2} \propto K$ at fairly small K, and $\varphi_{1,2}(0) = 0$. In addition, φ_1 and φ_2 have maximum values when the charge plane moves to the center of the crystal across its thickness and has the coordinate $z_0 = d_2/2$. In this position, the charge plane is spaced at the maximum distance from both crystal surfaces. Such a dependence of φ_1 and φ_2 on the position of the charge plane significantly distinguishes the transverse effect from the longitudinal one. Recall that the maximum modulation amplitude is observed for the longitudinal effect when the charge is on the crystal surface, and $\varphi_{1,2} = 0$ when the charge is in the center of the crystal. For convenience, we take the crystal without electrodes with a plane sinusoidal charge in the center across the thickness. The origin of coordinates along the z axis coincides with the charge plane. The solution of Poisson's equation (7.45), i.e., the potential of the electric field of the charge $\sigma(0) = \sigma_0 \sin Kx$, will be

$$\Phi(x,z) = \Phi_0 e^{Kz} \sin Kx \quad \text{at} \quad z \leq 0 ,$$

$$\Phi(x,z) = \Phi_0 e^{-Kz} \sin Kx \quad \text{at} \quad z \geq 0 , \tag{7.65}$$

where $\Phi_0 = \sigma_0 / 2\epsilon_0 \epsilon_2 K$, and ϵ_2 is the dielectric permeability of the crystal. By differentiating (7.65) over z and x, we find the longitudinal and transverse components of the electric field, respectively:

$$E_z(x,z) = E_0 e^{Kz} \sin Kx , \quad E_t = E_0 e^{Kz} \cos Kx \quad \text{for} \quad z \leq 0 ,$$

$$E_z(x,z) = - E_0 e^{-Kz} \sin Kx , \quad E_t = E_0 e^{-Kz} \cos Kx \quad \text{for} \quad z \geq 0 , \tag{7.66}$$

where $E_0 = \sigma_0 / 2\epsilon_0 \epsilon_2$, which coincides with (3.7).

Let us recall once more what are the reasons for major differences in the dependences of the modulation on charge position in the crystal in the longitudinal and transverse electro-optic effects. As seen from (7.66), the transverse and longitudinal field components are equal in moduli at the points in the crystal symmetric with respect to the charge plane that is at $(x,-z)$ and (x,z). The longitudinal components are of the opposite sign. As a result, on integration of (7.56), the phase shift of the light wave due to the longitudinal effect to the left of the charge plane is totally compensated for by the phase shift of the light wave to the right of this plane. Thus $\varphi_{1,2} = 0$ for the longitudinal effect when the charge is at the center of the crystal. Because of the same sign of the transverse components to the right and left from the charge plane, such a compensation does not take place in the transverse effect. Therefore, by integrating (7.62), we obtain $\varphi_{1,2} \neq 0$.

From (7.66) the amplitude of the electric field decays exponentially with separation from the charge plane and the grating spatial frequency $\nu = K/2\pi$. The electric field has a pronounced spatial modulation in a small layer immediately adjacent to the charge plane, rather than in the whole crystal. The thickness of this layer is $\propto 1/\nu$ and is the same for both the longitudinal and transverse field components. Therefore, the thickness of the charge layer that modulated the readout light decreases with increasing spatial frequency. This is the reason for a reduction in the amplitude at high frequencies of the recorded gratings.

The $\varphi_{1,2}(K)$ dependence of the transverse effect in the limit of high spatial frequencies can be obtained from (7.63). For this purpose, we ignore the $e^{-K(d_2 - z_0)}$ and e^{-Kz_0} values in (7.63). This is equivalent to assuming that the grating spacing is sufficiently small as compared with the separation between the charge and any of two electrodes with coordinates $z = 0$ and $z = d_2$. As a result, we have

$$\varphi_{1,2}(K) = \frac{2\pi A_{1,2} \sigma(z_0)}{\lambda K \epsilon_0 \epsilon_2} . \tag{7.67}$$

129

Thus, in the case under discussion, the amplitude in the high-frequency limit does not depend on the position of the charge plane in the crystal volume and is inversely proportional to the spatial frequency. Comparing (7.67) with a similar expression (7.59) for the longitudinal effect shows that these equations bear a close resemblance. In both cases, $\varphi_{1,2}$ have similar dependences on spatial frequencies when $z_0 = 0$ in (7.59), i.e., when the charge is on the crystal surface. As noted above, no compensation

of the light modulation in different parts of the crystal, with which e^{-Kz_0} is associated in (7.59), occurs with such a charge position in the longitudinal effect. For the transverse effect, compensation does not take place for an arbitrary charge position in the electro-optic-crystal bulk.

We now discuss the volume charge distribution for the transverse effect. As before, we assume that the electro-optic crystal is incorporated in the two-layer structure (Fig.7.4) and that the charge is in the crystal bulk in a layer interface. The charge is in the form of a sinusoidal grating with the amplitude ρ, which is constant across the charge-layer thickness. The $\varphi_{1,2}(K)$ dependences are obtained by summing the contributions of all layers of the volume charge to the light modulation. To this end, we integrate (7.63) over the variable z_0 and obtain [7.11]

$$\varphi_{1,2}(K) = \frac{2\pi A_{1,2}\rho}{\lambda\epsilon_0\epsilon_2 K}[1 - C(K)],$$

where

$$C(K) = \frac{\epsilon_1[chKd_2 + chKd_a - chK(d_2-d_1) - 1] + \epsilon_2 thKd_1 shKd_a}{K(\epsilon_1 thKd_2 + \epsilon_2 thKd_a)chKd_2}. \qquad (7.69)$$

Figure 7.6 shows the $\varphi_{1,2}(K)$ dependences calculated from (7.68,69) for different charge-layer thicknesses d_a. The other parameters entering (7.69) are typical of the PRIZ SLM, which uses the transverse effect in the

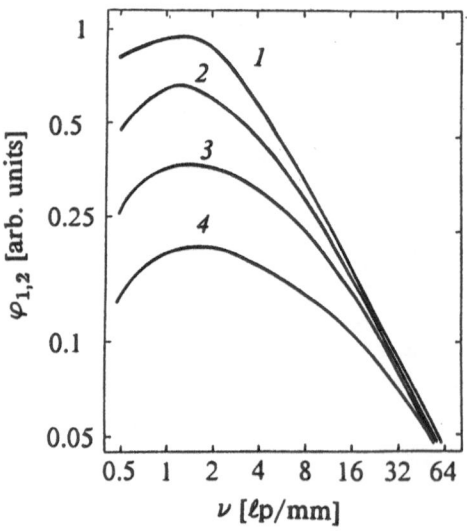

Fig.7.6. Phase modulation $\varphi_{1,2}$ of the readout light as a function of spatial frequency ν for the transverse electro-optic effect ($d_1 = 0$, $d_2 = 500\,\mu m$, $\epsilon_2 = 56$, d_a – 1: 200, 2: 100, 3: 50, 4: 25 [μm])

BSO-type crystals: $d_1 = 0$, $d_2 = 500$ μm, $\epsilon_2 = 56$ [7.6]. For sufficiently low spatial frequencies ($\nu < 1$ ℓp/mm in Fig.7.6), $[1-C(K)] \propto K^2$. As a result, at $K \rightarrow 0$, $\varphi_{1,2}(K) \rightarrow 0$ and $\varphi_{1,2}(0) = 0$. Thus, in the transverse effect, the low spatial frequencies in the recorded image are suppressed, and the zero spatial component - i.e., information on the average level of intensity in the image - is not reconstructed. The latter means in fact that the average value of the transverse field throughout the plate is zero for any charge distribution. The function $\varphi_{1,2}(K)$ peaks at the spatial frequencies $\nu \simeq 1/d_2$ and then decreases with increasing spatial frequency ν, with $C(K) \rightarrow 0$ at $K \rightarrow \infty$. Therefore, at fairly high ν [for the typical parameters included in (7.69), at $\nu > 10$ ℓp/mm], $\varphi(K) \propto 1/K$.

Figure 7.6 shows that the maximum amplitude increases with increasing charge-layer thickness at low spatial frequencies ranging from 1 to 6 ℓp/mm. Along with this, the amplitude does not change with thickness at high frequencies. If we regard $\varphi_{1,2}(K)$ as a transfer function of the SLM, then, according to the data given above, a decrease in the charge-layer thickness leads to an increase of the SLM resolution with an accompanying reduction of the maximum achievable modulation amplitude. The differences in the $\varphi_{1,2}(K)$ dependences for the transverse and longitudinal effects are evident from a comparison of Figs.7.5 and 6. The major difference is that the increase of the charge-layer thickness leads to a lower modulation amplitude at all spatial frequencies for the longitudinal effect. The SLM resolution also decreases, because the light modulation in the longitudinal effect is ensured by the charge located near the crystal surface. The thickness of the charge that efficiently participates in the light modulation decreases.

7.6 Sensitivity of Electro-Optic SLMs

According to the definitions given in Chap.3, we shall regard the sensitivity S^{-1} [J/cm^2] of an optically addressed SLM as the density of the recording light energy needed for the difraction efficiency to reach the magnitude of 1% when a sinusoidal grating is recorded. The diffraction efficiency is spatial frequency dependent. Therefore, to obtain unambiguous estimates, the spatial frequency at which measurements were carried out, should be stated. As a rule, the literature specifies the SLM sensitivity for the spatial frequencies corresponding to the transfer function maximum, where S^{-1} is the lowest. (To avoid misunderstanding, we note that the higher the SLM sensitivity, the lower the numerical value of sensitivity measured).

The results from Sects.7.4,5 allow calculation of the diffraction efficiency of the electro-optic SLM if the volume density of the charge $\rho(x, y, z)$ formed in the bulk of the modulator during grating recording is known. To estimate the maximum possible theoretical sensitivity of such SLMs (the minimum magnitude of S^{-1}), we assume that the recording light generates the charge with absolute efficiency; i.e., all the light is absorbed, and an absorption of a light quantum gives rise to excitation and removal of one electron from the modulator structure. The overall positive-charge density throughout the thickness is given by

$$\sigma(x, y, z) = \int_0^d \rho(x, y, z)dz = \frac{eW(x, y)}{\hbar\omega} , \tag{7.70}$$

where $\hbar\omega$ is the energy of the light quantum, and e is the electron charge.

Further, we need to make an assumption on how the charge is distributed throughout the modulator thickness, since, as shown above, this affects the readout-light modulation amplitude and, hence, the SLM sensitivity.

7.6.1 The Longitudinal Electo-Optic Effect

If the longitudinal effect is used, the maximum modulation amplitude for a given charge density is achieved if all the charge is on the electro-optic crystal surface insulated by a dielectric layer from the electrode. The diffraction efficiency is obtainable from (7.43,58), taking $z_0 = 0$. To use (7.58) to estimate S^{-1} at the transfer function maximum at $\nu = 0$, we substitute (7.70) into (7.58) and use the limiting transition $K = 2\pi\nu \to 0$. By using (3.4) and m = 1 we obtain

$$S^{-1} = \frac{\hbar\omega}{e} \frac{0.2}{\pi} \frac{U_{\lambda/2}\epsilon_0(\epsilon_2 d_1 + \epsilon_1 d_2)}{d_1 d_2} . \tag{7.71}$$

As (7.71) reveals, S^{-1} decreases and, hence, the SLM sensitivity increases with (1) an increasing half-wave voltage $U_{\lambda/2}$ of the crystal, (2) decreasing dielectric permeabilities of the materials used, and (3) increasing thickness of the layers in the modulator.

Let us estimate the sensitivity of the PROM and Phototitus. The PROM uses the electro-optic BSO crystal with $U_{\lambda/2} = 3900$ V, $\epsilon_2 = 56$, and $d_2 = 0.5$ mm, and a dielectric layer with $d_1 = 10$ μm and $\epsilon_1 = 3.1$ [7.16]. Blue light with the quantum energy of 2.8 eV is used for recording. By substituting these values into (7.71), we obtain $S^{-1} = 1.6$ μJ/cm^2. The larger of the two terms ϵ_1/d_1 and ϵ_2/d_2 in (7.71) for a PROM is determined by the parameters of the dielectric layer. To increase the sensitivity we must therefore make the dielectric layer thicker or reduce its dielectric permeability. One of the factors responsible for a higher experimentally measured S^{-1} of a PROM as compared with our estimate is that a volume rather than a surface charge is formed in the modulator during recording. The SLM sensitivity decreases in this case. By comparing (7.58 and 60) at K = 0 we obtain the ratio of $(d_2 - d_a/2)/d_2$ between the sensitivites S^{-1} for the surface charge and the charge across the crystal-layer thickness d_a.

The Phototitus devices employ the DKDP crystal, which operates near the phase-transition temperature and has a relatively low half-wave voltage $U_{\lambda/2} = 200$ V, a high dielectric permeability $\epsilon_2 = 600$, and thickness $d_2 = 200$ μm. The other layer is an amorphous photoconductor of thickness $d_1 =$

10 μm with $\epsilon_2 = 2$. The maximum theoretical sensitivity of the Phototitus estimated from (7.71) using the above parameters is $S^{-1} = 0.64$ $\mu J/cm^2$. The experimental S^{-1} is found above the theoretical value. The difference is evidently attributed to the inefficient charge formation during image recording, which is due to short drift lengths of the photoexcited carriers in the amorphous photoconductor.

7.6.2 The Transverse Electro-Optic Effect

The maximum theoretical readout-light modulation amplitude is here attained if the plane charge grating is in the volume of the electro-optic crystal. If there are electrodes on the crystal surface ($d_1 = 0$), the charge must be located in the middle of the crystal across its thickness ($z_0 = d_2/2$) (Sect. 7.5). Analysis of (7.63) reveals that the transfer function maximum is at $\nu \simeq 3/2\pi d_2$. Equations (7.43, 63, 70) yield

$$S^{-1} \simeq \frac{\hbar\omega}{e} \frac{0.6}{\pi} \frac{U_{\lambda/2}\epsilon_0\epsilon_2}{d_2} \qquad (7.72)$$

for the (111) cut cubic crystal at the transfer-function maximum. If we take into account the electro-optics, S^{-1} for the (110)-cut crystal is about 1.5 times as low as from (7.72). As with the longitudinal effect, (7.72) indicates that S^{-1} decreases with increasing thickness d_2 and decreasing dielectric permeability ϵ_2 of the crystal.

According to (7.72), the sensitivity S^{-1} of the PRIZ SLM, which uses the transverse effect in photorefractive BSO crystals with $d_2 = 500$ μm, is 1.4 $\mu J/cm^2$. Thus the theoretical estimate of S^{-1} for the transverse effect differs only slightly from S^{-1} for the longitudinal effect in the crystals of the same type (the PROM SLM). However, the experimental sensitivity of the PROM is an order of magnitude lower than that of the PRIZ. The difference is due to the fact that a volume photo-induced charge distribution, rather than a surface one, is formed in BSO during image recording. As a consequence, the transverse effect proves to be much more efficient for spatial light modulation.

7.7 Noise and Phase Distortions

When information is recorded and then read out from the SLM, the signal-to-noise ratio becomes lower due to noise introduced by the SLMs and other elements of the optical system. The SLM noise can arise from a variety of factors. Among them are such modulator defects as scratches and dust on the active surfaces and optical inhomogeneities of the volume, including mechanical stresses, which give rise to birefringence through the elasto-optic effect. If an analyzer is used, the birefringence leads to amplitude and phase noise. The listed defects do not distinguish the SLMs from the passive elements of optical systems, such as lenses, mirrors and prisms.

In addition, the photoelectric parameters can vary throughout the SLM structure, to result in different sensitivities of SLMs in various parts of the active surface. This also reduces the signal-to-noise ratio.

The listed defects do not typically vary during operation and can therefore be regarded as nonuniform background during information recording. The effect of the background on the image readout can be eliminated by taking special measures for each separate SLM sample. However, it is a complicated task in practical situations. Therefore, we consider here only such defects as a noise signal.

Such typical defects of optical elements as dust and scratches scatter light mainly without depolarization. The use of the linear electro-optic effect for the readout light modulation allows attenuation (rather strong one in some cases) of noise arising from these defects, defects of not only the SLM itself, but also of other elements located in the optical system before the SLM was included. To clarify the situation, let us consider the signal-to-noise ratio in the frequency plane of a coherent optical processor. We assume that an image of a sinusoidal grating is recorded on the SLM. The light scattered from defects of the optical elements and the SLM produces a halo around the zero diffraction order in this plane, thereby reducing the signal-to-noise ratio. If there is no analyzer, the signal-to-noise ratio is given by

$$S/N = I_1/I_f , \tag{7.73}$$

where I_1 is the light intensity in the diffraction spot, and I_f is the intensity of the halo in the region of the diffraction order.

As shown in Sect.7.4, the diffracted light has a linear polarization differing from the initial one if the linearly polarized readout light is used. For instance, the light polarization in the diffraction order can be orthogonal to the initial one. If a crossed analyzer is placed behind the modulator, it completely transmits the diffracted light. The zero order and halo are elliptically polarized in the general case. The degree of ellipticity depends on the phase difference φ_0 between the light wave's eigenmodes in the crystal averaged over the cross section of the readout beam. The intensity transmittance of an ideal crossed analyzer for the zero order and halo is $T = \sin^2 \varphi_0$. If $\varphi_0 = 0$, their polarizations do not differ from the initial ones, and a total suppression of noise arising from scattering (halo) must occur. A real analyzer does not perform the total suppression, and we can write $T = T_0 + \sin^2 \varphi_0$, where T_0 is the transmittance for the light of "crossed" polarization. If the diffraction order is polarized orthogonally to the initial one, the signal-to-noise ratio in the presence of the analyzer is similar to (7.73),

$$S/N = \frac{I_1}{I_f(T_0 + \sin^2 \varphi_0)} . \tag{7.74}$$

Thus, the analyzer allows a $(T_0 + \sin^2 \varphi_0)$-fold increase of the signal-to-noise ratio as compared with the situation when no analyzer is employed.

To achieve the maximum increase, the average birefringence of the SLM crystal itself and other optical elements through which the light passes before reaching the analyzer must be at the minimum. Experiments have been designed to suppress the halo by a factor of $10^2 - 10^4$. Special measures are taken in the Titus, Phototitus and PROM to compensate for the average birefringence. In the PRIZ, the compensation is automatic.

One more factor that deteriorates the noise properties of the SLMs is the unwanted reflections from the interfaces in the modulator structure. The reflections can give rise to additional reflexes in the operating frequency range in the frequency plane and complicated interference patterns (a stripped structure of the readout images) in the image plane of a coherent optical processor. This is a significant factor for SLMs having a complex multilayered structure with many interfaces, for instance the Phototitus, which will be discussed in Sect.8.3. Besides depositing antireflection coatings, a simple and efficient method to eliminate the effect of reflections, partly or completely, is the fabrication of modulator elements as wedges. This allows reflexes to be removed from the operating frequency range and to be filtered. Moreover, this phenomena is attenuated by using a polarizer, as discussed above, since polarization is mainly preserved on reflection.

To eliminate phase distortions in SLMs, they should first of all be fabricated with sufficiently flat surfaces to prevent strong distortion of the readout light wave front as it passes through the device. Typically, the wave front should be set accurate to $\lambda/4$ to $\lambda/10$. Since SLM resolution increases with decreasing crystal thickness, crystal plates of small thickness (100-500 μm) and large area ($1-10\text{cm}^2$) are required. Fabrication of such plates can cause certain difficulties.

Since the electro-optic crystals used in the SLMs are piezoelectrics, the crystal plates can deform because of the action of the external electric fields and changes in the internal fields during image recording. These can also cause phase distortions. As experiments indicated, the deformations do not introduce strong phase distortions during the transmissive readout. However, if the readout is performed in the reflection mode, where the readout light is reflected from the dielectric mirror on the surface of the deformable crystal plate, appreciable phase distortions can result.

8. Spatial Light Modulators

Before proceeding to a discussion of particular SLMs, it is worth summarizing their salient parameters (Table 8.1), which were discussed in Sect.2.3. Note that in some specific applications the requirements on SLM parameters may lie beyond the ranges indicated here, because we give values for the sake of reference. However, if a real device has a set of parameters inferior to the lowest values, the extent of its applicability will be restricted first of all because the optical information processing system using such an SLM will be unable to compete with the electronic processing units. We notice also that the list is not complete, since it does not include such parameters of practical significance as cost, reliability, energy consumption, and ease of handling, which may be deciding factors in the choice of a modulator. Besides the photorefractive SLMs, we shall consider the Titus and Phototitus devices that use electro-optic crystals. The readout light is modulated in these SLMs through the electro-optic effect, image recording occurs due to nonuniform charge accumulation. These devices can, therefore, be analyzed in the framework common to all SLMs discussed in this monograph.

Table 8.1. Reference requirements to SLM parameters

1. Space–bandwidth product	$\geq 10^6$
Resolution	$15 - 100$ $\ell p/mm$
Active area	$1 - 10$ cm^2
2. Speed of operation	$10 - 100$ cycles/s
3. Sensitivity	$10^{-4} - 10^{-7}$ J/cm^2
4. Dynamic range (in the Fourier plane, sinusoidal test)	≥ 60 dB
5. Nonlinear distortions	$1 - 10$ %
6. Phase distortions of the readout light wavefront	less than $\lambda/2$

8.1 Pockels Readout Optical Modulator

The Pockels Readout Optical Modulator (PROM) device [8.1-7] was the first SLM which made use of PRCs.

136

8.1.1 Description of the PROM

The first version of a PROM device employed epitaxially grown PRCs of ZnS [8.1-2]; later, modulators made of ZnSe and $Bi_{12}SiO_{20}$ (BSO) were suggested [8.3-4]. The highest PROM performance is obtained with the BSO crystal because of its relatively low half-wave voltage $U_{\lambda/2} = 3.9$ kV ($U_{\lambda/2} = 13$kV for ZnS and 11.3kV for ZnSe [8.4]) and the possibility of growing large crystals (more than 50mm in diameter). Devices using $Bi_{12}GeO_{20}$ (BGO) with $U_{\lambda/2} = 5.5$ kV [8.8] have also been fabricated. All the PRCs mentioned above are cubic in nature.

The crystal plate for the PROM is cut so that the normal to its surface coincides with the [100] axis. With this cut, the device modulates light through the longitudinal electro-optic effect (Sect.7.3). The plates, which are typically 0.25 to 1 mm thick, are made in the form of wedges with an angle of 1° to prevent reflection of light from the crystal surfaces into the image plane. The surfaces are polished to the quality of optically flat and coated with insulating dielectric poly-dichloro-para-xylylene (parylene) layers 3 to 10 μm thick, and transparent electrodes on both large faces. The electrodes are prepared by vacuum deposition, typically of In_2O_3.

In the modulator, depicted schematically in Fig.8.1a,b, the images are read out in the transmission and reflection modes, respectively. In the latter case, a dielectric mirror that reflects the readout light is sandwiched between the crystal surface and the dielectric layer. The modulation amplitude here is twice as high as in the readout via transmission, because of a double passage of the readout light through the crystal. Hence the diffraction efficiency is four times as high. In spite of this, application of a PROM, employing the BSO-type crystal and operating in the reflection mode in coherent-optical systems, is difficult. The piezoelectric effect exhibited by these crystals causes deformation of the crystal plate and, hence, of its mirror surface. The shape of the deformations depends on the field distribution, and also on how the crystal is supported in the SLM structure. This leads to almost uncontrollable phase distortions of the readout wave front, which can be higher than the distortions arising in the readout via transmission by several orders of magnitude.

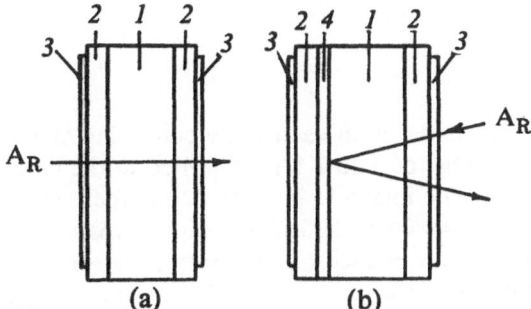

(a)   (b)

Fig.8.1. PROM modulators performing readout in the transmission (a) and reflection (b) modes. (*1*: a cubic photorefractive crystal, *2*: insulating parylene layer, *3*: transparent electrodes, *4*: dielectric mirror; A_R: readout light)

The choice of insulating layer defines the PROM quality in many respects. It has been shown that BSO plates are of high optical quality [8.7]. The optical noise of the PROM device due to light scattering from the imperfections of the device structure is mainly determined by the quality of the dielectric layers. In addition, the insulating layers affect markedly the information storage time and cause variations of the PROM parameters over the active surface, which contribute to the noise [8.9]. The organic polymer parylene has a set of parameters that makes it suitable for PROMs, namely, a good adhesion to the crystal surface, transparency in a broad wavelength range, a high breakdown voltage, low dielectric losses, and elasticity and vapor deposition gives thin uniform layers. Elasticity is important because of the deformation of the crystal plate during operation. The shortcomings of parylene involve its sensitivity to UV illumination as well as humidity, which reduce the breakdown voltage [8.10].

Besides parylene, other dielectrics were used in PROMs as insulating layers. For instance, a BGO modulator employing glass plates has been reported [8.8]. An optical glue used to attach the glass plates with transparent electrodes (Fig.8.2a) can also serve as an insulator [8.11]. In the SLM described in [8.12, 13], the electrodes were insulated from the crystal by mica (Fig.8.2b). Since mica is a birefringent material, the SLM was fabricated with two insulating layers of equal thickness with the optical axes at 90°. In this case the birefringence of one insulating layer compensates for that of the other.

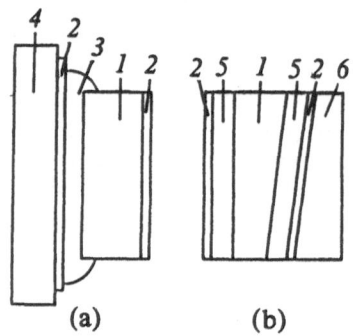

(a)   (b)

Fig.8.2. PROM modulators (a) with insulating layer made of optical glue; (b) with an insulating layer made of mica. (*1*: PRC, *2*: transparent electrodes, *3*: optical glue, *4*: glass plate, *5*: mica slice, *6*: PRC plate compensating for optical activity)

8.1.2 Operation of the PROM

The device is operated in the cyclic mode, with each cycle consisting of recording, readout, and erasure. A voltage of 1 to 2 kV is applied to the electrodes for the purpose of recording. The image to be recorded is focused on the PROM in blue or violet light (λ = 400 to 500nm), which gives rise to photoconductivity in BSO. Since the capacitance of the crystal plate is tens of times lower than that of the dielectric layers, initially all of the voltage applied drops across the crystal. The free carriers of the electric charge generated in the illuminated regions drift in the external electric field and are trapped both in the crystal volume and on the surface. As a consequence, a nonuniform charge distribution corresponding to the recording-light intensity is created in the crystal. The photoinduced charge formation in BSO in the longitudinal electric field was discussed in Sects.4.6 and 7.1.

The stored image is usually read out with linearly polarized light whose polarization direction is along one of the crystal's [100] axes (Fig.8.3). An analyzer, oriented to have minimum transmittance before recording when no voltage is yet applied, is placed behind the modulator in the readout-light path. The polarization modulation is thus converted into amplitude modulation. In the bright regions of the recorded image, the external field is screened by the photoinduced charge in the crystal volume, and the bire-fringence is lowest. An inverted image thus appears after recording. The BSO crystal exhibits optical activity [8.14]; i.e., it rotates the polarization plane of the light passing through the crystal by an angle proportional to the crystal-plate thickness (typically 15° to 20°). The analyzer orientation differs by the same angle from the normal to the direction of the initial polarization of the readout beam (Fig.8.3).

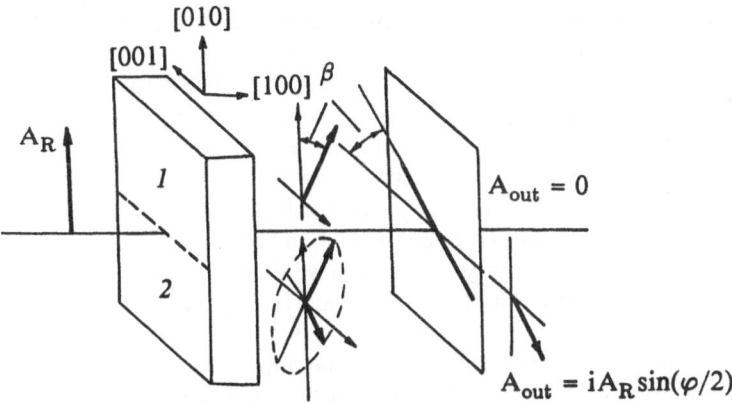

Fig.8.3. Mutual orientations of the readout light polarization direction A_R, crystal axes and axis of the polarization analyzer; β: angle of polarization rotation due to optical activity of the crystal, A_{out}: light amplitude after the analyzer. (1: light travels through the crystal part where the external field is screened, 2: external field is not screened)

The BSO photoconductivity in the yellow-red spectrum is reduced by several orders of magnitude, as compared with that in the blue. Therefore, images are read out from the PROM by yellow or red light to prevent rapid erasure. The energy of the readout light that can be transmitted through the device after an image has been recorded may be three to four orders of magnitude higher than the energy for recording.

The images are erased after readout by uniformly flooding the device with light that ensures the required excitation of free carriers in the crystal. A xenon flash lamp is typically used. The energy density needed for a com-plete erasure is on the order of 1 mJ/cm^2. Before flooding the crystal the electrodes are short-circuited or a voltage, with a polarity opposite to that used for recording, is applied. Under the erase illumination, the field is fully and uniformly compensated for in all parts of the crystal. If a reversed voltage is applied during erasure, the field in the crystal during image recording in the next cycle proves to be twice as high as that in erasure with short-circuited electrodes, since the field of the compensating charge accumulated during erasure is added to the external electric field. As a result, the sensitivity and diffraction efficiency of the device, which increase with increasing voltage, are superior to those observed in the case of short-circuited electrodes [8.6].

8.1.3 Frequency Dependence of Diffraction Efficiency and Sensitivity

Most experimental studies of PROM resolution employ holographic techniques [8.6, 15–18]. Figure 8.4 exhibits diffraction efficiency as a function of the recording exposure $W = I_0 t_{ex}$, where I_0 is the average intensity of the recording light in the modulator plane, and t_{ex} is the recording time. Note that the higher the spatial frequency of the recorded grating, the higher the exposure energy at which the diffraction efficiency peaks. Beyond the peak, strong nonlinear image distortions are observed. This means that the second and higher diffraction orders appear in the readout light diffraction plane when a sinusoidal grating is recorded. By measuring their intensity we can control and study the nonlinear distortions introduced by the device. The holographic PROM sensitivity can be derived from the data given in Fig. 8.4. Evidently the sensitivity differs for different spatial frequencies: the higher the spatial frequency, the lower the diffraction efficiency and, hence, the poorer the sensitivity. The PROM sensitivity at low spatial frequencies (1 to 2 ℓp/mm) is 200 μJ/cm^2 [8.17]. The sensitivity of the modulator, theoretically estimated on the basis of (7.71), is two orders of magnitude higher. Such a discrepancy with the experiment may be attributed primarily to the fact that in [8.17] SLMs with platinum electrodes were studied. These electrodes have a low transparency and thus lead to losses in the readout of the recording light. In addition, (7.71) does not take into account the fact that the charge in a PRC is not in the plane, but in the volume layer.

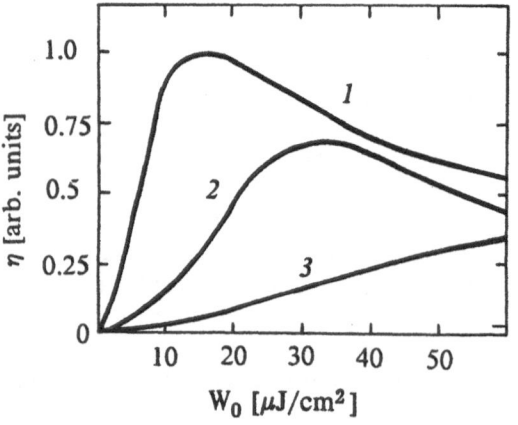

Fig. 8.4. Diffraction efficiency η of the PROM with BSO crystal as a function of exposure W_0 at different spatial frequencies. (*1*: 1.6, *2*: 5.1, *3*: 9.8 lines/mm) [8.31])

Figure 8.5 shows the PROM diffraction efficiency η as a function of spatial frequency for different exposures (5 and 70 μJ/cm^2). A modulator with 600 μm BSO plates and 2 to 3 μm parylene layers was used in the measurements. The curves are characterized by a reduction in η proportional to $1/\nu^4$ for $\nu > 10$ ℓp/mm (line pairs per mm). Since $\eta \propto \varphi^2$, we may infer that the modulation amplitude of the readout-light phase modulation φ is inversely proportionally to $1/\nu^2$ at these spatial frequencies. As we know (Sect. 7.5), φ decreases proportionally to $1/\nu$ at high ν if the charge is accumulated on the crystal surface and proportionally to $1/\nu^2$ if the charge is accumulated in the volume. The results measured for the PROM diffraction efficiency against spatial frequency were explained by the volume nature of the charge layer [8.16, 19-21], as discussed in Sect. 7.5.

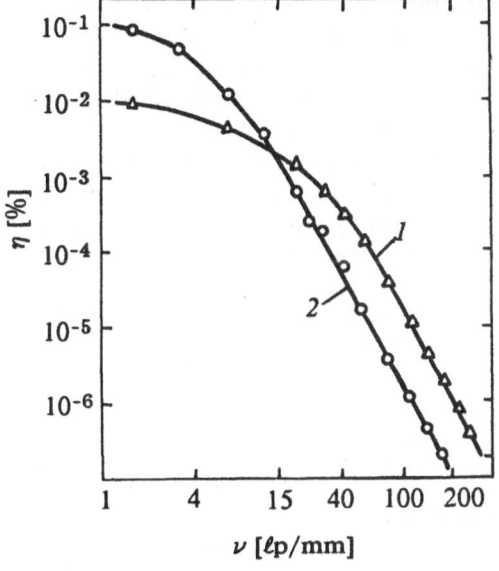

Fig.8.5. Diffraction efficiency η of the PROM as a function of spatial frequency ν for different exposures W_0 (1: 5, 2: 70 μJ/cm^2) [8.31])

The PROM diffraction efficiency estimated from (7.60) is in reasonable agreement with the experimental data if the charge-layer thickness is assumed to be several tens of micrometers (Fig. 8.5). As shown by experiment, the charge-layer thickness decreases with increasing exposure [8.16]. For an exposure of 5 μJ/cm^2, the values of the diffraction efficiency fit well with the theoretical curve for the charge-layer thickness $d_a = 20$ μm. The portion of the experimental plot for an exposure of 70 μJ/cm^2 where $\eta \propto \nu^{-4}$ nearly coincides with the theoretical curve for $d_a = 10$ μm.

Figure 7.5 also shows the PROM diffraction efficiency for an infinitesimally thin charge layer on the crystal surface calculated from (7.60). The volume nature of the charge layer leads to a reduction in the diffraction efficiency and resolution. The diffraction efficiency at low spatial frequencies (1 to 2 ℓp/mm) is on the order of 0.1%, and the resolution R defined as the spatial frequency at which $\eta = 0.1\eta_{max}$ is 6 ℓp/mm. Reduction in diffraction efficiency, by a factor of 100, is observed at a spatial frequency of 30 ℓp/mm. The maximum diffraction efficiency and resolution depend on the magnitude of exposure. For high exposure energies, when the nonlinear distortions of the recorded grating are strong, η_{max} decreases and R increases [8.16]. The values given above are obtained for exposures such that the nonlinear distortions can be neglected.

8.1.4 Speed of Operation

To estimate the minimum cycle time, i.e., the speed, characteristic times for the exposure with recording, readout, and erasure should be considered. Limitations on these times are similar to those to be discussed in Sect.8.2 for the PRIZ SLM. According to *Horwitz* and *Corbett* [8.7], the reciprocity law is fulfilled for recording light intensities ranging from 10^{-5} to 1 W/cm^2. This implies that in this intensity range the same exposure W = $I_0 t_{ex}$ is needed to obtain a given modulation of the readout light. Hence, as the intensity increases, the exposure time t_{ex} decreases. However, beginning from 1 W/cm^2, the PROM sensitivity decreases, i.e., W increases. The exposure required to record the image during 10^{-8} s is 10 times as high as that needed for the recording time t = 10^{-3} s. Estimates of the minimum durations of the main periods of the PROM cycle yield a maximum frame rate of hundreds of cycles per second. In experiments, the PROM was operated at 30 cylces/s.

The salient parameters of the PROM SLM employing the BSO crystal and parylene insulating layers are summarized in Table 8.2.

Table 8.2. PROM parameter and performance specifications

Resolution	
at $\eta = 0.1\eta_{max}$	5 – 6 ℓp/mm
at $\eta = 0.01\eta_{max}$	12 – 20 ℓp/mm
Holographic sensitivity (exposure at $\eta = 1\%$)	200 μJ/cm^2
Active area	up to 10 cm^2
Operating temperature	room temp.
Wavelength of recording light	400 – 500 nm
Readout light wavelength	600 – 800 nm
Required voltage	1 – 2 kV
Maximum diffraction efficiency	0.1 – 1 %

8.1.5 Contrast Modification and Image Subtraction

One of the operations that can be performed by the PROM is a modification of the image contrast before readout. The capacitance of the crystal plate in the PROM is less than the capacitance of all other layers. Therefore, the major part of the applied voltage drops across the electro-optic crystal. By varying the voltage after image recording we can efficiently change the average voltage across the crystal and, hence, the base-line readout-light amplitude A_0, whose magnitude determines the image contract. Thus, by changing the voltage during readout, the image contrast can be enhanced. The voltage can be chosen so that $A_0 = 0$; i.e., the zero component of the Fourier spectrum of the image being read out will here be zero. Experiments have been performed for suppressing the zero-component intensity of the image being read out by a factor of 10^4 as compared with the initial image. To achieve strong suppression, a precise adjustment of the voltage is necessary. The difficulty arises from the fact that the magnitude of voltage depends not only on conditions of recording, such as the recording light intensity and exposure time, but also on the type of image itself. Therefore, the voltage must be chosen individually for every image. If this is not done, strong suppression of the zero order cannot be achieved.

The PROM's ability to alter continuously the image contrast from positive to negative (or, to be more precise, to alter the zero component of the image Fourier spectrum) has been implemented in the Image Detail Enhancement System (IDES) developed by the Itek Corp. (USA) [8.7]. The device is intended for visual detection and detail enhancement of low-contrast images.

Image subtraction by PROM was reported in [8.22]. If two images are recorded in succession with different polarities of the voltage applied to electrodes, subtraction involves a selective erasure of the first image by the second one. The minimum size of the image region that can be selectively erased is not less than 0.5 to 1 mm. The value is governed by the spreading of free charge carriers by the internal electric field in the crystal. This technique allows subtraction of images with a low spatial resolution. The shortcoming can be overcome by using an optical configuration with two PROM devices (Fig.8.6). The modulators (1) and (2) are placed between the crossed analyzer (6) and polarizer (5). Lenses (3) and (4) focus the image of the first modulator onto the plane of the second device with the same scale. The polarity of the voltage applied during recording (Fig.8.6) is chosen so that the phase differences impressed on the readout light, as it passes through the first and second modulators, are of different signs. Thus, the

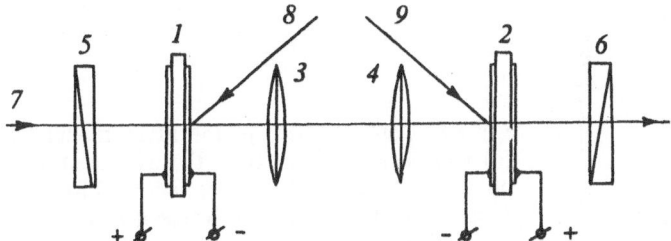

Fig.8.6. Arrangement for image subtraction using the PROM. (1 and 2: PROM modulators, 3 and 4: lenses, 5: polarizer, 6: analyzer, 7: readout light, 8 and 9: recording light)

143

Fig.8.7. Subtraction of two parts of the resolution chart

amplitude distribution of the readout light behind the analyzer corresponds to the difference between the images recorded on the modulators. Figure 8.7 demonstrates subtraction of two parts of the image of the resolution target obtained using two PROM devices.

The PROM SLM has attracted a great deal of attention because its performance parameters make it suitable for studies of optical information-processing techniques under laboratory conditions [8.23-30].

8.2 The Spatial Light Modulator PRIZ

The major distinction between a PRIZ[1] and the majority of SLMs using electro-optic crystals is that the readout light is here modulated through the transverse rather than the longitudinal electro-optic effect. The optically addressed PRIZ typically employs a plate of a cubic PRC - for instance, of the BSO type - as an active element. The transverse effect peaks when the plate of the cubic crystal has the (111) or (110) orientation.

Two modulator configurations have been studied (Fig.8.8). The first one is similar to that of the PROM and differs only by the crystal plate

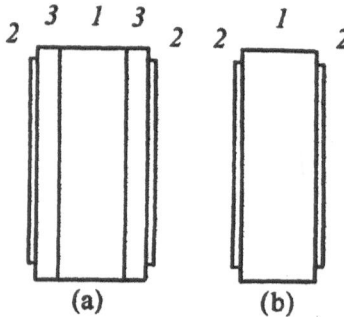

Fig.8.8. Configurations of the PRIZ modulator (a) with insulated electrodes, (b) with electrodes on the crystal surface. (*1*: cubic PRC, *2*: transparent electrodes, *3*: dielectric layer)

[1] The PRIZ device was proposed in [8.31-33]. The name of the device was derived from the Russian acronym that translates as "image transformer".

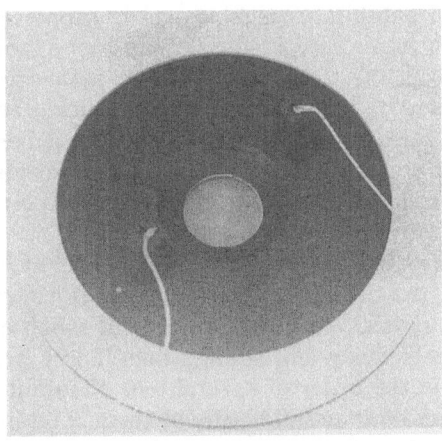

Fig.8.9. Photograph of the PRIZ modulator fabricated at A.F. Ioffe Physical Technical Institute

orientation. In the second one there are no dielectric insulating layers, and the electrodes are deposited directly on the crystal surface. Here the PRIZ has the simplest structure of all presently known SLMs. The modulators are typically fabricated from 300 to 600 μm thick crystal plates and 2 to 6 μm thick dielectric layers, which can consist of, for instance, parylene, as in the PROM device. Besides $Bi_{12}SiO_{20}$ [8.31-36], the PRIZ employs $Bi_{12}GeO_{20}$ [8.37] and $Bi_{12}TiO_{20}$ [8.38]. Figure 8.9 displays photographs of the PRIZ modulator fabricated at A.F. Ioffe Physical Technical Institute of the Academy of Sciences of the USSR.

As pointed out in the PRIZ description, the external longitudinal electric field is created by a pair of electrodes in the PRC. The field cannot change the readout-light polarization, since only the transverse effect is observed with the (111)- and (110)-oriented plates. The external field has the sole purpose to cause a drift of the electrons excited by the recording light. The photoinduced charge formation in an external longitudinal field was discussed in Sects. 4.6 and 7.1. After the image is recorded, the nonuniform space charge induces an internal electric field within the crystal, which has both longitudinal and transverse components. The transverse components governing the spatial readout-light modulation result from the image recording and are associated exclusively with the photoinduced space charge. To emphasize the specific nature, we may say that the readout light is modulated in the PRIZ through the internal transverse electro-optic effect.

Utilizing this effect leads to significant differences in the functional potential and the operational characteristics between the PRIZ and the longitudinal-effect devices. One difference is the transfer function of the PRIZ, which is unusual for photosensitive recording media. In this case the transfer function is a two-dimensional function with a zero at the origin of coordinates (Sect.8.2.1). As a result, the image is read out in a coded form, with a suppressed zero component in the Fourier spectrum. The zero-order suppression is extremely attractive for PRIZ applications in some optical information-processing systems. The ability to perform image transformation automatically is reflected in the name of the device. Moreover, in a certain mode of operation, the PRIZ exhibits unusual dynamic properties referred to as dynamic image selection (to be discussed below in Sect.8.2.6).

The modulator can be operated in the cyclic mode, with each cycle consisting typically of recording, readout, and erasure stages. During recording, a voltage of 1.5 to 3 kV is applied to the electrodes, readout is performed in the transmission mode, and erasure is accomplished by uniformly flooding the active surface by light from a flash lamp, the electrodes being short-circuited.

The two PRIZ configurations illustrated in Fig.8.8 differ primarily in dynamic characteristics. In operation, the dielectric layers prevent injection of electrons from electrodes into the crystal, to increase one and a half to two times the maximum diffraction efficiency and storage time [8.39]. The dark storage time without readout is in the order of several tens of minutes for modulators with dielectric layers and one minute without them. The insulating layers do not practically affect such parameters as sensitivity and resolution. The PRIZ without dielectric layers can be operated not only in the framed mode, but also in the continuous mode where dynamic image selection is observed. As will be shown later (Sect.8.2.6), the injection from electrodes into the crystal plays an essential role in the dynamic image selection. Devices with insulated electrodes do not exhibit this feature.

8.2.1 Transfer Function

As shown in Sect.7.5, the transfer function $\chi(\mathbf{K})$ of the transverse-effect SLM is anisotropic; i.e., the light modulation depends on the direction of the recorded grating wave vector $\mathbf{K}$. The anisotropy arises from properties of the transverse effect and is independent of the wave-vector length $K = |\mathbf{K}| = 2\pi\sqrt{\nu^2 + \xi^2}$. On the other hand, the dependence of χ on K is governed by the electrostatic relation between the charge formed during image recording and the internal field, and is isotropic in a cubic crystal, i.e., independent of the direction of $\mathbf{K}$. Here $\chi(\mathbf{K})$ can be represented in polar coordinates as a function of separable variables, i.e.,

$$\chi(\mathbf{K}) = \chi_I(\mathbf{K})\chi_A(\psi), \qquad (8.1)$$

where ψ is the angle which defines the direction of wave vector $\mathbf{K}$, and $\chi_I(\mathbf{K})$ and $\chi_A(\psi)$ are the isotropic and anisotropic multipliers of the transfer function, respectively. Note that we have to account for the analyzer when considering the SLM transfer function.

Let us inspect the isotropic part of the transfer function $\chi_I(\mathbf{K})$. In holographic experiments $\chi_I(\mathbf{K})$ is measured with different K and a fixed wave-vector $\mathbf{K}$ direction. By using an auxiliary coherent illumination in the observation plane, it was found that $\chi(K) = -\chi(-K)$ for the PRIZ. Figure 8.10 presents the experimental data obtained in the range of exposures where the SLM can be regarded as a linear element.[2] The grating was re-

---

[2] Linearity of the grating recording is essential for the unambiguous and correct measurement of the SLM transfer function.

corded with a He-Cd laser (λ = 442nm) and read out by linearly polarized light of a He-Ne laser (λ = 663nm). The experiments employed a modulator without dielectric layers and with a BSO plate of 450 μm thickness, as well as a linear analyzer placed behind the device. A voltage of 2 kV was applied to the electrodes during recording. The experimental data are found to be in a fair agreement with $\chi_I(K)$ calculated from (7.68), where the positive charge-layer thickness d_a was taken to be 100 μm, consistent with the results of investigation of the photoinduced charge (Sect.7.1). It was experimentally found that $\chi(0) = 0$, and at fairly high spatial frequencies ($K/2\pi >$ 15ℓp/mm) $\chi_I(K) \propto 1/K$, which is consistent with the results of calculations.

The spatial-frequency bandwidth R for $\eta(K) > 0.1\eta_{max}$ (which corresponds to $\chi_I(K) > 0.3\chi_{Imax}$), i.e. the resolution of the modulator, can be estimated from the $\chi_I(K)$ dependence. Here η_{max} and χ_{Imax} are the peak values of $\eta(K)$ and $\chi_I(K)$, respectively. The curve in Fig.8.10 yields R = 15 ℓp/mm.

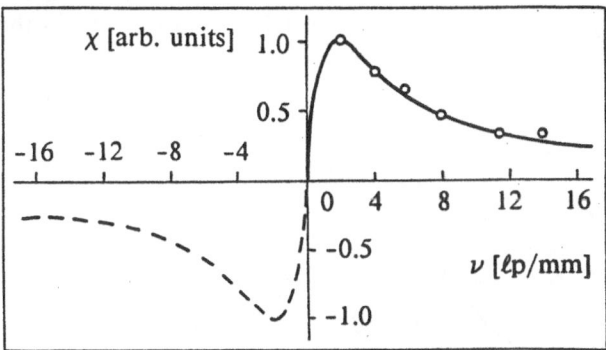

Fig.8.10. Transfer function of the PRIZ (450μm-thick BSO crystal, experimental points are obtained by the holographic technique, additional experiments showed that $\chi(\nu)$ is an odd function)

The anisotropic part of the transfer function $\chi_A(\psi)$ has been studied by measuring the diffraction efficiency as a function of ψ for the (111) and (110) cut PRIZ [8.40]. Using (7.27,43) $\chi_A(\psi)$ can be calculated for the (111)-cut modulator and linearly polarized light. Because the modulation amplitude of the readout-light, achieved in experiments with the PRIZ, is fairly low, (7.43) can be replaced, with sufficient accuracy and allowance made for (7.25), by

$$\eta(\psi) = \frac{1}{16}\varphi_1{}^2(K)\cos(\alpha' + \gamma) . \tag{8.2}$$

The angles α' and γ are shown in Fig.7.2; $\varphi(K)$ is γ-independent for the (111)-cut crystal. Accordingly, for the (110) cut, (7.43) yields

$$\eta(\psi) = \frac{1}{16}\varphi_1{}^2(K)\sin^2[2(\alpha' - \psi)] , \tag{8.3}$$

147

where ψ is related to γ through (7.22), and $\varphi(\mathbf{K})$ depends on ψ and γ according to (7.23).

In the measurements of the $\eta(\psi)$ dependence reported by *Petrov* and *Khomenko* [8.40], a grating image with $K/2\pi = 5$ ℓp/mm was focused onto the modulator from a photographic plate. The plate could be rotated about the optical axis of the imaging system, thereby changing the grating orientation relative to the crystal axes. The experimental data for the device with a 700-μm (111)-cut plate are shown in Fig.8.11 for readout with circularly and linearly polarized light. The shape of the $\eta(\gamma)$ dependence remains the same when the polarization direction is changed, but according to (7.25) and (8.2), the curve is rotated about the origin by an angle twice as large as the angle of rotation of the polarization plane of the readout light. The experimental data are found to be consistent with calculations only when voltage of negative polarity is applied to the electrode illuminated by the readout light. If a positive polarity is applied, the experimental $\eta(\gamma)$ curve is rotated by nearly 30° from the theoretical one (Fig.8.11). This is due to the optical activity of BSO, which was not taken into account in the calculations of $\eta(\gamma)$.

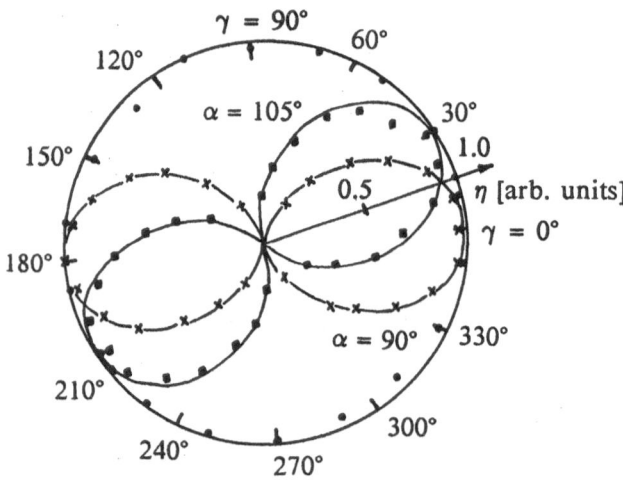

Fig.8.11. Theoretical and experimental values of diffraction efficiency as functions of the grating wave vector direction for the (111) oriented crystal [8.40]. (700μm-thick BSO crystal; × and ■ : experimental points for readout by a linearly polarized light, • : by circularly polarized light)

As pointed out above, both the nonuniform positive charge and the electric field responsible for the readout light modulation are formed near the negative electrode. As the readout light passes through the crystal, its polarization direction changes by about 15° (the crystal thickness is 700μm and the coefficient of the optical activity of BSO at $\lambda = 663$nm is 22°/mm [8.41]. Because light modulation occurs near the negative electrode and the $\eta(\gamma)$ dependence is rotated by an angle twice as large as that of the polarization plane, the difference arising from a reversal of the voltage polarity can be attributed to the optical activity.

Of particular interest is the case when images are read out from the (111) modulator by a circularly polarized light and a circular analyzer of polarization. The circular analyzer is a combination of a $\lambda/4$ plate and a linear analyzer. As calculations reveal, η is independent of γ in this case, as verified by experiments (Fig.8.11). However, as follows from (7.32), γ affects the phase of the diffracted wave. As a consequence, the complex transfer function of the SLM is here anisotropic as well, though its modulus is independent of the recorded grating wave vector.

Figure 8.12 depicts $\eta(\gamma)$ measured for (110) devices [8.40] with a positive polarity applied to the illuminated electrode. The experimental data are consistent with calculation. As the direction of the polarization plane of the readout light changes, the $\eta(\gamma)$ dependence is rotated as for the (111) modulator, but now, according to (7.22), there is a nonlinear relation between the angles of rotation of the polarization plane and the $\eta(\gamma)$ dependence. In addition, the peak value of $\eta(\gamma)$ undergoes a variation on rotation. The absolute maximum of η is observed when the grating vector is along the (110)-type axis and the readout light is polarized orthogonally to this direction. In this orientation, the theoretical value of η is 1.5 times as high as the maximum diffraction efficiency of the (111) modulator. Such a relation, within an experimental error, was indeed observed in experiments.

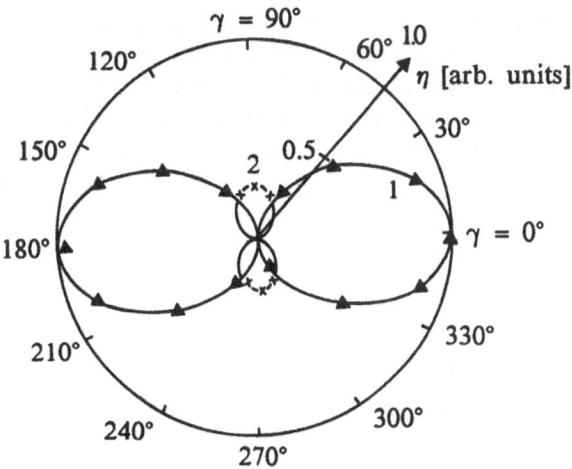

Fig.8.12. Theoretical and experimental values of diffraction efficiency for the (110) crystal orientation [8.40]. (*1*: $\alpha'=0$, *2*: $\alpha'=45°$)

Thus, the transfer function is anisotropic, and its shape depends on the crystal cut and the types of the readout-light polarization and the analyzer of polarization. However, in all cases $\chi(0) = 0$. Thus, the baseline amplitude of the image readout from the modulator is zero. The anisotropic properties of the modulator can be used to advantage for processing images. The angular dependence of $\eta(\gamma)$ for the linearly polarized readout light can be employed to enhance structures in the image oriented in a certain manner, for instance, only horizontal or vertical lines. This operation is equivalent to the directional filtering of the image spectrum, in which the spectral components oriented in selected angular directions are suppressed; it can be performed both in coherent and incoherent readout light. As mentioned above, directional filtering is absent in the (111) modulator and for circularly polarized light; i.e., the intensity of the spectral components of the readout image is independent of the wave-vector orientation. Therefore, the (111) modulator and circularly polarized light should be used when directional filtering is undesirable.

As noted above, any cross section of the PRIZ transfer function passing through the origin is represented by an odd function, i.e., $\chi(K) = -\chi(-K)$. The image transformation that the modulator with such a transfer function performs is somewhat similar to the Hilbert transformation [8.42, 43]. As is known [8.44], an ideal system that would produce the Hilbert transformation should also have an odd transfer function of the type sign (K). A theoretical treatment of the transformation performed by the PRIZ and an analysis of its similarities to and differences from the Hilbert transformation have been given by *Bryksin* et al. [8.45].

8.2.2 PRIZ Sensitivity

The PRIZ sensitivity to recording light at different wavelengths has been studied [8.46]. Sensitivity was determined from (3.34) during recording of the grating with $\nu = 5$ $\ell p/mm$ and a voltage of 2 kV applied to electrodes [8.46]. Figure 8.13 presents measurements for modulators with different thicknesses of the crystal plates. Note that the wavelength at which the

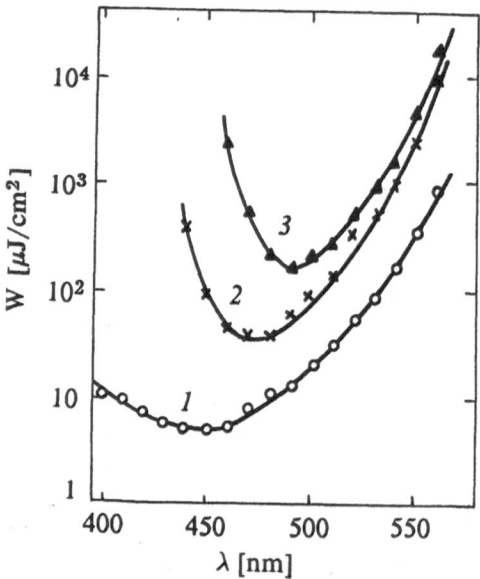

Fig.8.13. Sensitivity of the PRIZ employing BSO versus wavelength of the recording light for crystal thickness d_c (*1*: 610, *2*: 990, *3*: 1490 μm) [8.46]

maximum sensitivity is observed shifts to shorter waves as the plate thickness decreases. The PRIZ sensitivity at all wavelengths grows with decreasing crystal thickness, at least to 250 μm, which was studied in the experiments. The sensitivity increase is primarily attributable to the increase of the electric field strength in the crystal with decreasing thickness. This leads to a larger thickness of the positive charge layer responsible for the light modulation. The optimum crystal thickness for the SLM recording sensitivity equals the positive charge-layer thickness, that is, when the positive charge occupies the entire crystal thickness. As estimates show [8.47], the optimum thickness for BSO at a recording wavelength of 442 nm and an applied voltage of 2 kV is 150÷180 μm. However, fabrication of such thin plates with a sufficiently large (more than 2cm in diameter) active surface is a complicated technical problem.

8.2.3 Inherent Noise

As measurements reveal [8.48] the PRIZ has an extremely low level of inherent noise, probably the lowest for presently known SLMs. This is a result of the simplicity of the PRIZ structure and also the fact that BSO crystals with high optical and photoelectric parameters can be produced. Thus, inherent noise of the PRIZ is typically lower than total noises of other elements of the optical system in which it is incorporated.

Besides a low noise level, the PRIZ has the advantage of allowing an automatic and strong reduction of noise of other optical elements if the SLM is used in combination with an analyzer. As discussed in Sect. 7.7, the reduction is the largest when the electro-optic crystal birefringence averaged over the active surface vanishes ($\varphi_0 = 0$). For the PRIZ where the readout light is modulated through the transverse electro-optic effect under a longitudinal external field, this condition is met automatically in recording any image, since the transmittance of the SLM and the analyzer is zero before recording everywhere, and the average transmittance is zero after recording. In this respect, the PRIZ is superior to the longitudinal-effect SLMs, where special measures should be taken to compensate for φ_0. Such measures include a compensator of birefringence [8.6] or a variation of the voltage applied to electrodes during recording [8.49]. Here certain difficulties arise because φ_0 depends on the type of the recorded image, since $\chi(0) \neq 0$ for such SLMs.

Figure 8.14 demonstrates an improved signal-to-noise ratio of a coherent optical spectrum analyzer with the PRIZ at the input. Figure 8.14a shows an original object (a screw), and Fig.8.14b exhibits the Fourier spectrum of the object obtained by coherent illumination of the original object. The signal-to-noise ratio is low near the zero order because of scattering from the imperfections of the optical elements, and the low-frequency components are not distinguished. Figure 8.14c depicts the image after recording on the PRIZ. Note that only the edges of the original image appear during readout, which is a result of suppression of the zero component in the image spectrum. The photograph of the image spectrum (Fig.8.14d) de-

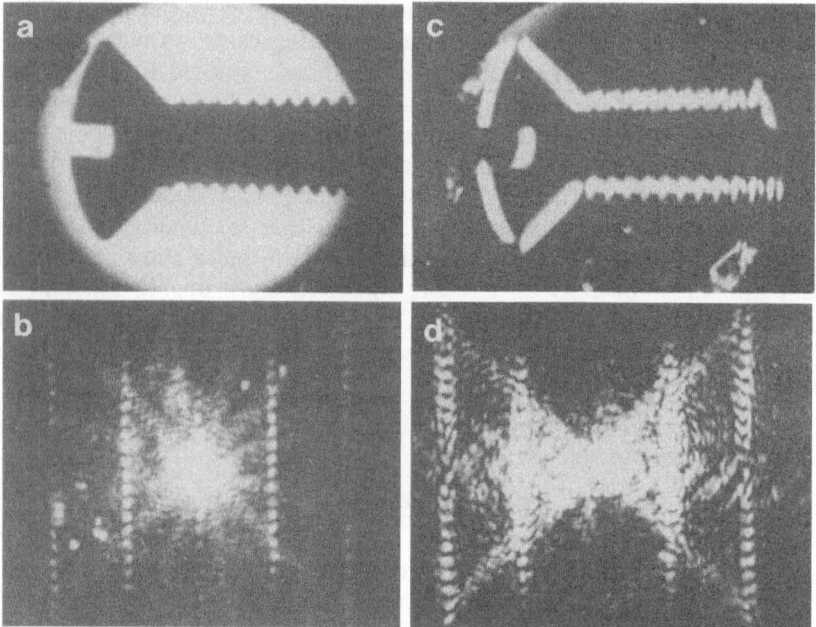

Fig.8.14. Illustrating the improved signal-to-noise ratio of a coherent optical spectrum analyzer: **a**: original object, **b**: Fourier spectrum of the object obtained by illuminating the original object by coherent light, **c**: the image after recording on the PRIZ, **d**: Fourier spectrum of the image readout from the PRIZ

monstrates an improved signal-to-noise ratio, which is especially pronounced at low frequencies. Thus, the low level of inherent noise and also the possibility to suppress effectively noise of other elements ensure a wide dynamic range at the output of the coherent-optical spectrum analyzer, in spite of the low diffraction efficiency of the device ($\eta < 1\%$). The measured signal-to-noise ratio in the Fourier plane on recording a sinusoidal grating is about 60 dB.

8.2.4 Phase Distortion

Elimination of phase distortion of the readout wave front in the PRIZ SLM is achieved by fabricating photorefractive plates of fairly small thickness ($0.3 \div 0.6$ mm) with flat faces of large area. In practice, the technological difficulties encountered in fabrication of the devices grow with increasing active surface. An SLM with an active surface of $20 \div 40$ mm in diameter with phase distortions not higher than $\lambda/4$ can be made from the BSO crystal.

8.2.5 Speed of Operation

The speed of operation is governed by the minimum duration of a cycle and the removal of heat released during the SLM operation due to charge currents and the absorption of recording and erase light. Estimates indicate that heating does not limit the speed of operation at least up to 10^3 Hz. The exposure time is as short as several nanoseconds if a fairly powerful recording light is used. The readout image is not immediately visible at such a short recording period, because time is needed for the charge excited in the crystal volume to separate. This time is commensurable with the electron transit time through the plate and amounts to about 10 μs for BSO. The readout time is determined by such factors as the readout-light intensity, sensitivity, and speed of the image detector. The duration of the erase-light pulse can be as short as that of the recording one. The erase pulse energy is an order of magnitude higher than that of the recording pulse. Thus, after the erase pulse and compensation of the internal fields within the crystal occur, a certain time is needed to allow relaxation of free charge before applying voltage to record a new image. As evidenced by experiments, this time varies from one BSO crystal to another, and reaches hundreds of milliseconds. This is an essential limitation on the SLMs' speed. The devices made of crystals with short relaxation times can be operated up to 30 Hz.

8.2.6 Dynamic Image Selection

Besides the cyclic mode of operation involving successive recording, readout, and erasure cycles, the PRIZ SLM without insulating layers can be operated in the continuous mode. Then there is no division into cycles, a fixed voltage is applied to the electrodes, recording and readout are continuous, and no erasure is accomplished. It was shown that, when operated in this mode, the PRIZ allows selection of time-varying images or parts of images on the fixed spatially nonuniform background. For instance, the device can respond to a moving object, while suppressing the images of fixed objects. This feature is termed Dynamic Image Selection (DIS). The DIS effect was first observed in the PRIZ [8.50]. Later, a liquid-crystal modulator was proposed that also exhibits the DIS feature [8.51].

Figure 8.15 shows temporal dependences of the response of the PRIZ (with no insulating layers) on switching on and off the recording light. Note that as the light goes on, the PRIZ response peaks and then decays almost to zero; i.e., the image disappears. However, as the light goes off, it appears again. The readout-light amplitude is of the opposite sign when the light is switched on or off, as shown in Fig.8.15. If the recorded image or parts of it move, the device will respond only to the moving object, because here the recording intensity will vary in that part of the SLM active surface where the moving object is found at that moment. It has been demonstrated [8.52] that, using the PRIZ in the DIS mode, a 20-fold increase of the intensity ratio between moving and fixed object images can be obtained. The PRIZ in the DIS mode was also regarded as a nonlinear element since, for instance, the magnitude of the response to a moving object depends on the intensity of the fixed-image background.

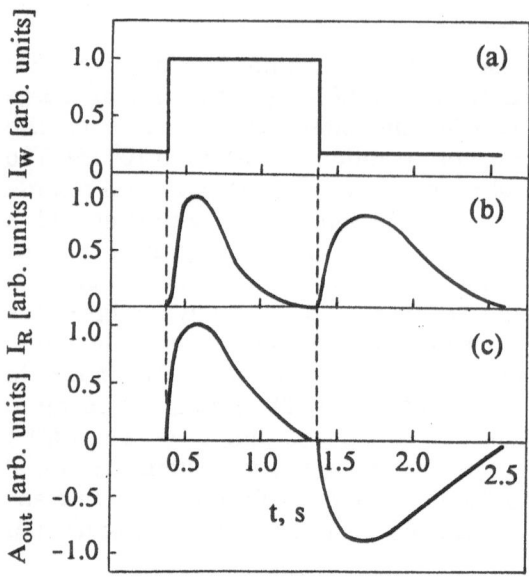

Fig.8.15. Temporal dependences of recording light intensity (a), readout light intensity (b), and readout light amplitude (c)

Studies of the physical mechanism responsible for the DIS effect in the PRIZ employing BSO reveal that the dominating role in formation of the response to switching off the recording light is played by injection of electrons from the negative electrode [8.53]. Indeed, as shown in Sects.4.6 and 7.1, the magnitude of the electric field near this electrode depends on the local exposure with the recording light. A strong electric field in the region adjacent to the electrode induced by exposure increases injection of electrons. Up to the moment when the light is switched off, the magnitude of the injection current is locally exposure-dependent. After the light goes off, the injection continues until the field in the near-electrode region decreases by a sufficient degree. The magnitude of the injected charge depends on the recording intensity and is different for different parts of the crystal. The charge field modulates the readout light after the recording light is switched off, and is opposite in sign to the positive charge field, which is formed when the recording light goes on. This causes an inversion of sign of the modulation amplitude of the readout light. Since the electrons are mainly captured by shallow traps and can be excited to the conduction band by, for instance, the red readout light, the negative charge relaxes with time.

The DIS mechanism has been verified by experiment. For instance, the DIS feature is not observed in modulators with insulating layers; the rate at which the response to switching off the recording light grows is determined by the intensity of this light, while the rate at which the response decays after the peak, is governed mainly by the readout-light intensity. Moreover, this mechanism is consistent with the results of studies of the photoinduced charge formation given in Sects.4.6 and 7.1.

Note that the time and space responses of the PRIZ in the DIS mode are interrelated. For instance, the transmitted spatial frequency band is dependent on the rate at which the images to be recorded change. Therefore, even in the linear approximation, we should introduce the transfer function $\chi(\nu,\xi,f)$, which depends on three parameters, i.e., spatial frequencies ν,ξ and temporal frequency f [8.51]. Here f describes not the light frequency, but the frequency of the intensity variation in the recorded image. The nonstationary image must here be represented by a superposition of running waves with differing amplitudes, frequencies, and propagation directions. Then each running wave will have a corresponding wave of the variation of the crystal birefringence modulating the readout light.

8.2.7 Recording of "Latent" Images

The PRIZ offers one more means of image recording unusual for SLMs. It can be called recording of "latent" images. In this mode, the image induced by the recording light exposure cannot be read out immediately and requires additional visualization.

To realize this mode, the modulator is pre-illuminated by uniform red light, and then the input image is focused into the device with blue light. Voltage is not applied to electrodes during recording. It has been shown that, though the image cannot be read out after the recording-light exposure is over, a "latent" pattern is stored in the crystal [8.54]. For its visualization a voltage should be applied to the electrodes and, in addition, in some cases a uniform illumnination of the crystal is needed. Figure 8.16 shows how the diffraction efficiency of the PRIZ using BSO grows on visualization of the sinusoidal grating image. Point $t = 0$ corresponds to the moment the recording light is switched off, and the voltage is switched on. Visualization begins when the crystal is uniformly illuminated with red light ($\lambda = 633$nm), which is used as readout light as well. It can be stimu-

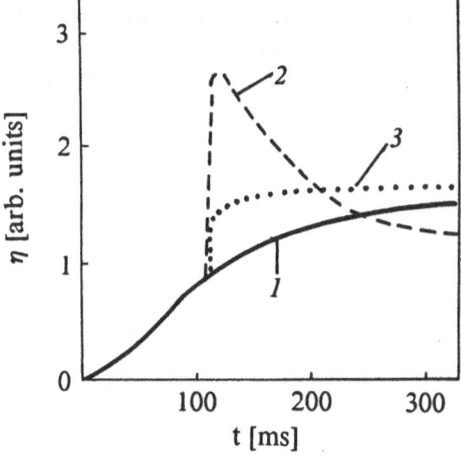

Fig.8.16. Diffraction efficiency as a function of time for visualization of a "latent" image of a sinusoidal grating in a BSO crystal. (*1*: visualization by readout light at $\lambda = 663$ nm, *2*: visualization using an additional pulse at 530 nm, *3*: pulse at 1060 nm) [8.54]

lated not only by the red, but by green light (λ = 530 nm), too. In several PRIZ samples, although made from nominally pure BSO, visualization of latent images occurs only through applying the external field to the sample; no auxiliary illumination is needed.

The mechanism of latent recording can briefly be described as follows [8.54, 55]. When the crystal is illuminated with the recording light, electrons are excited from donor levels into the conduction band. A fraction of them is captured by the traps emptied before recording by pre-illuminating the modulator with red light. Since the external field is not applied during the recording-light exposure, the spatial redistribution of electrons in the conduction band can occur through diffusion alone. The efficiency of the diffusion mechanism for charge formation during recording of the image with low spatial frequencies is, however, well below that of the drift mechanism. The diffusion-induced internal fields are small and cannot modulate the readout light noticeably. Thus, we can assume that there is no spatial-charge redistribution during the recording-light exposure. However, the latent image recording, which involves the spatially nonuniform electron redistribution in the bandgap, does occur. During visualization, the electrons in higher energy states are excited by light and drift in the external field. As a result, an uncompensated positive charge of donors whose field modulates the readout light spatially, is induced.

Studies of the latent image recording revealed that SLMs made from different samples of BSO crystals can have different latent image storage times, sensitivities to visualization with red and green light differing by orders of magnitude; some samples even do not require illumination for image visualization. Visualization in this samples is likely to occur through thermal excitation of electrons. This is evidence of the appreciably differing trapping level structures in various samples of BSO.

Nonlinear distortions during image recording in the PRIZ have been shown to arise from the nonlinearity of the photoinduced charge formation. In the latent image recording mode, charge is formed after recording and can be controlled. This allows recording and reconstruction of images free of nonlinear distortions over a wide range of exposure. Moreover, in certain PRIZ samples, image visualization occurred after a delay of 10 minutes following the recording-light exposure. This delay allows an appreciable increase in the information storage and recording times of the PRIZ as compared with an ordinary operating mode.

8.2.8 Photoinduced Piezoelectric Phase Modulation

Crystals of the BSO type typically modulate the readout light by means of the electro-optic effect. Yet, along with electro-optic properties, these crystals exhibit pronounced piezoelectric properties [8.56] that can also be utilized to modulate light spatially. Indeed, it has been shown that the internal field of the photoinduced charge gives rise to deformation of the BSO plate [8.57, 58]. Displacements of a plate's surface associated with deformation produce a marked phase modulation of the readout light reflected from the crystal.

This phenomenon was studied in experiments using the PRIZ with round crystal plates in the (110) orientation. Unlike in the conventional technique of image readout with the light passing through the crystal, here one observes diffraction of the light reflected from the crystal surface on recording a sinusoidal grating. Diffraction is observed only when a negative potential is applied to the electrode deposited on the reflecting surface.

It was shown that the diffraction efficiency depends on spatial frequency ν and the recorded-grating wave-vector direction with respect to the crystal axes. The $\eta(\nu)$ dispersion curve has a peak typical for PRIZ, and $\eta(0) = 0$. The position of the peak is determined by recording conditions in a range of 1-10 ℓp/mm. The maximum η does not exceed 0.15%. We can infer from this that the periodic displacement of the crystal surface is not more than 70 Å. Figure 8.17 presents η as a function of the recorded-grating wave-vector direction. The symmetry of the dependence indicates that the crystal surface deformation is mainly due to the transverse components of the internal electric field, since the longitudinal field would give an isotropic orientation dependence.

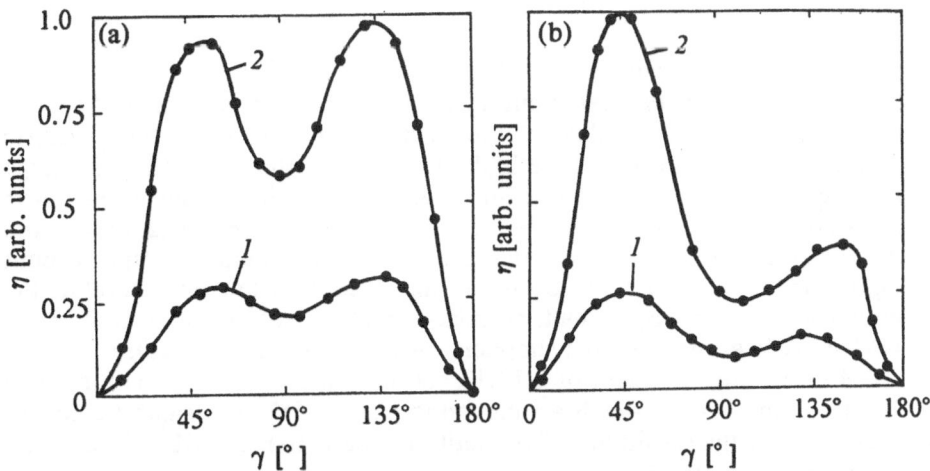

Fig.8.17. Diffraction efficiency as a function of orientation of the sinusoidal grating wave vector at $\nu = 4$ ℓp/mm for BSO (a) and $\nu = 15$ ℓp/mm (b). (*1*: recording light exposure corresponds to η_{max}, *2*: a low exposure when $\eta \simeq W^2$) [8.57]

The basic difference of the photoinduced piezoeffect from the electro-optic one in the PRIZ is that the magnitude of displacement of a given point on the surface depends upon the mechanical deformation field throughout the crystal and upon conditions at its boundaries. But the changes in birefringence induced by the electro-optic effect are local, i.e., they are independent of the field magnitude at other points with a sufficient degree of accuracy. That is why, for instance, the orientation dependences $\eta(\gamma)$ in the two cases differ appreciably, in spite of identical tensors describing the symmetry of the piezoelectric and electro-optic effects.

In conclusion, we summarize the salient parameters of the PRIZ SLM using BSO (Table 8.3). The parameters were measured in the cyclic operating mode.

Table 8.3. PRIZ parameter and performance specifications

Wavelength of recording light	400 – 500 nm
Sensitivity (for 1% η at 3 ℓp/mm)	5 μJ/cm^2
Wavelength of readout light	550 – 700 nm
Resolution at SHD = 1 % at SHD = 10%	15 ℓp/mm 30 ℓp/mm
Maximum diffraction efficiency	1 %
Active area (diam.) for phase distortions not more than $\lambda/4$	30 mm
Speed of operation	30 cycles/s

The estimates indicate that the parameters listed in Table 8.3 are consistent with the theoretical, limiting parameters for a modulator based on the crystal with given values of the electro-optic effect and the dielectric permeability. This is due to the high efficiency of the photoinduced charge formation in BSO caused by a high quantum efficiency and a long drift length of the electrons. Moreover, it is significant that the PRIZ utilizes the transverse electro-optic effect that most efficiently modulates the readout light with the photoinduced charge in the volume. A low noise level of the modulator is, to a large extent, associated with the technological ease of the BSO-crystals growth and the simplicity of the PRIZ SLM structure.

Further improvement of a PRIZ can be achieved by butt-joining it to an image amplifier through a fiber-optic plate [8.58]. As demonstrated experimentally, the sensitivity of the unit may be better than 10^{-9} J/cm^2, and visible and near-IR light can be used for recording.

8.3 Titus and Phototitus

The Titus and Phototitus SLMs mainly employ the electro-optic DKDP (potassium dideuterium phosphate) crystal. This crystal is operated close to the phase-transition temperature and therefore cooling to about -50° C is required.

8.3.1 Titus

Titus is an electron-beam-addressed SLM [8.59-61]. The input electric signal is used to modulate the current of the electron beam that scans the crystal surface and forms a raster (like in standard TV), thus inducing the nonuniform space charge corresponding to the input data (Fig.8.18). This figure illustrates two CRTs with an electro-optic crystal as a target, i.e., the Titus SLM. In the device shown in Fig.8.18a, the input electric signal is applied to the electrode, which modulates the current of the electron beam that records the image. The electron energy of the recording beam is 6 kV. The secondary emission coefficient of the DKDP crystal is less than unity and thus the crystal surface is charged negatively. The stored data are erased with an electron gun that floods the crystal surface with electrons simultaneously and uniformly. For the accelerating potential of 500-1000 V provided by the gun, the secondary emission coefficient is more than unity and the crystal surface is charged positively by loosing electrons. The surface potential becomes uniform; i.e., the image is erased and the device is ready for a new recording.

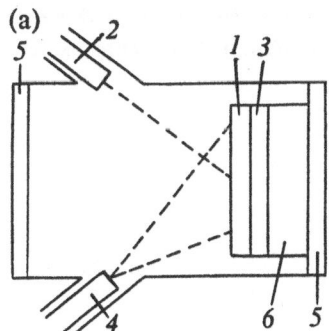 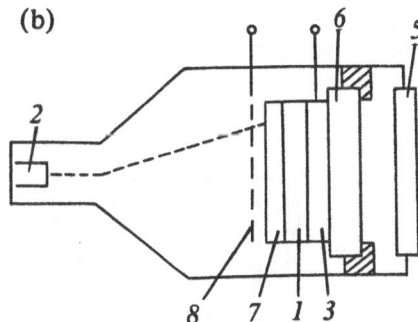

Fig.8.18. Versions of electron beam addressed Titus modulator (*1*: DKDP crystal, *2*: recording electron gun, *3*: transparent electrodes, *4*: erasing electron gun, *5*: optical window, *6*: CaF$_2$ substrate, *7*: dielectric mirror, *8*: control grid)

The SLM with the structure shown in Fig.8.18a operates with readout in the transmission mode. The dimensions of the device are large, and the crystal cannot be placed perpendicular to the axis of the recording system, leading to image defocusing at the crystal edges. These shortcomings are eliminated in the Titus scheme operating in the reflection mode (Fig.8.18b), where the light passes twice through the electro-optic crystal, with a resulting increase of the modulation depth by a factor of 2 and, hence, a decrease in the density of charge to be deposited on the surface of the crystal.

The Titus SLM described in [8.59] operates in the reflection mode and has a length of 40 cm, a maximum diameter of 10 cm, and target crystal size of 30×40×25 mm^3. The crystal surface facing the electron gun is coated with a dielectric mirror that reflects the readout light. The opposite surface has a transparent electrode. The crystal is positioned on a CaF$_2$ substrate, which has a high thermal conduction, and its coefficient of expansion is close to that of DKDP. The substrate and crystal are cooled by a two-stage Peltier cell cooling system.

The data are recorded by scanning the crystal surface with an electron beam of constant current. The input signal is applied to the grid mounted in front of the crystal at a distance of 210 μm. The electron beam short-circuits the grid and crystal surface, thereby causing a decrease or increase of the surface potential due to the secondary electron emission. As a consequence, each point of the crystal surface is charged to the potential corresponding to the video signal at the instant of time when the electron beam is deflected to this particular point.

8.3.2 Phototitus

Phototitus is an optically addressed SLM [8.59, 62–66] that uses a photoconductive layer consisting of, for instance, selenium on the electro-optic crystal surface as a photosensitive element (Fig.8.19). During image recording, the external longitudinal electric field created in the photoconductor and crystal by means of a pair of transparent electrodes gives rise to a drift of charge excited by light in the photoconductor, which in turn charges the crystal surface.

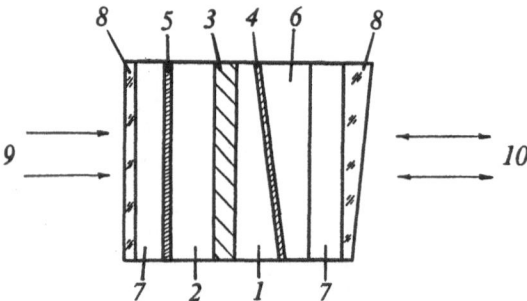

Fig.8.19. Structure of the Phototitus modulator (*1*: DKDP crystal, *2*: photoconductor, *3*: dielectric mirror, *4*: transparent electrode, In_2O_3, *5*: Au electrode, *6*: CaF_2 substrate, *7*: vacuum, *8*: optical windows, *9*: recording light, *10*: readout light)

The structure of one of the device versions operated with readout in the reflection mode is shown in Fig.8.19 [8.59]. The dielectric mirror is coated with a 12-μm amorphous Se photoconductor layer. The modulator is placed into a vacuum enclosure to protect the hygroscopic DKDP crystal from humidity during cooling. The crystal plate, CaF_2 substrate, and optical window for the readout light are made in the form of wedges to prevent light reflection from their surface into the image plane. The dielectric mirror provides nearly total reflection of the readout light, and hence the light does not reach the photoconductor. Therefore, readout can be accomplished in a wide wavelength range including the range of photoconductor sensitivity.

A voltage of 100÷200 V is applied to electrodes during recording. Because of a high dielectric permeability at the operating temperature, the capacitance of the crystal plate is at least five times as high as the capacitance of the photoconductor layer, and almost all of the applied voltage drops across the Se layer before recording. The electric field induced in this layer is fairly high (on the order of 10^5 V/cm), and the maximum operating voltage of the Phototitus is determined by the breakdown voltage of the photoconductor. Recording is performed with blue-green light (440÷520nm) when a Se photoconductor is used. The charge excited by the recording light passes through the photoconductor layer and charges the crystal surface. Since Se is a bipolar photoconductor, the crystal surface can be charged either negatively (by electrons) or positively (by holes), depending on the polarity of the applied voltage. In the illuminated regions, the field in the photoconductor is screened by the separating charge, and the voltage turns out to be applied to the electro-optic crystal. As a consequence, the readout light changes its polarization in these parts of the crystal, allowing readout of the positively recorded image. To enhance the contrast, electrodes are short-circuited during readout. The image is erased by uniformly flooding the photoconductor with light, which excites free carriers. During erasure, electrodes are either short-circuited [8.59], or voltage with polarity opposite to that used for recording is applied [8.65].

8.3.3 Parameters of Titus and Phototitus

The DKDP crystal is optically uniaxial, colorless, and transparent in the 0.2 to 2.4 μm range. The plate is cut normal to the optical axis (the so-called 0-z cut). The natural birefringence for the beams incident normal to the surface of a 0-cut crystal is not observed. At room temperature, the crystal is paraelectric with a tetragonal-scalenohedral symmetry, point group 42m [8.67]. Near -50° C DKDP undergoes phase transition and becomes a ferroelectric. Near the transition temperature, the crystal parameters crucial for use in SLMs - i.e., half-wave voltage $U_{\lambda/2}$ and dielectric permeability - experience drastic changes. Here $U_{\lambda/2} = 2200$ V, and components of the dielectric permeability tensor $\epsilon_x = 58$, $\epsilon_z = 48$ (the z axis coincides with the optical axis of the crystal) at room temperature change to $U_{\lambda/2} = 200$ V, $\epsilon_x = 66$, and $\epsilon_z = 650$ at -50° C.

When used in Titus and Phototitus, the DKDP crystal is cooled to a temperature near the phase transition, markedly improving its resolution and reducing inherent noise. Let us see how cooling affects resolution. Equations (7.58, 60) describing the longitudinal-effect SLM transfer function were derived for an isotropic cubic crystal. Since the DKDP crystal is anisotropic, the resolution R is determined from a more general expression given in [8.68]. On that basis, we obtain

$$R = \frac{\sqrt{10}}{2\pi} \frac{(\epsilon_1/d_1 + \epsilon_z/d_2)}{\epsilon_1 + (\epsilon_x \epsilon_z)^{1/2}} \tag{8.4}$$

for the charge grating on the electro-optic crystal surface, where, like in (7.58), d_2 is the crystal thickness, and ϵ_1 and d_1 are the dielectric permeability and thickness of the insulating layer, respectively.

As seen from Fig.8.18, the Titus has an electrode on one surface of the electro-optic crystal. We can assume that the second electrode is far from the crystal, so that $1/d_1 \simeq 0$ and the main term in the denominator of (8.4) is ϵ_z/d_2. The Phototitus employs the photoconducting layer to insulate the electrode from the crystal. Its thickness and dielectric permeability are such that ϵ_1/d_1 is also small as compared with ϵ_z/d_2. As is evident from (8.4), because of such a specific feature of the modulator, the variation of the dielectric permeability of the crystal on cooling gives rise to changes of resolution. If we take the above data on the dielectric permeability variation on crystal cooling, (8.4) yields a resolution improvement by a factor of 3. In Titus experiments, a four-fold resolution improvement was observed [8.60].

The SLM sensitivity is governed not only by the dielectric permeability, but also by the half-wave voltage of the crystal, which also changes on DKDP cooling. The amplitude of the charge grating to be formed on the anisotropic crystal surface for the modulation amplitude of the readout-light phase difference at low spatial frequencies to reach a specified value, for instance π [8.68], is given by

$$S_\pi = \epsilon_0 U_{\lambda/2}(\epsilon_1/d_1 + \epsilon_z/d_2) . \qquad (8.5)$$

As discussed in Sect.8.6, the charge grating amplitude is proportional to the recording-light exposure in the linear approximation. Since $U_{\lambda/2}$ reduces and ϵ_z increases on crystal cooling, the readout-light modulation amplitude is nearly unaltered. Hence the recording sensitivity of the Phototitus also remains unchanged. For the Titus, no changes in the beam current of the scanning electron gun are required to retain the readout-light modulation amplitude.

Noise in SLM employing DKDP at room temperature is mainly attributable to a nonuniform charge leakage over the crystal plate area, which gives rise to a nonuniform erasure of the recorded image. At room temperature the charge is stored on the surface for only 0.1 s, and image erasure results in noises even in short-term storage. This storage is required, for instance, for successive (point-by-point) recording of images. At -50°C the charge is stored for two hours [8.60]. Thus, operation at -50°C not only leads to a longer storage time, attractive in many practical situations, but also improves the noise performance of the SLMs using DKDP [8.60, 69].

Images are read out in Titus and Phototitus by linearly polarized light. Light emerging from the crystal is spatially polarization modulated. The polarization modulation may be converted to amplitude modulation by passing the light through an analyzer. Since the DKDP crystal is birefringent, the readout light should be collimated sufficiently. The linearly polarized light passing through the crystal at a nonzero angle to the optical axis emerges with an elliptic polarization, even when there is no electric field in the crystal. Thus poorly collimated readout light reduces the contrast of the readout image by increasing the reference amplitude (background), which, as discussed in Sect.7.7, can deteriorate the noise performance of a coherent optical system. For instance, a contrast ratio of 100:1 was obtained for Titus using a 10° angle of incidence for the incoherent readout light. The maximum contrast ratio of 1 000:1 may be achieved with coherent light, for which the beam can be collimated better.

Figure 8.20 shows a frequency dependence of the contrast of the image produced by recording an electric signal on the Titus [8.69].[3] Equation (8.4) describing the resolution does not take into account the sizes of the electron beam, an additional factor that reduces resolution. Curves 1 and 2 in Fig.8.20 plot calculations for the infinitesimally thin and 80-μm beams, respectively (crystal thickness is 220μm) [8.69]. The experiments agree well with calculated values. The data yield resolution of 3-5 ℓp/mm for the contrast level of 0.5 of the maximum level. Other Titus parameters are summarized in Table 8.4.

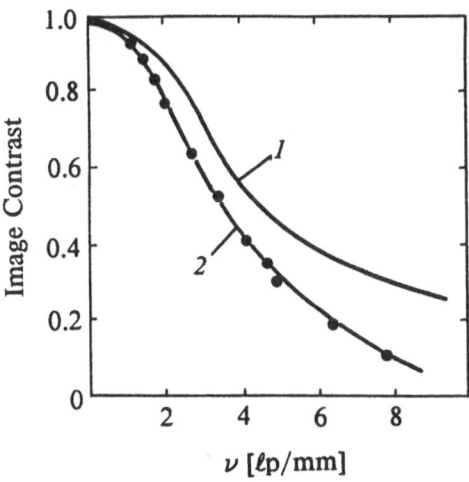

Fig.8.20. Contrast vs. spatial frequency for Titus [8.73]. (220μm-thick crystal plate, crystal temperature is -53°C, 1: electron beam thickness is neglected, 2: calculations and experimental points for 80μm-thick electron beam)

Figure 8.21 presents contrast as a function of spatial frequency for the Phototitus with 150 and 180 μm thick crystals [8.64]. The resolution of 5-10 ℓp/mm at the 50% contrast level is consistent with calculations carried out on the assumption that the plane, sinusoidal charge grating is at the crystal-photoconductor interface. The published data on resolution of the Phototitus differ markedly, for instance, 16 ℓp/mm [8.65] and 85 ℓp/mm [8.59]. The difference is probably due to the techniques and criteria used for estimates. The resolution of 85 ℓp/mm was obtained [8.59] as the maximum spatial frequency of the grating that could be visually observed on readout

---

[3] When used in a coherent system, the SLM is characterized by a transfer function. However, because of the absence of necessary data in the literature, we use the frequency dependence of the contrast to discuss the resolution of the Titus and Phototitus, and also a definition for the Phototitus sensitivity differing from those used above.

Table 8.4. Titus parameter and performance specifications

Bandwidth	10 MHz
Target size	50×60 mm^2
Resolution at 0.5 contrast level	20 ℓp/mm
Space–bandwidth product	10^6
Cycle time	33 ms
Contrast	100:1, 1000:1
Storage time	more than 1 hour
Operating temperature	– 50° C

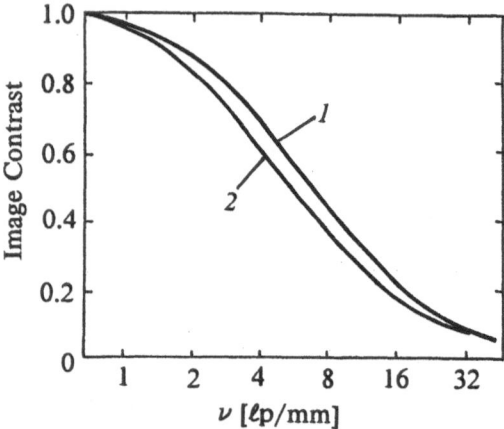

Fig.8.21. Contrast vs. spatial frequency for Phototitus [8.64]. (The crystal thickness was *1*: 150 μm, *2*: 180 μm)

with incoherent light. As supposed in [8.64], the grating can be visually detected at a contrast level of 0.05.

The experimentally obtained recording sensitivities of the Phototitus (10μJ/cm^2 [8.59] and 75 and 290μJ/cm^2, depending on the voltage polarity [8.65]) differ by more than an order of magnitude. This can be explained by differences between the modulators tested and the measurement procedures used.

Table 8.5 summarizes salient parameters of the Phototitus. The sensitivity is defined [8.65] as the exposure required to achieve a 80% intensity transmittance with respect to the maximum value, which amounts to 25% in this case. The light losses are a result of the voltage drop across the crystal not reaching the half-wave voltage.

A drawback of the Titus and Phototitus using the DKDP crystal is the need for the crystal cooling, which complicates device usage. The parameters exhibit a sharp dependence near the operating temperature. Therefore uniform and stable crystal cooling is necessary, which requires a precise thermal stabilization. These limitations can be avoided by substituting DKDP with crystals that do not require cooling and thermal stabilization,

164

Table 8.5. Phototitus parameter and performance specifications

Resolution	5 - 7 ℓp/mm
Active area	27×38 mm²
Sensitivity	75 μJ/cm²
Storage time (under readout)	30 min
Required voltage	200 V
Operating temperature	-50° C

such as ferroelectric ceramics [8.70], LiNbO$_3$ [8.71], BGO or BSO [8.72, 73], which were used for the Titus-type device.

8.4 Optically Addressed Microchannel Spatial Light Modulator

The optically-addressed SLMs discussed above (PROM, PRIZ, and Phototitus) have recording sensitivities on the order of 10^{-4} to 10^{-6} J/cm² (10^{-11} to 10^{-12} J/pixel), which allows recording of signals and images in real time if, for instance, a laser scanning device or a TV tube with an enhanced brightness is used. However, this sensitivity is too low to record naturally illuminated objects in real time, even in daylight. To accomplish this task, the brightness of the images should be pre-amplified, which markedly complicates the recording system.

One way to increase the sensitivity of electro-optic SLMs is to employ a Microchannel Array Plate (MCP). This approach was brought into practice in the so-called microchannel SLM proposed by *Warde* et al. [8.74, 75] and modified by a group of Japanese researchers [8.76, 78]. Two versions of a vacuum-sealed device are depicted in Fig. 8.22. The modulator consists of a photocathode which serves as a photosensitive element, a MCP which amplifies the electron image, and an electro-optic crystal which modulates the readout light.

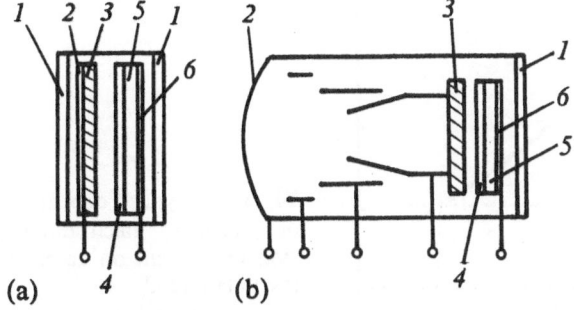

Fig. 8.22. Versions of microchannel SLMs. (a) proximed focusing device, (b) device with an electron focusing system (*1*: optical windows, *2*: photocathode, *3*: microchannel array plate, *4*: dielectric mirror, *5*: electro-optic crystal, *6*: transparent electrode)

The MCP [8.75] is a semiconducting glass plate with channels (pores) typically 10 μm in diameter, with 10^5 pores per cm^2. The plate is coated with electrodes, leaving the pores open. A voltage of about 1 kV is applied to electrodes during operation. Each pore functions as a continuous dynode electron multiplier. Electrons in the pore are accelerated by the external field and grow in quantity because of multiple collisions with the pore walls, as in an ordinary electron multiplier. The positive charge formed at pore walls is compensated, since the material has a noticeable conductivity. Gains of 10^4 with a single plate and of 10^7 with two cascaded plates were reported. The active area of the plate may be made more than 10 cm^2; the gain variation is less than 5%.

Microchannel SLMs employ plates of uniaxial birefringent $LiNbO_3$ [8.74-76] and $LiTaO_3$ [8.75],[4] with a thickness ranging from 500 to 50 μm. The plates have a so-called 55° cut, at which the natural birefringence is not observed for light waves incident normal to the crystal surface, and light is modulated through the longitudinal electro-optic effect. Images are read out through light reflection from the dielectric mirror on the crystal side facing the MCP. The devices use photocathodes sensitive in the visible spectrum. The configuration shown in Fig.8.22b has an electrostatic system focusing the electron image from the photocathode onto the MCP surface. To improve the image fidelity, the photocathode is placed on a fiber-optic plate whose shape is chosen to compensate for aberrations.

The energy of primary electrons arriving at the crystal surface is determined by the potential at the accelerating grid. In turn, the energy of the primary electrons affects the secondary-electron-emission coefficient δ, i.e., the ratio between the number of electrons leaving the crystal to that of incident (primary) electrons. Figure 8.23 shows a plot of δ as a function of primary electron energy typical of insulators. If $\delta < 1$, the surface collecting electrons is charged negatively; if $\delta > 1$, the surface is charged positively. As the charge is accumulated, the surface potential varies, thus leading to a change of the primary electron energy. The surface is charged until the

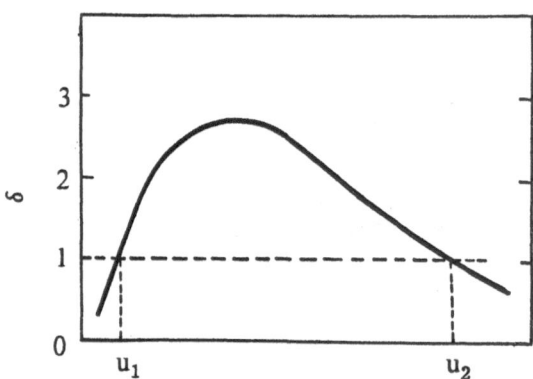

Fig.8.23. Typical dependence of secondary electron emission coefficient on accelerating potential U

---

[4] Though $LiNbO_3$ and $LiTaO_3$ are photorefractive crtystals, microchannel SLMs use only their electro-optic properties.

electron energy reaches the value at which $\delta = 1$ (eU_1 or eU_2 in Fig.8.23).
Gradually the surface charge density becomes uniform, and the image is
erased.

To record a new image, the potential at the accelerating grid must be
changed. Then the image can be recorded through a nonuniform accumulation of positive or negative surface charge. Thus, both negatives and posi-
tives of a recorded image may be produced. By successively recording two
images (with different potentials on the accelerating grid), image addition,
subtraction, and logic operations can be performed [8.77]. In addition, the
electrostatic focusing system allows zooming (with a magnifying ratio of 0.5
to 1.5) and deflection of the readout images. An auxiliary magnetic system
provides image rotation by angles approaching 180° [8.76]. These operations
can be useful for SLM application in coherent optical systems, though
attention should be paid to the fact that the electrostatic system can give
rise to nonlinear distortions of images, especially severe for the geometrical
operations mentioned above.

Microchannel SLMs have the highest recording sensitivity among the
electro-optic SLMs. A sensitivity as high as $2.2 \cdot 10^{-9}$ J/cm^2 ($\lambda = 655$nm)
was reported for the modulator employing MCP with a gain of 10^3 and a
500-μm LiNbO$_3$ plate [8.75]. Resolution of such an SLM is relatively low:
the thinner the plate, the higher the resolution and the lower the SLM sensitivity. Figure 8.24 shows the MTF curve for microchannel SLMs with
300- and 500-μm crystal plates [8.77]. The resolution at 50% modulation is
10 and 3 ℓp/mm, respectively (the active surface of the SLM is 15 mm in
diameter). The sensitivity is lower than that given above and amounts to (5
and 50)$\cdot 10^{-9}$ J/cm^2 [8.78]. A further increase of the SLM sensitivity might
be achieved by using MCPs with higher gains and/or special materials for
coating the dielectric mirror surface to ensure a more efficient secondary
electron emission [8.78].

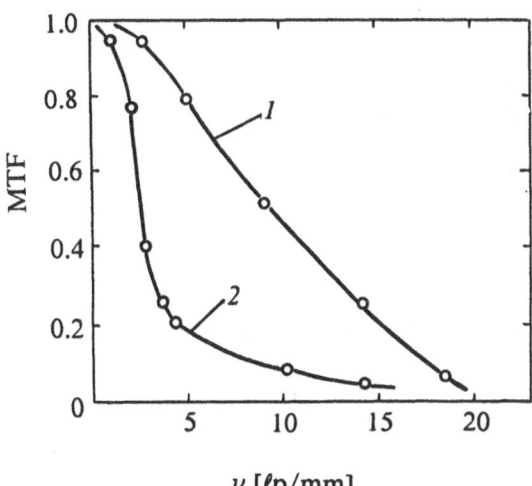

Fig.8.24. MTF curve for a
microchannel SLM. (*1*: crystal
thickness $d_c = 50\,\mu$m, *2*: $d_c =$
300$\,\mu$m)

ν [ℓp/mm]

8.5 Photorefractive SLMs with a Spatial Carrier Frequency

All the electro-optic SLMs discussed above employ the longitudinal electric field to record images. In this section we shall briefly consider the SLMs where image recording occurs in cubic crystals of the BSO type in the transverse external electric field and readout is accomplished through the transverse electro-optic effect. The recording and readout geometries - i.e., mutual orientations of the external field, crystallographic axes, and the light propagation directions - are similar to those used for holographic recording. (Chap.4).

The incoherent-to-coherent image conversion may be performed as follows [9.79, 80] (Fig.8.25): During recording, two plane coherent beams A_1 and A_2 create a sinusoidal interference pattern with spatial frequency ν_c in the crystal and, in addition, incoherent recording light $I_S(x, y)$ focuses an image to be converted with spatial bandwidth $\Delta\nu < \nu_c$. The crystal may be exposed with the coherent and incoherent light simultaneously, or in succession. In either case, a sinusoidal charge grating with ν_c, whose amplitude is spatially modulated in accordance with the image intensity $I_S(x, y)$, is formed. For instance, when the coherent and incoherent light act simultaneously and the amplitude of the pattern of interference between the coherent beams R_1 and R_2 is well below the incoherent light intensity, the contrast of the resulting intensity distribution within the crystal is

$$m(x, y) \simeq \frac{I_C}{I_C + I_S} \simeq \frac{I_C}{I_S(x, y)} . \tag{8.6}$$

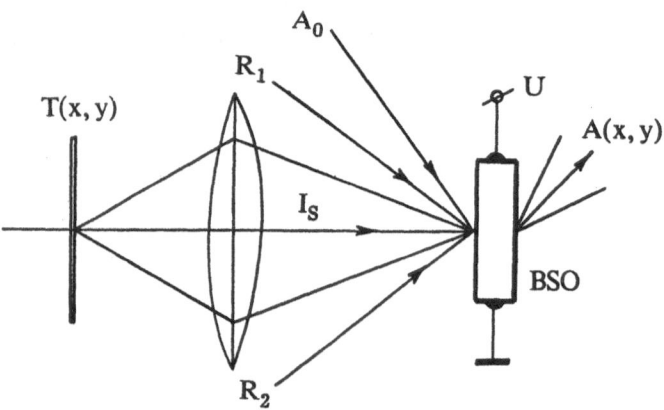

Fig.8.25. Photorefractive incoherent-to-coherent image conversion (PICOC). (R_1 and R_2: coherent light beams, I_S: incoherent recording light, A_0: readout coherent light)

The diffraction efficiency of the hologram recorded in BSO-type crystals in the steady-state regime of recording is determined by $m(x, y)$ and is independent of the recording intensity (Sect.4.2.2). Therefore, the light diffracted from the sinusoidal grating with the spatial frequency ν_c during readout forms a light field with amplitude

$$A(x,y) \propto \sqrt{\eta(x,y)} \propto m(x,y) \propto 1/I_s(x,y) \tag{8.7}$$

in the image plane of the PRC. Thus we obtain amplitude modulation of the readout light that yields a coherent "replica" of the initially incoherent image, which resembles its inverse, since its darkest parts correspond to the brightest parts of the image.

Different operating modes of the PRC performing photorefractive incoherent-to-coherent image conversion were studied [8.81]. All the experiments used the BSO crystal.

In the transverse-effect SLM, PRCs were employed in a similar fashion [8.82]. The electric field induced in the crystal by the image recording was modulated with the spatial frequency ν_c by focusing the image through a periodic amplitude grating so that the recorded image and that of the grating were multiplied in the crystal plane. This obviated the need for coherent light during image recording. The coherent replica of the image was observed in the direction of the first diffraction order of the periodic grating.

Note that the sensitivity of such an optically addressed SLM must be commensurable with the sensitivity of PRCs to holographic recording at the spatial frequency ν_c; i.e., it must be well below the sensitivity of the PRIZ SLM employing crystals of the same type.

The fundamental limitation on resolution is related to two factors [8.79,81]: the Bragg diffraction from a volume hologram and a finite depth of focusing of the incoherent image in the crystal volume. It can be shown (Sect.5.2) that the Bragg diffraction limits the maximum bandwidth of spatial frequencies in the converted image by

$$\Delta\nu_{max} \leq \frac{2\pi n}{\lambda \nu_c d}, \tag{8.8}$$

where n is the refractive index of the crystal, λ is the readout wavelength, and d is the thickness of the crystal layer where the sinusoidal grating is recorded. From (8.8) $\Delta\nu_{max}$ increases with decreasing ν_c. The minimum ν_c can be estimated from the condition $2\Delta\nu_{max} = \nu_c$, which allows the image spectra formed around the first and zero diffraction orders to be spatially separated. Then

$$\Delta\nu_{max} = \sqrt{\frac{\pi n}{\lambda d}}. \tag{8.9}$$

From (8.9), $\Delta\nu_{max} = 100$ lp/mm for typical values of n = 2.5, $\lambda = 0.63$ μm, and d = 1 mm.

The other limitation on $\Delta\nu_{max}$ - a finite depth of focusing of the input incoherent imaging system [8.81] - yields

$$\Delta\nu_{max} \sim \frac{4nF}{d} \; , \qquad\qquad (8.10)$$

where F is the F-number of the incoherent imaging system. For F = 5, $\Delta\nu_{max} \simeq 50$ ℓp/mm [8.81]. Equations (8.9,10) predict that resolution increases with decreasing crystal thickness d. However, a smaller d will be accompanied by a reduction in the SLM sensitivity ($S^{-1} \propto d$) and diffraction efficiency ($\eta \propto d^2$).

Another version of the optically addressed SLM with the transverse external field, called EPOS (EPOS is the abbreviated form of "electro-optic converter of optical signals" in Russian), has been described [8.83]. Figure 8.26 shows the interdigital electrode structure used to create the external field in the BSO crystal for image recording. The electrode width is 30 μm; the interelectrode separation is 300 μm. A voltage of 2.5 kV is applied to the electrodes. To prevent electrical breakdown, a glass plate was attached to the crystal surface by an optical glue with high breakdown voltage.

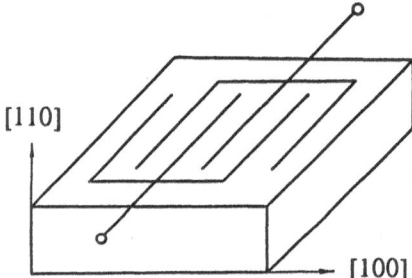

[110]

[100]

Fig.8.26. Electrode structure and orientation of crystal axes in the EPOS modulator [8.83]

The voltage applied to the electrodes induces a strong periodically alternating field within the crystal. Recording is performed with blue-green light (λ = 0.4 to 0.5 μm). The sensitivity for recording a sinusoidal grating with ν = 5 ℓp/mm was S^{-1} = 50 mJ/cm^2, and the maximum diffraction efficiency reached several percent.

The major specific features of the EPOS SLM arise from the periodic external field E(x). During recording, it is modulated by the recorded image $I_S(x,y)$ so that the readout-light amplitude in the linear approximation is

$$A(x,y) \propto I_S(x,y)E(x) \; . \qquad\qquad (8.11)$$

There is an analogy with the photorefractive holographic image converter discussed above, where image recording occurs in the external field modulated by recording a sinusoidal grating with coherent light. In the EPOS, the external field is also modulated, but using a system of electrodes on the crystal surface. However, a relatively low spatial frequency of the external field ($\nu = 1.5\,\ell p$/mm) does not allow readout of the recorded images (at least of two-dimensional images) in the first order of diffraction of the readout light from the external field grating. The task may be accomplished by decreasing the spacing between electrodes. However, because of the proportionality between the depth of penetration of the periodic external field into the crystal and the spacing between electrodes, the smaller spacing will lead to a reduction in the crystal layer where image recording takes place and, hence, to a lower sensitivity and diffraction efficiency of the SLM.

PRCs can be used for electrically addressed information recording [8.84]. In the simplest version, such a recording may be performed, for instance, on the PRIZ or PROM SLMs. Indeed, the charge density induced in the PRC is in the linear approximation $\rho \propto EW$, where E is the external field, and W is exposure with the recording light. Therefore, by scanning the crystal surface with recording-light spot of constant intensity and varying voltage applied to the electrodes, U(t), we can induce a nonuniform electric charge distribution in the crystal

$$\rho(x, y) \propto W E(x, y) ,\qquad (8.12)$$

where $E(x, y) = U(x, y)/d$, and x and y correspond to the instant of time at which the recording light illuminated this particular point of the surface. The image so recorded may then be read out as usual through the transverse electro-optic effect (as in the PRIZ) or longitudinal electro-optic effect (as in the PROM).

Figure 8.27 shows the modulator configurations allowing simultaneous recording of several electric signals. In the first case, the device consists of the BSO plate with electrodes in the form of transparent conducting strips on one face and a continous electrode on the other. The number of simultaneously recorded signals corresponds to that of strip electrodes. The electrodes create mainly the longitudinal electric field in the crystal. In the

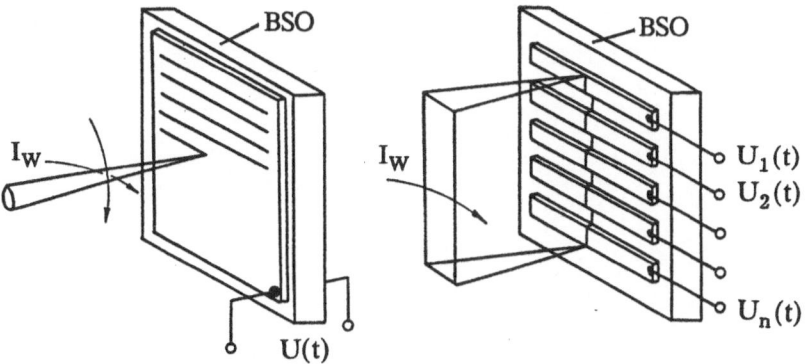

Fig.8.27. Electrically controlled recording on photorefractive crystals: (a) single-channel recording, (b) multichannel recording (I_W: blue light beam scanning the crystal surface)

171

second version (Fig.8.27) the electrical signals are applied to a pair of strip electrodes attached to one surface of the crystal plate. The electrodes for experimental samples were produced by the photolithographic technique, with the interelectrode separation of 125 μm. The magnitude of the input signals was limited by the breakdown voltage and did not exceed 200 V, which limited the diffraction efficiency of the SLM. During recording, all electrodes were simultaneously scanned by a narrow line of the recording light. Readout was through the transverse effect. The electrically addressed recording was performed in the BSO crystal in the frequency range from 2 to 200 Hz at a scanning rate of 10 mm/s. This bandwidth was determined by a tenfold decrease of the SLM diffraction efficiency from its maximum value [8.84].

9. Applications of Photorefractive Crystals

This chapter reviews different suggestions for applications of PRCs. Due to the high sensitivity, high diffraction efficiency, reversibility, and the ease of handling, these dynamic photosensitive media attracted the attention of researchers working in different areas of coherent optics. Early proposals on the use of photorefractive crystals and SLMs for holographic interferometry and 2D image processing are still interesting. Recently, a number of new ideas on PRC applications based on various phase-conjugate geometries have appeared, too.

Before dwelling into our discussion it seems worth mentioning one of the most important applications of holographic recording in PRCs, which proved to be beyond of the scope of this chapter. This is the measurement of different parameters of crystals, in particular, electro-optic and photogalvanic coefficients, dark- and photo-conductivity, diffusion length, lifetime and sign of the photoexcited carriers, concentration and structure of impurity centers, etc. (see, e.g., Chaps.4 and 5, and Appendix). The holographic methods are contactless, nondestructive, and exhibit high spatial resolution. They seem to be especially attractive for the characterization of semiconductor PRCs, such as GaAs and InP.

9.1 Holographic Interferometry

Holographic interferometry [9.1-5] is, to date, one of the most efficient tools for remote nondestructive testing, which finds wide use in such areas as industry, science, medicine, and metrology. It involves the comparison of two or more wave fronts, with at least one of them being reconstructed from a hologram. The holographic recording techniques enable testing of the most complicated wave fronts, including those reflected from real diffusely scattering objects.

The typical measurement accuracy afforded by holographic interferometry is at the submicrometer level. It is determined by the technique used for the interpretation of interferograms [9.2] and the errors in localization of the interferogram fringes, which are typically about 0.5 to 0.1 of the fringe width. Since a $\lambda/4$ displacement of a tested surface results in a fringe shift of a distance equal to its width, the final measurement accuracy is $0.1 \div 0.02\ \mu$m (for $\lambda \simeq 0.5\mu$m).

The first successful attempts to use PRCs for holographic interferometry followed the discovery of the photorefractive $Bi_{12}SiO_{20}$ (BSO) crystal [9.6], which still remains one of the most promising material for such applications. Its holographic sensitivity in the blue-green region for the spatial frequency $\Lambda^{-1} \simeq 1000$ mm^{-1} is $S^{-1} \simeq 10^{-3}$ J/cm^2 ($S^{-1} \simeq 10^{-4}$ J/cm^2 at $\Lambda^{-1} \simeq 100$ mm^{-1}) and the diffraction efficiency on recording under an external electric field can reach tens of a percent.

The literature gives many examples for the application of BSO (as well as BGO and BTO, which are of the same family) in different holographic interferometry geometries. We list here only the most typical examples, with a brief description of the basic distinctive features of PRCs as photosensitive media.

The most promising PRC applications in this area seem to be a continuous interferometric monitoring of objects or processes, and also the control of their behavior under the influence of external factors (temperature, load, frequency of excitation, etc.) needed to reveal critical situations. These tasks require operation of PRCs in a continuous mode or a framed mode with a high-cycle repetition rate, without any processing procedures and noticeable degradation of the PRC itself.

9.1.1 Double-Exposure Holographic Interferometry

The technique of double-exposure holographic interferometry involves recording two holograms of one and the same object in sequence in a single photosensitive medium using a pulsed laser (Fig.9.1a). Under illumination of such a superposition hologram by an original reference wave, the image of the object covered by interference fringes is reconstructed. Fringe location, orientation, and spacing carry information about the changes experienced by the object during the time interval Δt between exposures.

Decoding the resultant interferogram - i.e., calculation of real changes in the object - is a separate and often fairly complicated task [9.2]. With some simplification, we can say that for a reflecting object the bright fringes are localized in the object region, which either were not displaced or were displaced at distances being multiples of $\lambda/2$ along the direction of the highest sensitivity k; i.e., along the bisector between the directions of the illuminating and the reconstructed beams [9.2]. The dark fringes correspond to the regions displaced by $\lambda/4+n\lambda/2$ (here n is any integer), where coherent subtraction of the wave fields reconstructed from two successively recorded holograms is observed.

Double-exposure holographic interferograms of a stressed transparent plastic blade and the thermal index gradient induced by a transistor in a radiator (Fig.9.1b) were recorded in a BSO crystal with an external DC field $E_0 \simeq 6$ kV/cm at $\lambda = 488$ nm [9.7]. The typical diffraction efficiency of the hologram η for the optimized crystal thickness $d = \alpha^{-1} = 0.3$ cm (α being the optical absorption) was 2%. The complete write-read-erase cycle time for the incident light intensity $I_0 \simeq 13$ mW/cm^2 was < 0.1 s.

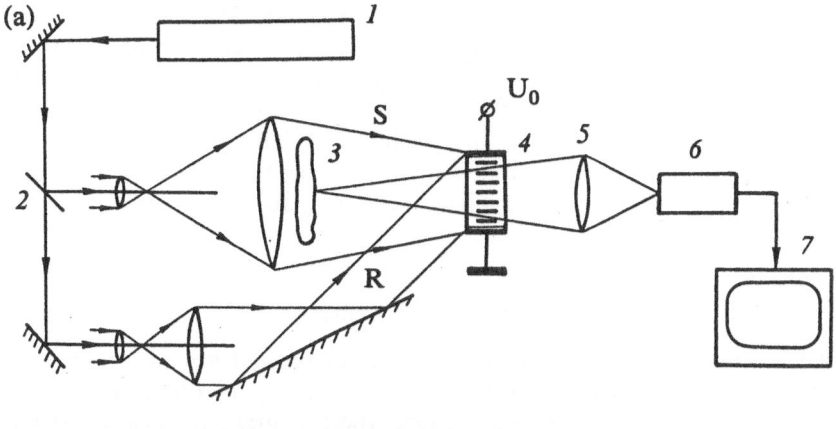

(a)

U_0

S

R

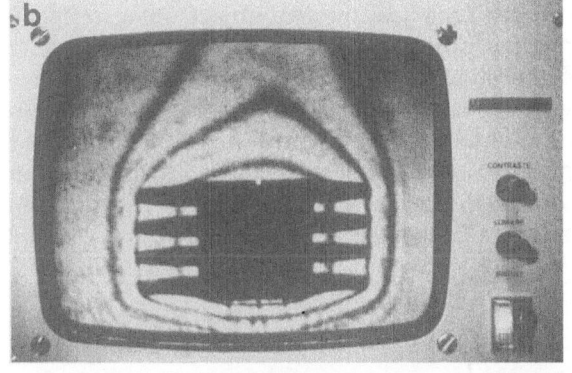

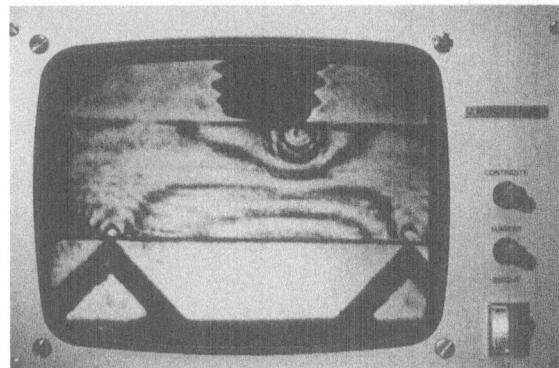

Fig.9.1. (a) Double-exposure holographic interferometry of transparent objects using PRC [9.7] (*1*: laser, *2*: beamsplitter, *3*: object under test, *4*: PRC, *5*: imaging lens, *6*: vidicon tube, *7*: TV monitor). (b) Typical double-exposure holographic interferograms obtained in [9.7] with BSO crystal

An important feature of PRCs that should necessarily be taken into account is the dynamic nature of the recorded holograms. Reconstruction of the resulting double-exposure hologram by the reference beam at the recording wavelength leads unavoidably to its optical erasure. Therefore, to make the observation period of the interferogram longer, either the intensity of the readout beam should be lowered or a TV monitor should be employed [9.6].

175

The erasing effect manifests itself not only during readout, but also during recording of the second hologram, when the first hologram is also subjected to erasure. To achieve maximum visibility of the interference fringes, the first hologram should therefore be recorded up to an amplitude approximately twice as large as the second one. This implies a two times longer exposure time for the first hologram recording [9.6, 7], if the initial linear stage of recording is used where the hologram amplitude $E_{sc} \propto t$, see (4.13, 15).

9.1.2 Real-Time Holographic Interferometry

In the technique of real-time holographic interferometry a single hologram of an object at some initial time t_0 is recorded. Then the object illuminated with the original beam is viewed through the hologram reconstructed with the same reference beam. As a consequence, the real object wave, reflected from (or transmitted through) the object at time t, and the reconstructed wave, yielding the image of the object at the initial time t_0, interfere behind the hologram plane (Fig.9.2a). In a way similar to the double-exposure hologram, the observer sees the object covered with fringes that show how the object has varied during the interval between t_0 and t. If the hologram is stable and illumination of the hologram and the object remains unchanged, a continuous comparison of the altered state of the object with the

(a)

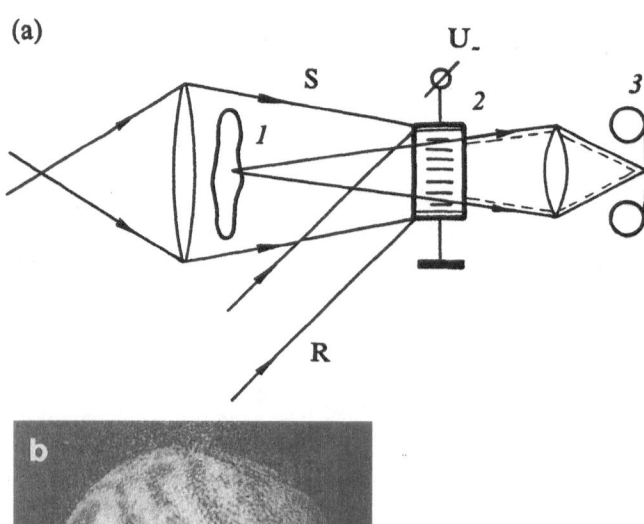

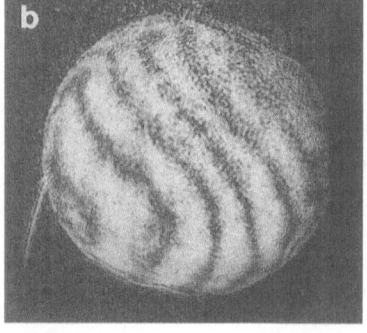

Fig.9.2. (a) Real-time holographic interferometry of transparent objects using PRC [9.8] (*1*: object, *2*: PRC, *3*: output plane, dashed lines show the light beam reconstructed from the hologram). (b) Examples of interferograms of thin crystalline plates of BSO obtained in [9.8] by real-time holographic interferometry using BTO crystal

176

initial one is performed - hence the name of the technique: *real-time holographic interferometry*.

The use of a PRC as a photosensitive medium prevents realization of this technique in a pure form. A continuous reconstruction of the hologram recorded in a PRC eventually leads to erasure of the hologram, which limits the maximum possible observation time. Moreover, because simultaneous illumination of the PRC by the object wave, the holograms of the whole set of object states during the observation period are additionally recorded within the crystal volume if special precautions are not taken.

Nevertheless, the technique is rather simple and, in certain cases, can be used in practice. Figure 9.2b demonstrates the result of its use [9.8] for measuring phase homogeneity of transparent crystal plates. The holograms were recorded at $\lambda = 633$ nm in a photorefractive $Bi_{12}TiO_{20}$ (BTO) crystal under an AC electric field with an amplitude of $E_\sim \simeq 10$ kV/cm. The complete recording cycle time was $\tilde{<}1$ s, at a total light power $P \simeq 1$ mW incident on the PRC area of 1×1 mm^2.

A somewhat modified form of real-time interferometry for studying transparent objects was used in [9.9]. In contrast to conventional configurations (Fig.9.2a), a counterpropagating plane reference wave reconstructs a phase-conjugate object beam from the hologram, which then travels back through the object under observation. Changes in the phase relief of the object are seen as fringes resulting from the interference of this wave with the original plane wave used to illuminate the object.

The experiments were carried out at $\lambda = 514$ nm and used BSO crystals with a thickness of $d \simeq 3$ mm. The diffraction effeciency of the hologram $\eta \sim 10^{-3}$ was observed under a DC field of $E_0 \simeq 6$ kV/cm.

9.1.3 Two-Wavelength Holographic Interferometry

Unlike the previously described techniques, a two-wavelength method is used for measuring the surface profile and positioning accuracy of diffusely reflecting objects. The double-exposure interferogram is also involved, with the only difference that instead of recording holograms at two different times, two slightly differing wavelengths are employed: $|\lambda_1 - \lambda_2| \ll \lambda_{1,2}$ (Fig.9.3a) [9.2]. The interferogram is read out at one of the wavelengths λ_1 or λ_2. As a result the phase relief of one and the same object but with different longitudinal magnification is produced due to interference. The final image will be superimposed with fringes indicating the surface profile, with the distance in depth between fringes:

$$\Delta h \simeq \frac{1}{2} \frac{\lambda_1 \lambda_2}{\lambda_1 - \lambda_2} . \tag{9.1}$$

The principal use of PRCs in this geometry is essentially the same as that discussed above for double-exposure holographic interferometry. Figure 9.3b presents the result of reconstruction of a two-wavelength holographic interferogram of a coin surface recorded using a BSO sample and a krypton laser [9.10].

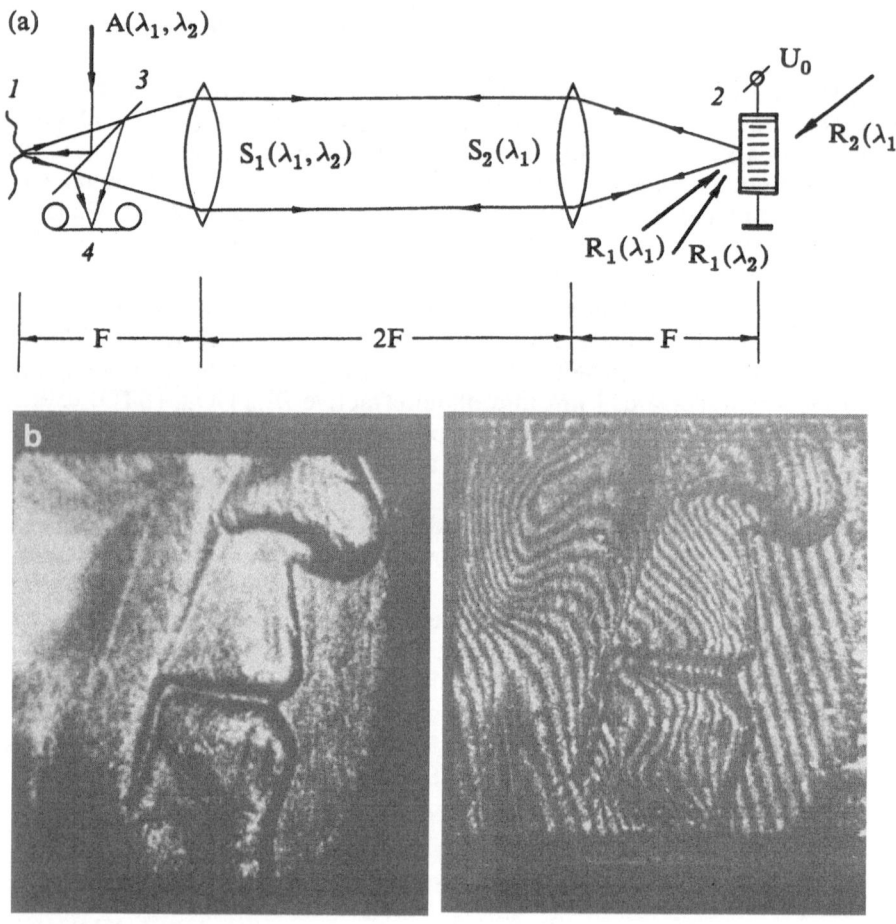

(a)

$A(\lambda_1, \lambda_2)$

$S_1(\lambda_1, \lambda_2)$ $S_2(\lambda_1)$

U_0

$R_2(\lambda_1)$

$R_1(\lambda_1)$ / $R_1(\lambda_2)$

1   3   2

4

F   2F   F

b

Fig.9.3. (a) Two-wavelength interferometry of diffusely scattering objects using PRC (*A*: illuminating light beam, *1*: object, *2*: PRC, *3*: beam splitter, *4*: output plane). **(b)** Example of a two-wavelength interferogram of a coin relief obtained in [9.10] using BSO ($\lambda_1 = 520.83\,\text{nm}$, $\lambda_2 = 530.87\,\text{nm}$, $\Delta h = 13.88\,\mu\text{m}$)

9.1.4 Time–Average Holographic Interferometry

Time-average holographic interferometry, a technique used to study vibrating objects, employs continuous recording of the hologram during a fairly long time interval $\Delta t \gg f^{-1}$, f being the vibration frequency of the test object. Reconstruction of such a hologram averaged over Δt gives the image of the initial object covered with fringes of differing brightness. The brightest fringes correspond to the lines of zero amplitude (or nodal lines) in the vibration distribution throughout the object. The holographic recording proceeds optimally for these regions, since their relevant interference structures turn out to be stable during Δt .

178

Those components of the recorded signal wave that correspond to the vibrating regions of the object are phase-modulate in accordance with

$$\propto \exp\left[i\frac{4\pi}{\lambda}\delta(\mathbf{r})\cos(2\pi ft)\right] . \tag{9.2}$$

Here $\delta(\mathbf{r})$ is the vibration amplitude at a given point $\mathbf{r}$ along the direction of the highest sensitivity $\mathbf{k}$ of the arrangement. The average contrast m (when averaged over the time interval $\Delta t \gg f^{-1}$) of the interference pattern and, hence, the amplitude of the corresponding component of the recorded hologram decays as

$$m(\mathbf{r}) = m(\mathbf{r},\delta=0)J_0(4\pi\delta(\mathbf{r})/\lambda) . \tag{9.3}$$

As a consequence, the brightness of the hologram-generated image turns out to be modulated by the same law that yields information on the spatial distribution of the vibration amplitude $\delta(\mathbf{r})$.

The PRCs seem to be best suited for vibration analysis through time-average holographic interferometry. Reconstruction of the hologram during its recording enables continuous monitoring (visually or using a TV monitor) of changes in the spatial distribution of vibration amplitudes caused by variations of the excitation frequency f, its intensity, and also external factors such as temperature, loads, etc.. Note that the time of averaging, Δt, in this case is the characteristic τ_{sc} time for hologram recording in PRC.

Several techniques for reconstructing time-average photorefractive holograms in a continuous manner have been proposed. Originally [9.7b, 11, 12], an auxiliary readout light wave R_2 was used. It travels strictly opposite to the plane reference wave R_1 participating in the hologram recording (Fig.9.4a). This means that the four-wave mixing geometry is used here. The reconstructed light wave is a complex conjugate replica of the recorded signal wave S_1, and therefore it forms a real image of the object. To separate the object and its reconstructed image spatially, a beam splitter is placed between the test object and the photorefractive sample.

Two somewhat different versions of this geometry have been considered [9.11, 12], namely, a configuration with a plane retroreflecting mirror for beam R_1 behind the crystal and the second one with an independently formed readout beam R_2 (Fig.9.4a). The first scheme is simpler and, hence, more easily aligned. Nonetheless, it places severe requirements on the phase homogeneity of the crystal and flatness of its faces. The second, a more complicated scheme, ensures an optimum intensity ratio between the recording and readout beams ($|S_1|^2+|R_1|^2 \simeq |R_2|^2$) nearly doubling the intensity of the reconstructed interferogram as compared with the former geometry.

A typical example of an interferogram of a vibrating diffuser [9.12] obtained in the geometry with a retroreflecting mirror is shown in Fig.9.4b. The hologram was recorded using the green line of an argon laser ($\lambda = 514$nm) in a 2-mm thick BSO sample under an external DC field of $E_0 = 7$

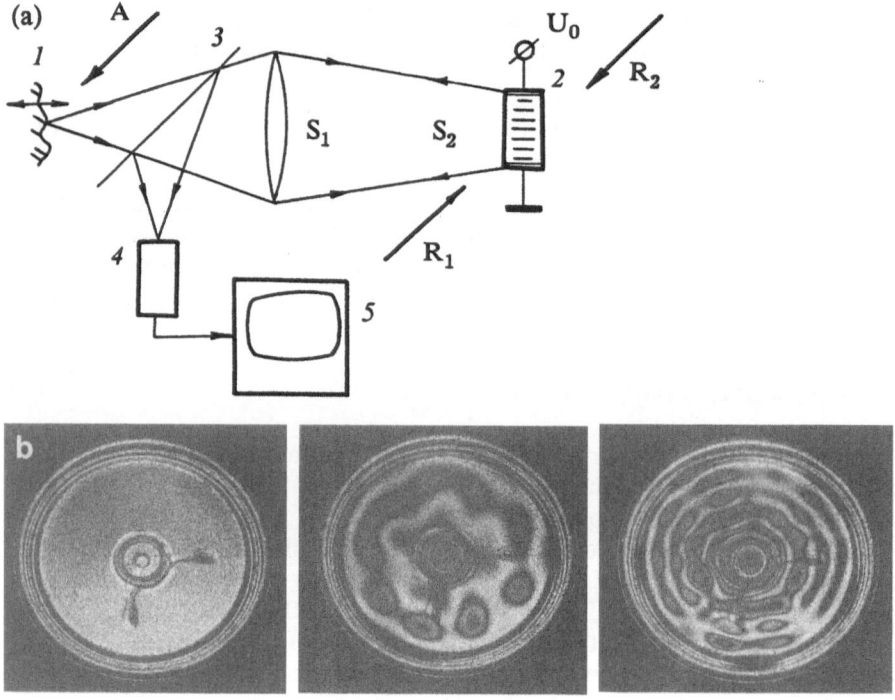

Fig.9.4. (a) Continuous reconstruction of a time-average interferogram by a counter-propagating readout beam R_2 [9.11, 12] (*1*: vibrating object, *2*: PRC, *3*: beam splitter, *4*: vidicon tube, *5*: TV monitor). (b) Time-average interferograms of a diffuser obtained with BSO for different excitation frequencies [9.7b]

kV/cm. For a total laser power of 800 mW, the intensity of the reference recording wave incident on the sample was $I_{R_1} \simeq 15$ mW/cm². This ensured a hologram recording time of $\simeq 0.5$ s at a spatial frequency of $\Lambda^{-1} \simeq 500 \div 300$ mm^{-1}. The vibration frequency range of the test object was 10 to $2 \cdot 10^4$ Hz.

An alternative way to record and to simultaneously reconstruct a time-average hologram of a vibrating object relies on two-wave amplification via a shifted hologram in a PRC [9.13]. Here a reduction in the contrast of the recorded interference pattern (9.3) caused by the object vibration changes the effective gain factor of the PRC (Sect.6.2). As a consequence, enhancement of the vibrating object depends on the amplitude of its vibrations at a given point **r**.

Besides its apparent simplicity, this scheme has no particular demand on quality of the light beams used, phase homogeneity of the crystal, and flatness of its faces. In addition, there is no additional readout beam, and the reconstructed interferogram experiences gain. A possibility to observe the higher-order interference fringes (for $\delta = \frac{1}{2} n\lambda$ with $n \gtrsim 1$) with a high contrast in this geometry is, however, doubtful. In fact, the two-wave mixing process is efficiently suppressed here as a result of a continous sinusoidal phase modulation of the object beam.

A third technique for continuous reconstruction of an interferogram exploits diffraction with polarization rotation [9.14]. The main idea is that the polarization of the original signal wave and that of the reconstructed wave can be orthogonal in a cubic PRC. This is achieved in the (110) cut sample when its [110] crystallographic axis lies in the incidence plane and the polarizations of the recording beams R and S are either in the incidence plane or normal to it (Sect.5.5). If a polarizer adjusted to suppress the signal wave is placed behind the PRC (Fig.9,5a), only a reconstructed signal wave is observed at the output.

Figure 9.5b demonstrates an application of this technique for the visualization of vibrations of a diffusely scattering ultrasonic transducer. Recording was performed at the wavelength of an HeNe laser ($\lambda = 633$nm) in a cubic BTO ($Bi_{12}TiO_{20}$) crystal with no external field applied. Use of an

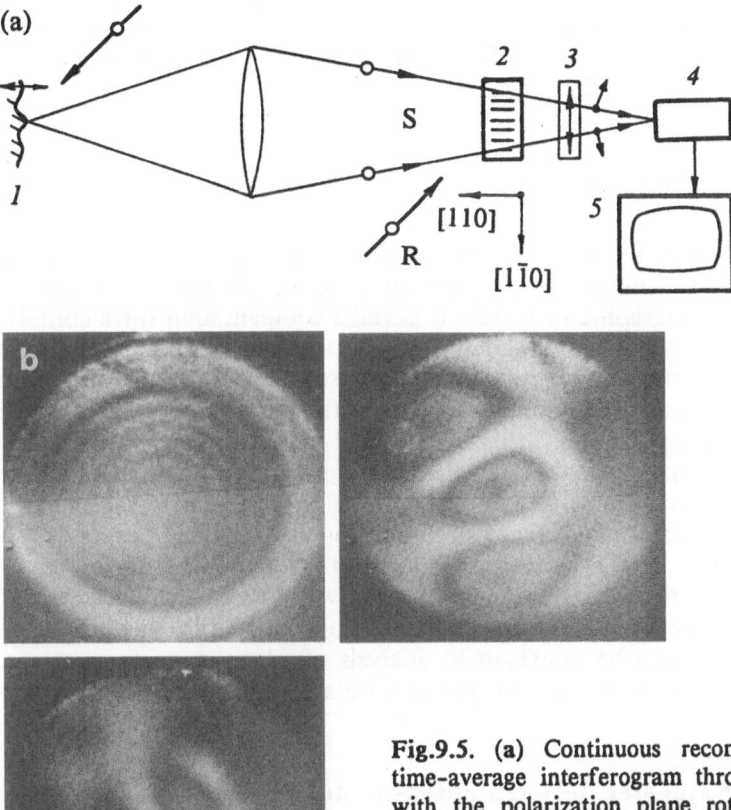

Fig.9.5. (a) Continuous reconstruction of a time-average interferogram through diffraction with the polarization plane rotation [9.14] (*1*: vibrating object, *2*: PRC, *3*: polaroid oriented to suppress the readout light polarization, *4*: vidicon tube, *5*: TV monitor). (b) Interferograms of vibrations of a diffusely scattering ultrasonic transducer obtained in [9.14] using the inteferometric arrangement shown in Fig.9.5a

externally applied electric field distorted drastically the polarization of the output light wave and made it nonuniform over the beam cross section, thereby precluding a total suppression of the original signal wave. Along with this, the diffraction efficiency of the recorded diffusion hologram in an 8-mm thick BTO crystal at a spatial frequency of $\Lambda^{-1} \simeq 500$ mm^{-1} was found to be high enough to ensure normal operation of a TV vidicon.

Without going into detail, we note that literature also reports on other applications of PRCs of the BSO type for interferometric monitoring of two-dimensional objects in real time. For instance, different geometries of speckle photography for measuring displacements and rotations of diffusely scattering objects were studied in [9.15]. An original technique for monitoring of vibrating objects based on holographic recording in BSO under an external alternating electric field was proposed in [9.16]. An interferometer for testing the phase homogeneity of transparent objects was also suggested in [9.17]. It has an increased sensitivity and operates through comparison of the wave transmitted through the object with its phase-conjugate replica. Suppression of speckle noises through incoherent summation of a series of interferograms recorded in a BSO crystal using a time-varying random phase mask was reported in [9.18].

9.2 Adaptive Interferometry

Generally speaking, the term "adaptive" is also applicable to conventional holographic interferometry relying on nondynamic photosensitive media, such as usual photographic materials. It permits compensation for a complicated relief of a test object (in other words, adaptation to it) and extraction of information exclusively about the changes experienced by the object. In this section we consider a continuous adaptation to slowly varying wave fronts. As will be shown later, this is necessary to perform an optimum detection of their rapid changes. The potential applications of this technique lie therefore in such fields as vibrometry and interferometric sensors (the fiber-optic ones included) of rapid time-varying or oscillating processes.

The evolution of this promising branch of holographic interferometry was motivated by the discovery of high-sensitivity PRCs. The discussion of the physical foundations of this promising application of dynamic holography will therefore be confined to analysis of phase dynamic gratings, with the idea that similar results are also obtainable for amplitude holograms.

9.2.1 Energy Transfer Between Phase-Modulated Beams

Let the photorefractive sample be exposed to an interference pattern of two intersecting plane coherent light beams of equal intensities I_0, one of them being phase modulated with a certain frequency Ω (Fig.9.6a):

$$I(x, t) = 2I_0[1 + \cos(Kx + \delta\cos\Omega t)] , \qquad (9.4)$$

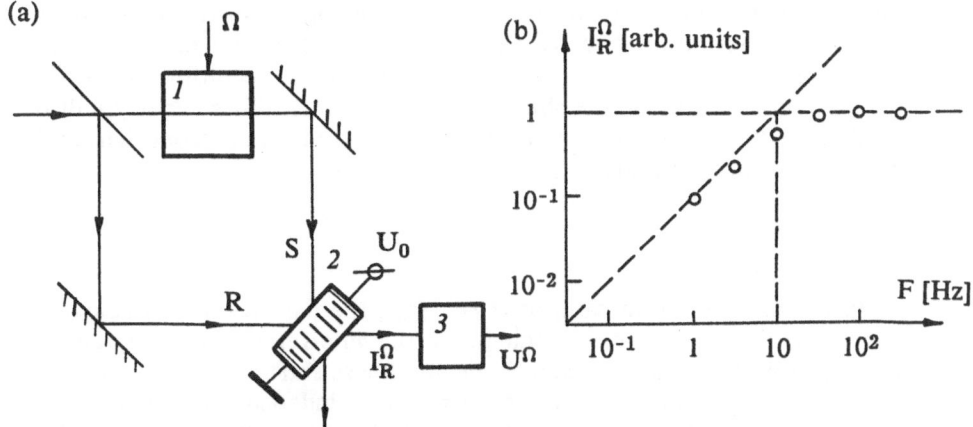

Fig.9.6. (a) Adaptive interferometer using PRC (1: the element which performs phase modulation of the signal beam, 2: PRC, 3: photodetector converting the beam intensity modulation into an electric signal). (b) Transfer function of the adaptive interferometer using photorefractive BSO (λ = 633nm, $I_0 \simeq$ 1mW/mm², Λ = 100μm, E_0 = 12 kV/cm)

where δ is the modulation depth. If the vibration frequency is high compared with the inverse characteristic time for hologram formation in PRCs under these illumination conditions ($\Omega \gg 1/\tau_{sc}$), the hologram fails to follow the displacements of the interference pattern. Its amplitude is determined by the average effective contrast m, (9.3), and may be expressed as

$$\Delta\epsilon^\omega(\delta) = J_0(\delta)|\Delta\epsilon^\omega|\exp(i\psi) , \tag{9.5}$$

where $|\Delta\epsilon^\omega|$ and ψ are the steady-state amplitude and phase shift of the hologram recorded by a stationary interference pattern ($\delta = 0$).

Direct substitution of (9.4,5) into (6.4a), which describes the energy exchange between the interfering light beams via the recorded phase grating, yields the temporal dependence of the beam intensity at the crystal output:

$$I_{S,R} \simeq I_0\left[1 \pm J_0(\delta)\left[\frac{\pi|\Delta\epsilon^\omega|d}{n\lambda}\right]\sin(\psi + \delta\cos\Omega t)\right] . \tag{9.6}$$

In this expression we ignored the light losses due to absorption and Fresnel reflections, considered the diffraction efficiency of the hologram to be small, and also neglected the possible effects of stationary energy exchange between the light beams. It demonstrates nevertheless the basic effect, consisting in conversion of the original phase modulation into the amplitude modulation of the output beams.

For a small modulation depth ($\delta \ll 1$), (9.6) yields the expression for the amplitude of the first harmonic in the output signal:

$$I^\Omega_{S,R} \simeq \pm I_0\delta\left[\frac{\pi|\Delta\epsilon^\omega|d}{n\lambda}\right]\cos\psi . \tag{9.7}$$

183

Thus, to achieve optimum phase-to-amplitude linear conversion, a phase grating of an unshifted type is required ($\psi=0$, $\pm\pi$). It can be readily obtained in PRCs via the drift mechanism of hologram recording. Note that with the phase grating of the shifted type ($\psi = \pm\pi/2$), this geometry will be operating in the regime of square-law detection, with the second harmonic amplitude at the output given by

$$I_{S,R}^{2\Omega} \simeq \pm I_0 \frac{\delta^2}{4} \frac{\pi|\Delta\epsilon^\omega|d}{n\lambda} . \tag{9.8}$$

In the other limiting case, at $\Omega \ll 1/\tau_{sc}$, the dynmaic phase hologram succeeds in following the movements of the interference pattern, i.e., in adapting to it. The hologram amplitude and phase shift between the grating and pattern, (9.4), turn out to be nearly time-independent and coincide with their steady-state values $|\Delta\epsilon^\omega|$ and ψ. If we use the analogy with a semitransparent mirror employed typically to observe beating between two coherent light beams, we can say that here we deal with a multilayer interference mirror. The latter, however, has a significant advantage in that the position of its reflecting surfaces follows the fringe movements in the interference pattern, thereby keeping constant the phase shift between the interfering beams at the output. For this reason the output beam intensities are nearly constant when one of the input beams is phase modulated with a frequency $\Omega \ll 1/\tau_{sc}$.

A more thorough analysis [9.19] reveals that if the record-erase process of the phase hologram is purely relaxational, the transfer function of such an adaptive phase-to-amplitude converter is

$$I_{S,R}^{\Omega}(\Omega) \propto \frac{\Omega\tau_{sc}}{(1 + \Omega^2\tau_{sc}{}^2)^{1/2}} . \tag{9.9}$$

Thus, it is identical with the transfer function of a conventional differentiating electronic RC circuit with a time constant RC = τ_{sc}. A more complicated form of transfer function can be observed for "moving" drift holograms in PRCs with a long photoelectron drift length (Sect.4.5.1).

9.2.2 Experiments with Photorefractive Adaptive Interferometers

The effect of energy exchange of phase-modulated beams as itself is an excellent means of exploring the PRCs and other dynamic holographic media [9.20-22]. The adjustment is easy, no auxiliary readout beams are needed, and the information on such parameters as the grating amplitude, its phase, and characteristic time of recording can be obtained.

The suggestion to use dynamic holograms in fiber-optic interferometric sensors was first made in [9.23]. *Hall* et al. pointed out that this allows use of a multimode optical fiber in the arms of the interferometer, simplifies alignment of the interferometer, and also ensures suppression of slow changes in the interference pattern caused by environmental changes.

It is known that temperature or pressure changes can induce an appreciable, slowly varying phase shift between the interferometer arms with amplitude $\gtrsim 10^3$ rad in high-sensitivity fiber-optic sensors with long arms [9.24]. Because of an essentially nonlinear operating regime of the photodetector, the spectrum of the useful high-frequency signal is broadened, due to the occasional time-varying phase shifts discussed, thus decreasing dramatically the threshold sensitivity of the sensor. The dynamic hologram allows compensation for the slow drift of the phase delay and transmission of the useful signal in the frequency range $\Omega \gtrsim 1/\tau_{sc}$ almost without attenuation. From a practical point of view, this method seems to be less complicated than the conventional techniques of heterodyne detection [9.25] or active stabilization of the operation point [9.24].

An adaptive interferometer of this type was experimentally investigated in a scheme of a laser vibrometer [9.26]. The dynamic phase holograms in SBN:Ce were recorded at $\lambda = 442$ nm under an external DC field $E_0 = 3.6$ kV/cm. For a total laser power $P_0 \simeq 20$ mW, the characteristic cutoff frequency was $f_0 = (2\pi\tau_{sc})^{-1} \simeq 20$ Hz. The minimum vibration amplitude detected was about 1 Å (in a frequency range from 10^2 to 10^5 Hz and for a bandwidth of $\Delta f \simeq 10^{-2} f$) and was mainly determined by the laser beam noise. The same measurement accuracy was also obtained with LiNbO$_3$ crystal at $\lambda = 475$ nm.

A similar scheme of energy transfer between phase-modulated beams vibrometric monitoring of real, diffusely scattering objects and in [9.28] for the stabilization of an interferometer used for the fabrication of conventional holographic diffraction gratings.

9.2.3 Other Geometries of Adaptive Interferometers

The two-beam geometry of adaptive interferometry discussed above is not the only possible one. In particular, a similar ability to compensate for a slowly varying phase delay is also displayed by the four-wave mixing geometry (Fig.6.4). This feature is one of the fundamental properties of nearly degenerate four-wave mixing scheme [9.29, 30] and relies on compensation for the phase shift after the backward propagation of the phase-conjugate wave through the perturbing medium. The dynamic properties of such an adaptive interferometer were studied in experiments using BaTiO$_3$ [9.31] and BSO [9.32]. The cutoff frequency in the BSO for a diffusion hologram with a spatial frequency $\Lambda^{-1} \simeq 300$ mm^{-1} at the argon-laser wavelength $\lambda = 514$ nm was $f_0 \simeq 22$ Hz.

The four-wave mixing geometry of the adaptive interferometer is, however, more complicated than the two-wave mixing scheme. The former needs precise alignment and phase stabilization of the reference beams. Self-pumped phase-conjugate mirrors are not applicable here since an absolute phase conjugation of the reflected wave is required. A similar interferometer using the multimode fiber and a self-pumped four-wave mixing geometry in a BaTiO$_3$ crystal was studied in [9.33]. It was shown to ensure almost ideal compensation for the complicated structure of the light wave after it backpropagates through the optical fiber, but it did not suppress the average slow phase drift. From this point of view it operates as an ordinary mirror.

In conclusion, note that similar adaptive properties are encountered by a photodetector for phase modulated signals, employing the phenomenon of non-steady-state photo-EMF [9.34,35]. It manifests itself in an alternating electric current through a short-circuited volume sample of a photoconductor illuminated in a conventional holographic arrangement (Fig.4.1) by a vibrating interference pattern (9.4). This current is also associated with the formation of a dynamic space-charge field grating $E_{sc}(x)$ in the sample volume.

9.3 Lensless Imaging and Compensation of Phase Distortions

One of the essential features of phase-conjugate optical systems pointed out in early works [9.36,37] is their ability to compensate for phase distortions after a two-way transmission of the wave front through a phase-aberrating medium [9.38-40]. In particular, it is possible to restore the light field distribution after a transparency or slide whose image, formed by a conjugate wave front, can be transferred to another plane by a simple beam splitter (Fig.9.7a).

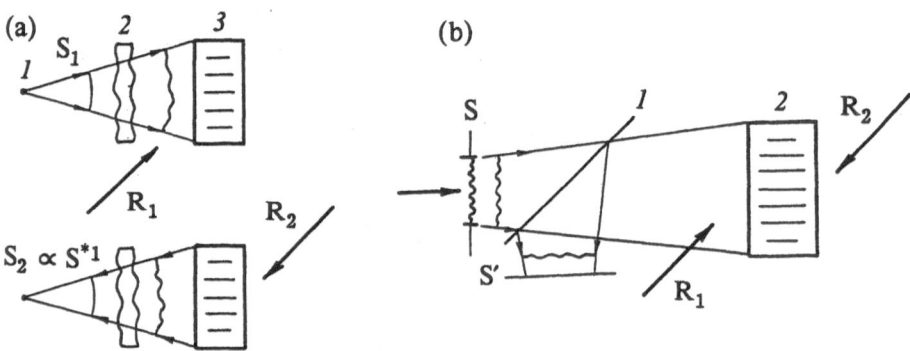

Fig.9.7. (a) Reconstruction of the wavefront shape after transmission of the phase-conjugate wave through the same phase aberrator (1: point source, 2: aberrator, 3: PRC). (b) Lensless image formation (S: original image plane, 1: beam splitter, 2: PRC, S': output plane)

The use of PRCs in the four-wave mixing geometry to implement this operation was first demonstrated for BSO [9.41] and for LiTaO₃ [9.42]. It was also shown in [9.42] that the time-varying phase distortions can be continuously compensated for. Different suggestions on PRC applications related to this fundamental property have recently appeared in the literature.

9.3.1 Lensless Imaging

The main task to be solved here is the spatial translation of a transparency image S to another plane S' (Fig.9.7b). The urgent demand for such an unconventional (lensless) imaging system has arisen in fine-line lithographic techniques. Standard objectives cannot provide the required submicrometer resolution for areas of $\gtrsim 10$ cm^2. The lensless imaging system in question would be able to cope with the problem if wave front conjugation of sufficient quality with a numerical aperture N.A. $\gtrsim 0.5$ for the area indicated could be performed.

The experimental study of such a device using four-wave mixing in a γ-irradiated LiNbO$_3$ sample was reported in [9.43, 44]. With a CW krypton-ion laser operating at the violet line ($\lambda = 413$nm) and a numerical aperture N.A. $\simeq 0.65$, the experiments succeeded in achieving a fairly high spatial resolution $\Lambda^{-1} \simeq 1000$ mm^{-1}.

A major shortcoming of this lithographic system is a low speed, likely due to the low sensitivity of the PRC chosen. With the average laser power $P_0 \simeq 0.4$ W, an area of the projected mask 6.8×6.8 mm^2, and sensitivity of the photoresist S$^{-1} \simeq 0.1$ J/cm^2, the exposure required four hours. As was pointed out in [9.44], the system needs a high-quality beam splitter and beam expanders, and a spatially homogeneous PRC sample. It also suffers from the speckle noise that is observed in any coherent imaging system.

9.3.2 Image Reconstruction by Two-Way Transmission Through a Multimode Optical Fiber

A specific case of an optically inhomogeneous medium is a multimode optical fiber. A major reason for image distortion is the modal dispersion, i.e., the difference in phase delays between the waveguide modes arising at the fiber exit. In particular, an additional phase delay of π radians between the zero mode and the highest-order mode is reached in a fiber of length

$$L \simeq (2n/N.A.)^2 \lambda \tag{9.10}$$

for a step-index fiber. For the typical numerical aperture N.A. $\simeq 0.2$ and $\lambda \simeq 1$ μm, it takes place at $L \simeq 0.2$ mm.

The use of phase conjugation for compensation of the modal dispersion in optical fibers, which is required to restore the transmitted image, was proposed in [9.45]. The phase of every mode transmitted through a length L of the fiber is reversed as a result of phase conjugation. This phase shift is entirely compensated for after the light transmission through another fiber section of length L with a similar mode structure. Precise compensation means that the input amplitude distribution is restored with high fidelity.

As far as we know, such a scheme consisting of two identical multimode fibers and a phase conjugator between them was not realized in practice because of the difficulty in fabrication of two identical fibers. Along tice because of the difficulty in fabrication of two identical fibers. Along with this, successful experiments on image reconstruction by two-way transmission through the same fiber in opposite directions were reported. The experiments employed BaTiO$_3$ in the conventional configurations of active [9.31] and passive [9.33] phase conjugators. Images were transmitted through sections of step-index fibers 1.75 and 0.75 m in length with up to 10^4 modes. The reflectivity of the phase-conjugate mirror obtained in [9.31] for weak signal waves was 150%.

The applicability of this configuration is still limited. Indeed, it is difficult to imagine a situation where the image is deliberately degraded by transmission through the fiber to be recovered at the input, i.e., at the place where it exists in the undistorted form initially. An important exception is, however, when the multimode fiber is used as one of the arms of the Michelson interferometer. The possibility to reproduce the Gaussian wave at the fiber input with high fidelity was demonstrated, in particular, in [9.33]. This enables multimode fibers to be employed in high-sensitivity fiber-optic interferometers, that simplifies construction and adjustment of the device.

9.3.3 One-Way Phase-Distortion Compensation

As the name of the technique indicates, the phase distortions introduced in the medium are compensated in a one-way transmission of the light from the object plane to the observation plane. In contrast to the two-way compensation geometry, these systems could find more extensive use for viewing real objects through a layer of a phase-aberrating medium.

Photorefractive devices of this type [9.46, 47] employ conventional holographic methods suggested earlier in [9.48, 49]. The techniques involve encoding of information about an a-priori unknown phase relief of an aberrator by transmitting the reference wave with a known wave front through it.

In the method suggested in [9.48], a lensless Fourier hologram is recorded in the plane immediately behind the aberrator (Fig.9.8a). If the aberrator is sufficiently thin, it does not disturb the recorded interference pattern. Therefore, when it is removed and the hologram is illuminated with the original reference beam, an aberration-free signal wave front is reconstructed.

In the technique described in [9.49], a focused image hologram of the aberrator is recorded (Fig.9.8b). The hologram so formed is read out by the beam passing through the aberrator, into which the transmitted image (a slide) is additionally introduced. The focused image hologram is nearly equivalent to the hologram of the phase front after the aberrator. Thus the distortions of the object wave caused by the aberrator are compensated for as it passes through the hologram, and an unaberrated object wave is observed in the first order of diffraction.

These two methods differ in that the aberration compensation in the object wave occurs at the recording stage in the former case and at the

188

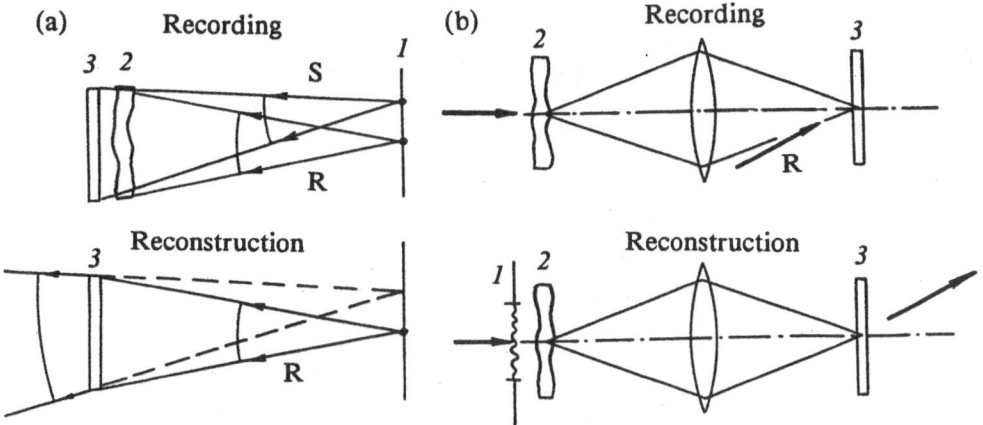

Fig.9.8. Phase distortion compensation using conventional (not dynamic) holograms suggested in [9.48] (a) and in [9.49] (b) (*1*: image plane, *2*: aberrator, *3*: recorded/reconstructed hologram)

reconstruction stage in the latter. In dynamic holography, where hologram recording and reconstruction are usually simultaneous, this difference is of no significance. A one-way compensation scheme using BaTiO$_3$ [9.46] is presented in Fig.9.9a. It uses recording of a focused image hologram of the aberrator, a change in the sequence in which the object beam passes through the hologram and the aberrator is not important. The presence of the reference beam R is not obligatory, since a passive (self-pumped) four-wave mixing geometry can also be employed here.

In a compensation scheme using BSO [9.47] a special beam illuminating the aberrator is absent (Fig.9.9b). It is formed automatically by a point reflector located in the center of the plane where the undistorted object image is formed. To initiate this process, a plane pump wave R$_2$ traveling opposite to R$_1$ is introduced into the scheme.

In [9.49] limitations on the information capacity of the transmitted images associated with the finite thickness of the hologram used in the device were estimated, too. Indeed, the maximum number of pixels in the image in the incidence plane (N$_x$) is determined by

$$N_x \simeq L_x \Delta\theta \frac{1}{\lambda} \simeq L_x \frac{\Lambda}{d}\frac{1}{\lambda} , \tag{9.11}$$

where L$_x$ is the linear transverse size of the hologram, and $\Delta\theta \simeq \Lambda/d$ is the angular width of the Bragg maximum. For the parameters used in [9.49] (L$_x$ $\simeq$ 10mm, d $\simeq$ 3mm, $\Lambda \simeq 5\,\mu$m, $\lambda = 0.5\,\mu$m), N$_x \simeq$ 30. Additional losses in the quality of compensation can also arise because of the difficulties in superimposing the hologram and the aberrator, which can also be of finite thickness [9.46].

Distortion-compensation geometries using four-wave mixing in PRCs in an essentially nonlinear regime of recording were also proposed in [9.50].

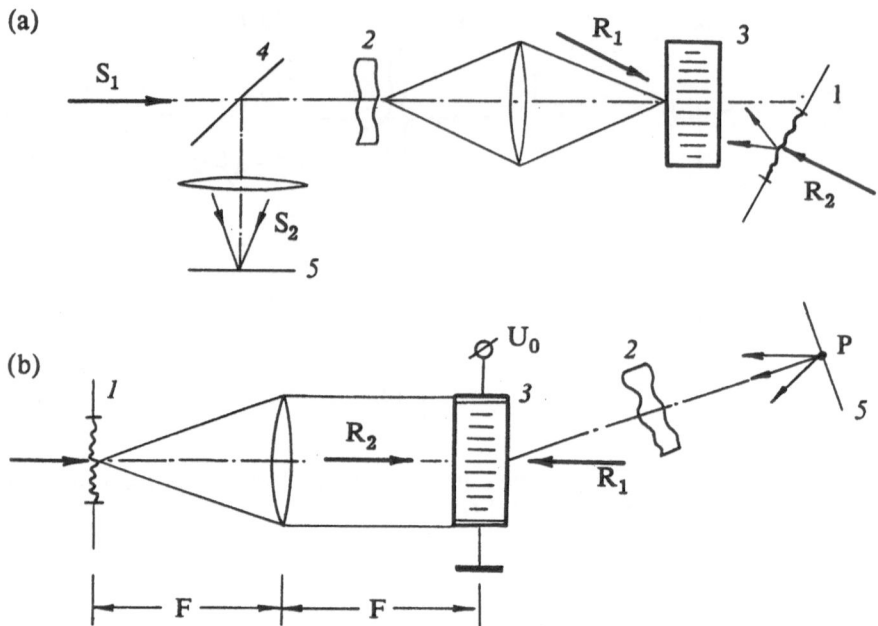

Fig.9.9. Phase distortion compensation using dynamic holograms in PRC suggested in [9.46] (a) and in [9.47] (b) (*1*: original image plane, *2*: aberrator, *3*: PRC, *4*: beam splitter, *5*: output plane, *P*: point reflector)

9.4 Lasers with Photorefractive Phase-Conjugate Mirrors

The use of a PRC in a conventional laser employing a traditional gain medium is motivated by the unique features of the resonators, which incorporate an ordinary and a phase-conjugate mirror (PCM) [9.51, 52].

1) The wave front radiated through the ordinary mirror of the resonator has the curvature of this mirror and is independent of the intracavity distortions (automatic compensation for intracavity distortions).
2) The resonator is always stable, even if the ordinary mirror is convex.
3) If the phase-conjugate mirror performs an absolute phase conjugation of the intracavity wave, there is no discrete spectrum of the longitudinal modes and the laser can generate any frequency within the bandwidth of the gain medium.

The standard geometry for phase conjugation by four-wave mixing in PRCs (Sect.6.3, 4) is unsuitable for the laser applications under discussion [9.52], since it requires auxiliary pump waves. For this reason the first experiments with PRCs as PCMs of the laser resonator were started only with the invention of passive (self-pumped) phase-conjugate configuarations (Sect. 6.5) which do not need external pump waves.

190

9.4.1 The First Experiments with Laser Resonators Using Photorefractive PCMs

The first experiments with $BaTiO_3$ in the configuration depicted in Fig. 9.10a were described in [9.53]. One of the mirrors of a powerful CW argonion laser was replaced by a passive photorefractive phase-conjugate mirror (with a linear or semilinear auxiliary resonator, Fig.6.8a,b). It was shown experimentally that this geometry offers the capability for efficient compensation of phase distortions induced by an intracavity aberrator. The power output of 500 mW was achieved as compared with 1 mW in a conventional resonator with the same aberrator. The presence of longitudinal modes corresponding to a conventional laser resonator was reported.

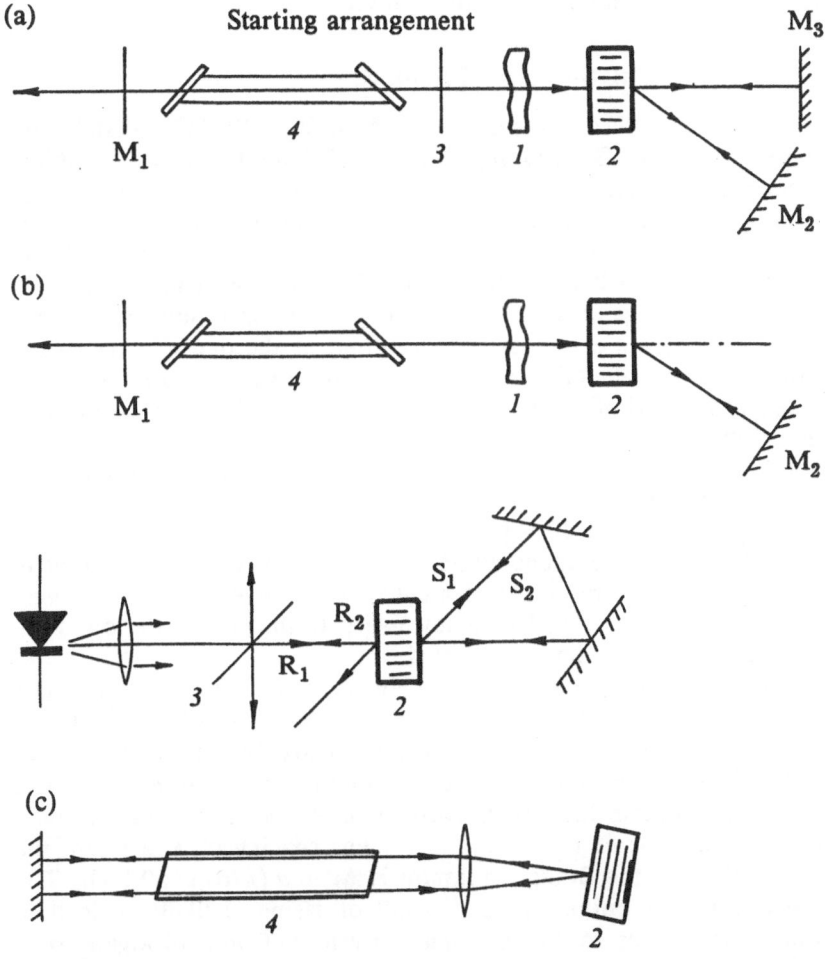

Fig.9.10. (a) Intracavity distortion compensation using four-wave mixing in PRC [9.53]. (b) Semiconductor laser with an external passive ring phase-conjugate mirror [9.57]. (c) Self-starting laser cavity using $LiNbO_3$:Fe as a dynamic mirror [9.60]. (*1*: aberrator, *2*: PRC, *3*: beam splitter, *4*: inverted laser medium)

191

Note that the self-pumped phase-conjugate configurations used provide only a relative phase conjugation of the different spatial components of the reflected wave and do not perform its absolute conjugation. As a result use of the passive PCM does not exclude the discrete structure of the longitudinal modes in the generated wave, with the first two properties (1 and 2) listed above being preserved.

The phase-conjugate resonator geometry discussed was found to need a starting procedure. The starting configuration includes a semitransparent mirror M_3, which formes an ordinary resonator with a front mirror M_1. After oscillation has been established between M_3 and M_1, a proper hologram is recorded in the PRC and then the mirror M_3 is removed. The initial coherence of fluorescence of the active medium was found to be sufficient to initiate operation of the photorefractive PCM.

9.4.2 Self-Sweeping of the Laser Frequency

Similar experiments with a self-pumped PCM using a $BaTiO_3$ crystal were performed for a dye (R6G) laser with pulsed [9.54] and CW [9.55] pumping. The geometry differs from that used in [9.53] by the PCM scheme (Fig. 6.7c). It also requires a starting arrangement consisting of a removable semitransparent mirror. Note that in both cases [9.54, 55] the output beam intensity increased remarkably and the linewidth narrowed to 2 to 5 GHz when the phase conjugator was switched on (with the removed auxiliary mirror). In addition, the effect of self-scanning of the dye laser from the initial value of the wavelength $\lambda \simeq 575$ nm mainly toward the red up to $\lambda \simeq 623$ nm, i.e., to the end of the dye tuning range where lasing extinguished, was observed in [9.55].

A similar behaviour - namely, an increase of the power output, narrowing of the generated beam spectrum, and frequency sweep - were also reported in an earlier work [9.56]. As distinguished from the previous geometry, the $BaTiO_3$ self-pumped PCM in this experiment was external to an ordinary (two-mirror) resonator of a conventional dye laser. In this configuration the lasing initiation, frequency sweep to the end of the dye tuning range, and lasing cessation were cycled.

The laser output increase can be explained in the following way. The reflectivity of $BaTiO_3$ PCM proved to be rather high (up to 50-60%) [9.55, 56], so the "swiching on" of the PCM reduced remarkably the light losses in the resonator. The line narrowing can be attributed to a fairly high spectral selectivity of the volume holograms formed in PRCs. In particular, for a typical crystal thickness of $d \simeq 1$ cm and fringe spacing of $\Lambda \simeq 1$ μm, the spectral half-width of the Bragg maximum is $\Delta \nu \simeq \nu (\Lambda/d) \simeq 50$ GHz. The positive feedback that builds up as a result of lasing is likely to lead to further line narrowing up to the experimental values of units of gigahertz.

The frequency sweep can evidently be caused by the frequency shift in the wave reflected from the photorefractive self-pumped PCM (Sect.6.5). Indeed, if frequencies of the incident wave and of the wave reflected from the PCM differ by $\Delta \omega$, the PCM is equivalent in some sense to a conventional mirror moving with a velocity of $v = c\Delta\omega/\omega$. The eigenfrequencies of the longitudinal modes of such an ordinary expanding ($\Delta\omega < 0$) or contracting ($\Delta\omega > 0$) resonator sweep at a rate

$$\frac{\delta\omega}{\delta t} = \frac{\omega v}{L} = \frac{\Delta\omega c}{L}. \qquad (9.12)$$

Here L is the resonator length.

Since, as indicated above, the absolute value and the sign of the frequency shift $\Delta\omega$ are influenced by a number of unpredictable factors, a sweep erratic in rate and direction should be expected. In fact, such a behavior of the effect was observed experimentally in [9.55,56]. The magnitude and direction of the sweep was affected by the cavity length, orientation of the crystal in the incidence plane, and even table vibrations.

9.4.3 Cavities with a Passive Ring PCM

The first experiments on a semiconductor single-mode GaAlAs laser ($\lambda = 0.85\,\mu m$) with an external photorefractive PCM were performed in [9.57]. Formation of the PCM in BaTiO$_3$ passive ring phase-conjugate geometry (Fig.9.10b) resulted in lowering of the lasing threshold and switching to multimode operation. In addition, noise spikes with a period equal to a round-trip time of the PCM ring cavity were observed in the output radiation. This phenomenon was attributed to phase locking in the external resonator, which is most promising for semiconductor lasers with small active areas.

A similar passive ring geometry with BaTiO$_3$ was used in experiments with semiconductor lasers in [9.58,59]. In the former paper a periodic frequency sweeping within a frequency range of about 10 nm was observed. In the latter the researchers reported on the measurements of the frequency width of the generated light, which proved to be lower than 100 kHz.

9.4.4 Laser Cavities with a Reflection Holographic PCM

A somewhat different geometry of the laser resonator with a volume hologram in a PCM as one of the mirrors was studied in [9.60] (Fig.9.10c). The nonlinear dynamic mirror was formed within a LiNbO$_3$:Fe crystal volume via recording a reflection hologram. The experiment employed an active copper-vapor element operating at the green line ($\lambda = 0.51\,\mu m$), which is characterized by a narrow luminescence spectrum ($\simeq 0.3\,cm^{-1}$). The scheme was self-starting; i.e., approximately a one-minute period of the superradiance regime of the active element was required to establish lasing between the front mirror M$_1$ and the reflection hologram in the PRC.

The mechanism responsible for the formation of the reflection hologram relies on the initial presence of a weak counterpropagating wave within the PRC volume, which is due to reflections from inhomogeneities within the sample and at its back face. As a result of the energy exchange via the reflection hologram being recorded [9.61] the reflected wave experiences gain. In such a way an effective multilayer mirror tuned in resonance with the generated wavelength is eventually formed in the volume of the PRC. An optimum shape of this dynamic mirror leading to the observed narrowing of the radiation angular spectrum is supposed to be chosen automatically due to the positive feedback through the active element and the front resonator mirror. The problem needs, however, more detailed studies, since as stated in [9.60], the holographic mirror in this case is not a purely phase-conjugate one.

9.4.5 Phase Locking of Independent Lasers

Investigations of phase locking of independent lasers started recently [9.62-68] and are considered encouraging. Dynamic holograms in PRCs are used for matching of two (or even more) laser cavities. This implies transformation of the transverse beam structure by the first laser into the phase conjugate replica of that generated by the second one. Two typical arrangements are presented in Fig.9.11.

In the the first one (Fig.9.11a) the laser L_1 is the master laser, and the second one L_2 (with one of the cavity mirrors removed) is a slave laser [9.62]. Lasing in the cavity of L_2 occurs only when a PCM in a volume of the PRC (BaTiO$_3$ in [9.62]) is formed. It is clear that the frequency detuning ($\Delta\omega$) between slave and master lasers for such an external pumping of the PRC cannot exceed the inverse characteristic time for photorefractive grating formation ($1/\tau_{sc}$).

Note that the geometry under consideration, as it is presented in Fig. 9.11a, is a conventional semilinear PRC oscillator (Fig.6.8b) pumped by master laser. An additional intracavity active medium is only introduced. As

(a)

(b)

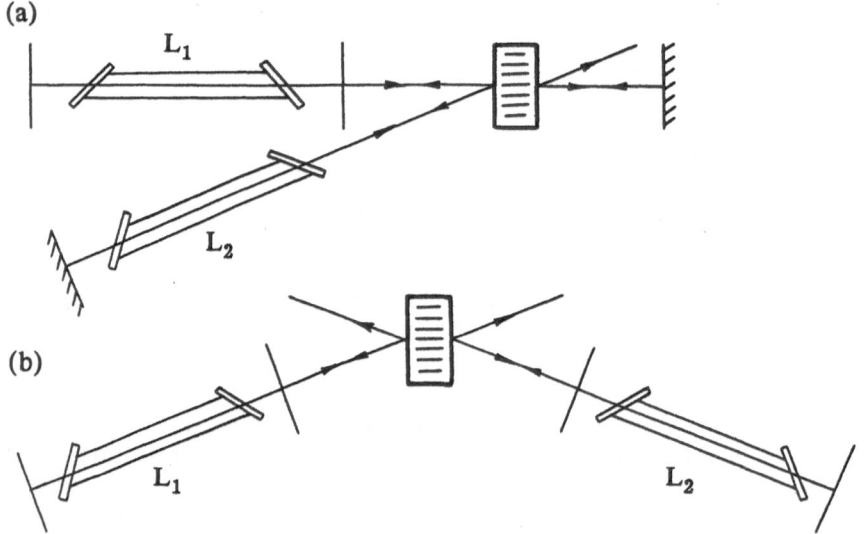

Fig.9.11. Phase locking of lasers by means of a PRC: (a) geometry with a master laser, (b) geometry with a double phase-conjugate mirror

194

was shown in [9.62], generation of a slave laser (L_2) can, however, be reached even with a hollow cavity without active medium.

Other geometries of multi-beam interaction in PCRs were employed in [9.63-65] for phase locking of independent lasers. Here a double phase-conjugate mirror geometry [9.66-68] is considered to be most promising (Fig. 9.11b). Interesting results with $BaTiO_3$ were obtained with linear arrays of GaAlAs laser diodes [9.66-68]. A frequency shift and narrowing of the generated line as well as a transformation of the transverse structure are worth mentioning.

9.5 Optical Gyroscopes

There are two major types of optical gyroscopes based on the Sagnac effect. One is the ring laser gyroscope [9.69, 71], which is actually a ring laser (Fig. 9.12a), i.e., a ring optical resonator, one or several arms of which are filled with a gain medium (typically a HeNe gas mixture operating at $\lambda = 633\,\text{nm}$). If the resonator is at rest, the cavity eigenmodes that are the clockwise- and counterclockwise-propagating light waves are degenerate in frequency. If the gyroscope is rotated, the counterpropagating modes experience a frequency shift due to the Doppler shift in the coordinate system related to the gyroscope:

$$\Delta\omega^\Omega \simeq 2\omega \frac{\Omega R}{c} , \tag{9.13}$$

where ω is the light wave frequency, c is the velocity of light, Ω is the rotation rate and R is the ring resonator radius. This frequency shift proportional to the rotation rate is registered as beats between the counterpropagating frequency-shifted light waves.

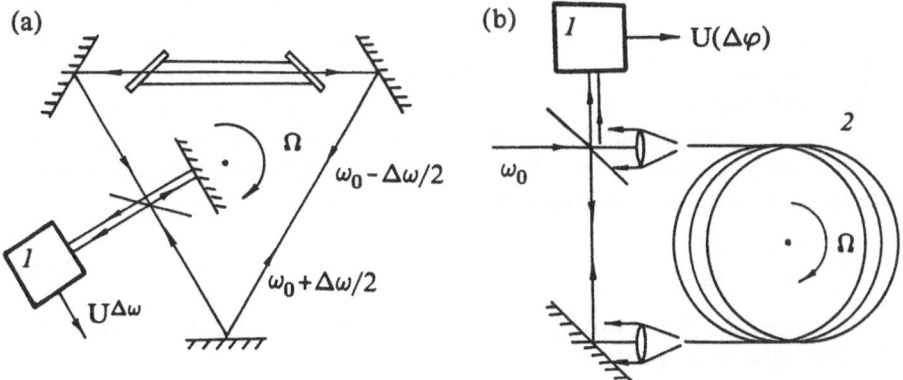

Fig.9.12. Conventional laser gyroscopes: (a) Ring laser gyroscope, (b) Fiberoptic gyroscope (*1*: photodetector converting the light beam intensity into an output electric signal, *2*: fiber coil)

The other type of optical gyroscope (i.e., a fiber-optic gyro [9.70]) is a fiber coil excited simultaneously by an external laser from its opposite ends (Fig.9.12b). If the gyroscope is at rest, the phase shift between the light waves traveling through the fiber in opposite directions proves to be the same. If the gyroscope is rotated at a constant rate Ω, the counterpropagating waves experience an additional nonreciprocal relative phase shift

$$\Delta\psi^\Omega = \frac{4\pi RL\Omega}{\lambda c} ,\qquad(9.14)$$

where L is the total fiber length. The magnitude of $\Delta\psi^\Omega$, which serves as a measure of the rotation rate, is registered using an ordinary interferometric technique.

These two types of optical gyros have their own merits and limitations and, as a result, somewhat different areas of application [9.70]. Both of them are now commercially available. The attempts to incorporate PRCs are aimed at overcoming one or another specific drawback of an existing device and rely primarily on the ability of PRCs to enhance and phase-conjugate wave fronts. The major distinguishing features of such systems are application of an external laser source, possibility to use multimode optical fibers, and, as a rule, frequency output of the device.

The simplest demonstration of such a device is a laser gyro where the gain medium is replaced by an externally pumped PRC (Fig.9.13a) [9.71, 72]. Two counterpropagating light waves shifted by $\pm \Delta\omega/2$ relative to the pump frequency ω_0 are excited in the resonator through four-wave mixing. The major problems here are the need for a pumping laser with a coherence length exceeding the resonator length L and precise tuning of the ring resonator to the pump frequency ω_0 [9.73]. Moreover, as was shown in this paper (also Sect.6.5), the real detuning $\Delta\omega$ in the waves at the output of this gyro should be remarkably inferior to the maximum value (9.13) because the PRC amplifier used is usually a narrow-band one.

9.5.1 Fiber-Optic Gyroscopes with Photorefractive PCMs

The fiber-optic gyroscope where optical radiation is launched only into one end of the fiber (Fig.9.13b) has been described in [9.74]. The radiation transmitted through the fiber is retroreflected by the phase-conjugate mirror and passes through the fiber in the opposite direction, to interfere with the original light wave. After the forward trip through the fiber in this arrangement, the wave experiences the phase shift

$$2\pi \frac{nL}{\lambda} - \frac{\Delta\psi^\Omega}{2} ,\qquad(9.15)$$

that changes the sign after the reflection from the PCM. After the return trip, the light beam phase is

$$- \left(2\pi \frac{nL}{\lambda} - \frac{\Delta\psi^\Omega}{2}\right) + \left(2\pi \frac{nL}{\lambda} + \frac{\Delta\psi^\Omega}{2}\right) = \Delta\psi^\Omega ;\qquad(9.16)$$

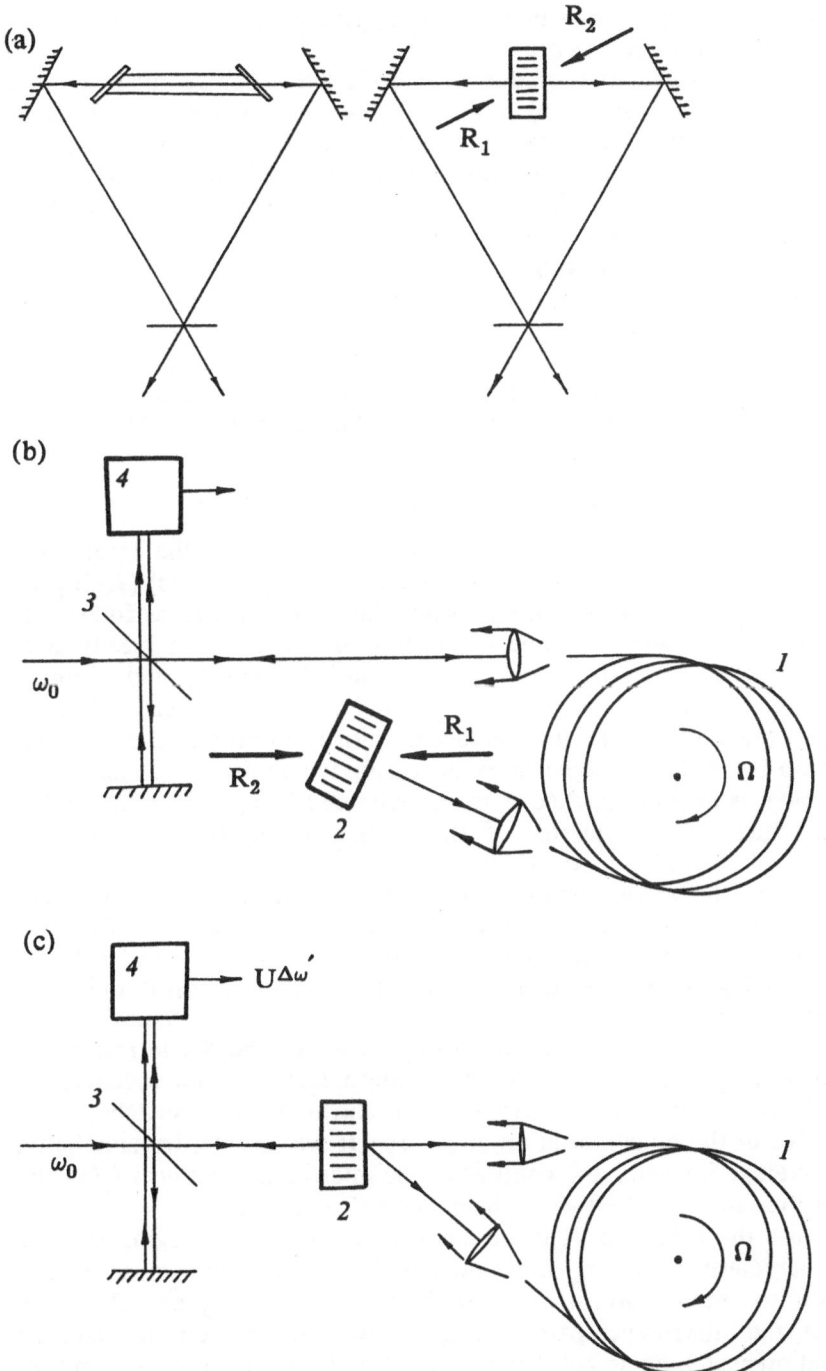

Fig.9.13. (a) Direct substitution of a gain medium by PRC in a ring laser cavity [9.71,72]. Fiberoptic gyroscopes with PRCs suggested in [9.74] (b) and [9.72] (c) (*1*: fiber coil, *2*: PRC, *3*: beam splitter, *4*: photodetector)

i.e., it equals the phase shift between counterpropagating waves in a conventional fiber-optic gyroscope.

An essential distinction of this fiber-optic gyroscope geometry from a conventional device (Fig.9.12b) is that it allows use of a multimode fiber with markedly reduced adjustment requirements. In fact, after the phase reversal and return trip, an original shape of the laser beam emerging from the fiber is restored. along with this, special, more complicated holographic geometries that restore the light polarization state are required to accomplish high-fidelity phase conjugation [9.75-77].

We emphasize that this gyroscope needs the absolute conjugation of the reflected wave phase. The feature is characteristic of the externally pumped phase-conjugate geometry, whereas the self-pumped (passive) phase conjugators do not possess it. In addition, a highly coherent laser is needed here; its coherence length is to exceed the doubled fiber length (2L).

9.5.2 Fiber-Optic Gyroscope with a Passive Ring PCM

An entirely different operating principle is employed in the fiber-optic gyroscope based on a ring passive phase conjugator (Fig.9.13c) [9.72b]. Its operation can be described in a somewhat simplified fashion as follows. If the necessary threshold condition of generation is fulfilled, a steady-state regime characterized by formation of a transmission grating and counterpropagating complex conjugate wave $S_2 \propto R_1^*$ within the crystal volume is established (Sect.6.5.4). The interference pattern $I_1(x)$ produced by the original pump beam R_1 and signal wave S_1 arising from self-diffraction of the former wave is always matched with the recorded hologram. In particular, they are shifted by a quarter of the spatial period for the diffusion mechanism of recording used. When the gyroscope is at rest, the phase shifts imposed on the waves passing through the fiber coil in opposite directions turn out to be equal. Therefore, the interference pattern $I_2(x)$ formed by the waves R_2 and S_2 within the crystal volume coincides with $I_1(x)$; i.e., it is also matched with the stationary hologram, which is at rest in this case.

If the gyroscope is set into rotation with a rate Ω, a nonreciprocal phase shift $\Delta\psi^\Omega$ (9.14) arises between the waves R_2 and S_2, to result in a corresponding shift of the interference pattern $I_2(x)$. In consequence, the net interference pattern $I(x) = I_1(x)+I_2(x)$ is shifted from its matched position relative to the hologram at an angle $\Delta\psi'$. If we assume, for simplicity, that the average intensity and contrast of the interference patterns $I_1(x)$ and $I_2(x)$ are the same and $\Delta\psi^\Omega \ll 1$, the angle $\Delta\psi' \simeq \Delta\psi^\Omega/2$.

Because the PRC is a dynamic medium, the hologram begins to move with a characteristic time τ_{sc} to a new position of the interference pattern $I(x)$. With the nonreciprocal phase shift $\Delta\psi^\Omega$ between S_2 and R_2 being preserved, the interference pattern will, however, also move to maintain an additional mismatch angle $\Delta\psi'$ between itself and the hologram. As a result, the hologram pattern will be continuously moving at a rate

$$v \simeq \Delta\psi'/\tau_{sc}K .$$
(9.17)

The frequencies of signal waves S_1 and S_2 arising due to diffraction of R_1 and R_2 from the moving hologram will be shifted by

$$\Delta\omega' \simeq 2\Delta\psi'/\tau_{sc} \simeq \omega \frac{RL\Omega}{c^2\tau_{sc}} = \left(2\omega\frac{\Omega R}{c}\right)\frac{L}{2c\tau_{sc}} . \tag{9.18}$$

So the nonreciprocal phase shift $\Delta\psi^\Omega$ in this fiber-optic gyroscope scheme is converted to the frequency detuning $\Delta\omega'$ between the signal waves S_1 and S_2. This value can be measured precisely enough by means of a conventional interferometric technique. The sensitivity of this device (9.18) differs, however, from that of a conventional ring-laser gyro (9.13) by a factor of $L/2c\tau_{sc}$. That is why to achieve the sensitivity of a conventional gyro, fast PRCs should be used (with $\tau_{sc} \simeq 10^{-6}$s for $L = 10^3$ m).

When estimating the real accuracy of such a gyroscope one should take into account the fact that the value of this shift $\Delta\omega'$ depends on the light intensity incident on the crystal ($\tau_{sc} \propto I_0^{-1}$), its sensitivity, and the spatial frequency K. It can also vary under the influence of an external electric field E_0 [9.78] and, hence, of internal effective fields of photovoltaic, pyroelectric and piezoelectric nature.

9.6 Holographic Memory

One of the first suggestions on the practical applications of PRCs concerned optical holographic memories. Researchers were attracted by two specific features of PRCs, namely, the possibility of reversible holographic recording and the volume nature of recorded holograms. In particular, the pronounced angular and wavelength selectivity of volume holograms (Sect.5.6) allows utilization of the third dimension (thickness) of the recording medium [9.79]. Retrieval of a needed hologram from this holographic memory unit can be accomplished by changing the wavelength of the plane readout beam (Fig.9.14a), its incidence angle (Fig.9.14b), or the refractive index of the medium the hologram is recorded in (Sect.9.7.1).

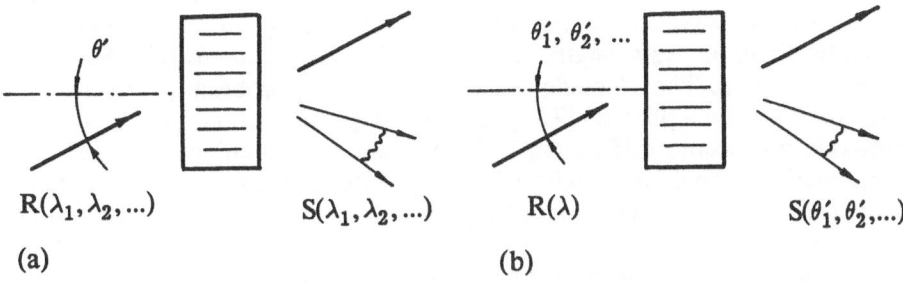

Fig.9.14. Retrieval of a set of holograms from a volume of PRC by changing the wavelength (a), or the incidence angle (b) of the readout beam

The limiting information capacity of the hologram with a volume $V = L_x \times L_y \times L_z$ (or to be more exact, the total number of pixels reconstructed from the volume hologram) can be estimated as [9.79]

$$N^{max} \sim \frac{n^3 V}{\lambda^3} . \tag{9.19}$$

For instance, if the hologram retrieval is performed by varying the readout incidence angle, it equals the number of pixels in one hologram times the number of holograms

$$N \simeq \left[\frac{(N.A.)L_x}{\lambda} \frac{(N.A.)L_y}{\lambda} \right] \frac{\delta\theta'}{\Delta\theta'} \simeq \frac{V}{\lambda^3} 2n\sin\theta \, (N.A.)^2 \, \delta\theta' . \tag{9.20}$$

Here N.A. is the numerical aperture of the imaging optics, $\Delta\theta'$ is the angular width of the Bragg maximum, and $\delta\theta'$ is the possible range of the readout incidence angles. For N.A., $\delta\theta' \sim 0.1$, and $2n\sin\theta \sim 1$, the information capacity of the volume holographic medium is about 10^9 pixel/cm^3.

9.6.1 Permanent Holographic Memory

For the first time multiple hologram storage was experimentally investigated in [9.80] for LiNbO$_3$:Fe at the argon-ion laser wavelength $\lambda = 488\,\text{nm}$. After recording an hologram the crystal was rotated by a certain angle ($\simeq 0.1°$), and the next hologram was recorded, etc. In this fashion, a series of 500 holograms with the final diffraction efficiency of 2.5–25% was recorded in a 1-cm thick sample.

The literature gives a fairly thorough discussion of the factors limiting the information capacity of real volume holograms in PRCs [9.1, 81–83]. Among them are limitations on the maximum value of Δn, concentration of trapping centers, and also cross talk between different holograms. On the other hand, the applicability of permanent volume holographic photorefractive memory is first of all restricted by the degradation of holograms during their reconstruction.

Different techniques for hologram fixing were proposed, such as "thermal" fixing in LiNbO$_3$ [9.80, 84–86] or "electrical" fixing in SBN [9.87]. Both of them provide long-term hologram storage and the nondestructive readout at the original wavelength. Methods for volume hologram readout at a remarkably different wavelength, to which this PRC is insensitive, have also been developed. They include readout by a diverging beam [9.88], recording through two-photon [9.89] or two-step absorption [9.90], readout through anisotropic diffraction [9.91, 92], and also nonlinear recording of combinational holograms [9.93-95]. All the techniques listed above have, however, fairly restricted areas of application and their own specific limitations.

Another important problem related to a permanent photorefractive holographic memory is the reproduction of holograms. Indeed, all the holograms must be recorded in succession into a photorefractive element on a vibrational-isolated optical table. For instance, when a series of 500 holograms mentioned above was recorded in LiNbO$_3$ [9.80], the required total light exposure was about 1400 J. Hence, when a laser with an average light power $\simeq 1$ W is used, up to one hour is needed for recording of the whole information.

9.6.2 Read-Write Holographic Memory

The capability for reversible recording, readout, and erasure of holograms is particularly attractive for read-write memory. In the literature [9.96], this type of memory is usually associated with the "flying spot" geometry [9.1, 97], where addressing of the recording and readout light beams is performed by an X-Y acousto-optic deflector (Fig.9.15). In addition to the reversible holographic medium and deflector, this memory system comprises many other complicated elements, such as a fairly powerful recording laser, a page composer, an array of photosensors, special optics (a lens array), and so on.

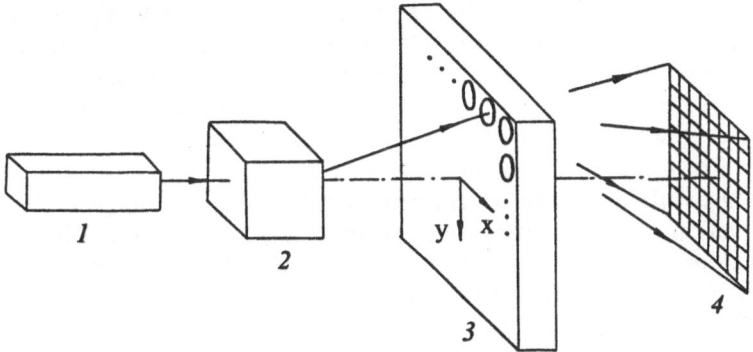

Fig.9.15. Simplified geometry of information retrieval from a holographic memory by a deflector ("flying spot" scheme) (*1*: laser, *2*: X-Y deflector, *3*: hologram matrix, *4*: matrix of photodetectors)

With the available deflectors, page composers and high-sensitivity PRCs, these read-write memories ensure a fairly high average speed of recording and retrieval of information (about 10^{10} bit/s). Their major limitation is apparently a small total storage capacity to which a high-speed access can be provided. It equals the capacity of a page-composer times the total number of resolvable points of the deflector, and typically does not exceed $10^7 \div 10^8$ pixels [9.96]. A further increase of the storage capacity by a mechanical replacement of matrices of holograms leads to a considerable increase of the access time [9.98].

9.6.3 Associative Holographic Memory

In contrast to conventional holographic memories, associative memory systems rely mainly on dynamic properties of the holograms recorded in PRCs. Proposals on these devices appeared only recently and the interested reader may consult the current literature [9.99-101].

9.7 Modulation and Deflection of Laser Beams

The literature describes a variety of methods for efficient modulation, deflection, and switching of laser beams by means of volume phase holograms in PRCs. This section discusses the basic concepts underlying these techniques.

9.7.1 Electric Control of the Bragg Condition

In a simplified manner, effect of the external DC electric field E_0 on the optical properties of the electro-optic PRC consists of a change in the refractive index. This affects both the wavelength of the readout light and its propagation direction in the crystal volume (the incidence angle being fixed). Therefore, we should expect variations in the conditions of Bragg diffraction as a result of application of the field E_0.

Suppose that initially (with no applied electric field) the Bragg condition for diffraction of a readout beam R from a particular hologram is satisfied. Application of a field E_0 can entirely "switch out" diffraction from this hologram. Also vice versa, if initially (at $E_0 = 0$) the diffraction condition for another volume hologram was not fulfilled, the external field E_0 can "switch on" diffraction from this hologram. A further change in the field can bring out this hologram from the readout conditions as well, but "switch on" the third hologram, and so on. Thus, a steplike variation of the external field E_0 results in switching (or an electric retrieval) of a series of holograms.

This series of holograms must be recorded in the PRC volume at the same fixed incidence angle of the reference beam and with exactly the same applied fields under which the holograms will afterwards be reconstructed. Needless to say, this technique requires relatively stable and non-degrading (during readout) volume holograms. This can be achieved by employing a fixing process, for instance, thermal [9.80, 84, 86] or electric [9.87] fixing.

The efficiency of the electric control is determined by the typical field variation ΔE_0 that causes noticeable violation of the Bragg condition [9. 102]. Let us assume that at $E_0 = 0$ the readout plane light beam R satisfies the Bragg condition (Fig.9.16a). Under an external electric field E_0 the wave-vector surface of the electro-optic crystal deforms. The component of the readout light wave vector K_R tangential to the sample surface remains unaltered. That is why application of E_0 results in a deviation of the grating

202

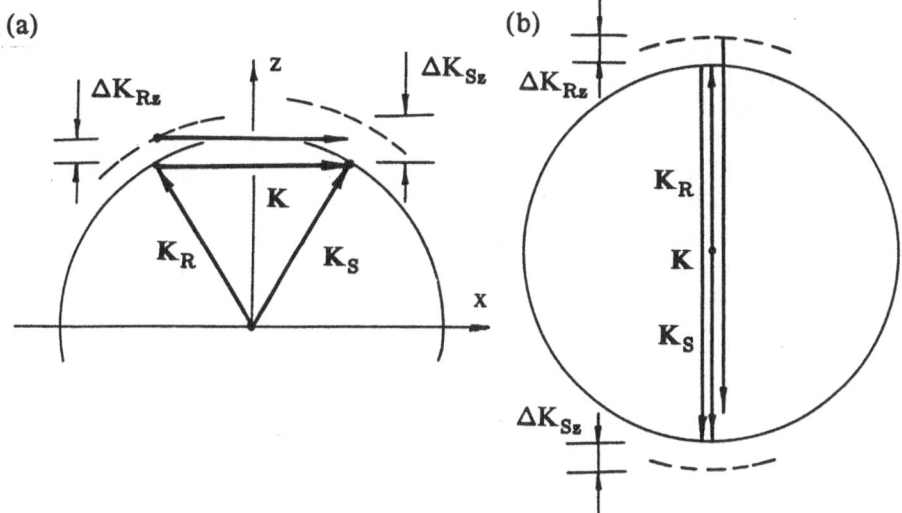

Fig.9.16. Vector diagrams illustrating violation of the Bragg condition as a result of application of external electric field E_0 to the electro-optic crystal: (a) Transmission hologram, (b) reflection hologram (dashed lines show shifts of relevant portions of the wave vector surface induced by field E_0, note that in the general case they are not the same for the readout and diffracted beams)

vector **K** tip from the wave-vector surface along the normal to the sample surface (Fig.9.16a). When the mismatch reaches the uncertainty of the grating wave vector ($\sim\pi/d$, Fig.5.2a) due to a finite grating thickness d, the intensity of the diffracted light beam S becomes half as large as its maximum Bragg value.

We can reasonably take $\Delta E_{0.5}$, which causes this twofold decrease of the diffracted intensity, as a quantitative measure of the effect of the electric field E_0 on the Bragg condition. Its value can be determined by the following simple rule: "Application of the electric field $\Delta E_{0.5}$ to the crystal must provide an additional π phase shift between waves R and S at the crystal output". This immediately follows from the above consideration if we take into account that deformation of the crystal wave-vectors surface is directly related to the additional phase delays for waves R and S transmitted through the crystal (Fig.9.16a).

A similar rule is also applicable to the reflection hologram (Fig.9.16b). Here, however, the sum, and not the difference, of additional phase shifts in the light waves R and S must be equal to π. This general consideration indicates that the typical values of the characteristic electric field $\Delta E_{0.5}$ (and hence the voltages $\Delta U_{0.5}$ applied to the crystal) correspond to the driving fields and half-wave voltages ($U_{\lambda/2}$) used in the conventional electro-optic modulators employing the same crystals in a similar geometry.

The electrically controlled switching of plane laser beams through the mechanism described above was first suggested and experimentally demonstrated for a $LiNbO_3$ crystal [9.103]. The effect of electric switching of plane beams in $LiNbO_3$ was also studied in [9.91, 104, 105], though somewhat modified geometries were employed.

Clearly this technique allows switching of complicated wave fronts as well [9.91, 106, 107]. This property of PRCs can be used in volume holographic memory cells with electrically controlled retrieval of information. An example of retrieval of a series of holograms of complicated images from $LiNbO_3$:Fe is presented in Fig.9.17a. Note that the experiment [9.91, 107] employed the most efficient "transverse" geometry of the field application (Fig.9.17b). The driving voltage $\Delta U_{0.5}$ required to perform switching of two successive holograms can have a minimum value of

$$\Delta U_{0.5} = \left(\frac{\lambda}{rn^3}\right)\frac{L_x}{L_z} = \frac{L_x}{d}U_{\lambda/2} \tag{9.21}$$

in this case. An example of use of the electrically controlled diffraction in $LiNbO_3$:Fe for a commutation in a fiber-optic net is given [9.102].

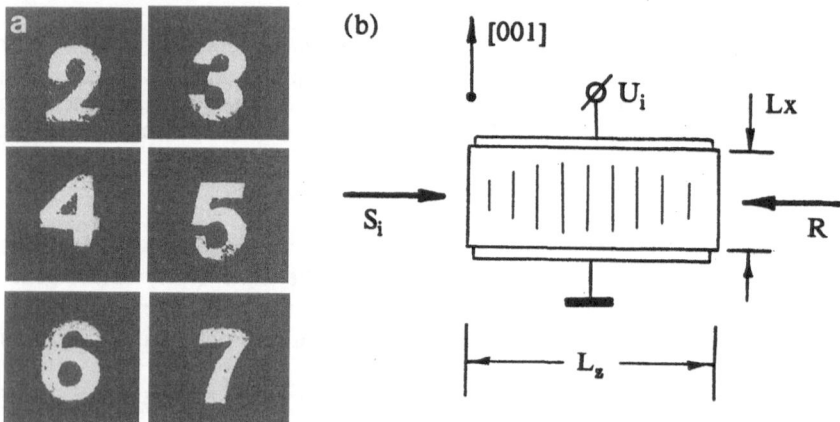

Fig.9.17 (a) Series of images reconstructed from volume reflection holograms recorded in $LiNbO_3$:Fe [9.91, 107] ($d_z/d_x \simeq 3$, U [kV]: 1: -0.5, 2: +0.5, 3: -1.5, 4: +1.5, 5: -2.5, 6: +2.5). (b) Transverse geometry of electrically controlled diffraction from a reflection hologram in $LiNbO_3$

9.7.2 Gratings with a Variable Spacing

This technique of plane laser beam deflection reported in detail in [9.108-110] exploits the same principle as the acousto-optic deflectors [9.111]. In both cases deflection of the laser beam results from diffraction off a volume phase grating with a variable spacing Λ. In acousto-optic deflectors variation of the grating spacing is achieved by changing the frequency of the acouststic wave generated in the acousto-optic cell volume. In the photorefractive deflectors the required diffraction grating is formed by a continuous illumination of the PRC by two intersecting plane laser beams (Fig.9.18a). The spacing Λ of the grating is varied by changing the wave-

(a)

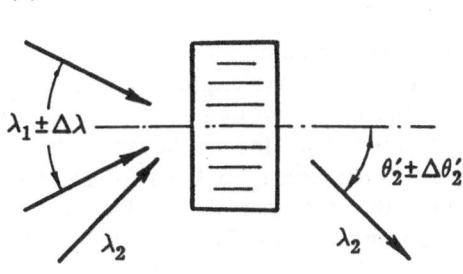

Fig.9.18. (a) Photorefractive deflector. (b) Increasing the angle of deflection in a photorefractive deflector by means of an additional diffraction grating (G) [9.108, 110]

(b)

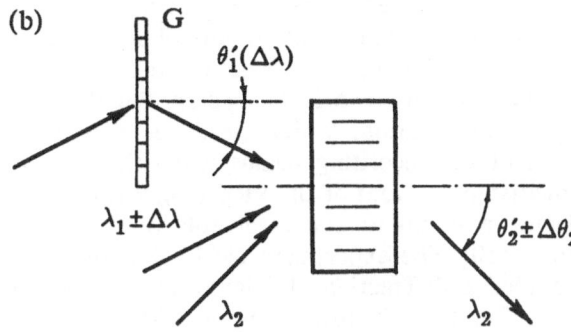

length of the recording light beams, their incidence angles θ' remaining fixed:

$$A = \frac{\lambda}{2\sin\theta'} \cdot \tag{9.22}$$

Clearly, the operation speed of such a deflector is determined by the minimum time required to switch from one grating to another, i.e., by the characteristic time τ_{sc} for hologram recording.

Because of similarity of the operating principles, the deflectors we are discussing (photorefractive and acousto-optic) share a basic limitation, i.e., a limited scan angle due to the Bragg diffraction from the volume phase grating. This problem is usually solved in acouso-optic deflectors by using anisotropic (intermode) diffraction in a special "wideband" geometry first proposed in [9.112]. (Fig.5.12a). This geometry provides an increase in the scan angle of the diffracted beam up to

$$\Delta\theta' \simeq 2\sqrt{\frac{n\lambda}{d}} \cdot \tag{9.23}$$

The angular divergence of a plane light beam diffracted from an aperture with a linear size L_x is $\simeq \lambda/L_x$. Hence deflectors with the following maximum number of resolvable positions

$$N_x \simeq \Delta\theta' \left[\frac{\lambda}{L_x}\right]^{-1} \simeq 2L_x\sqrt{\frac{n}{\lambda d}} \cdot \tag{9.24}$$

205

can be fabricated. For the typical values $L_x \simeq 1$ cm, $n \simeq 2.5$, $\lambda \simeq 0.5~\mu m$, and $d \simeq 3$ mm, (9.24) yields $N_x \simeq 10^3$.

A similar geometry of a photorefractive deflector was investigated in a biaxial photorefractive $KNbO_3$ for different wavelengths of recording argon-ion laser (457.9nm $< \lambda <$ 514.5nm) [9.113]. *Voit* et al. managed to demonstrate scanning of a HeNe-laser beam within the angle $\Delta\theta' \simeq 5.67°$, with the characteristic time for hologram formation $\tau_{sc} \simeq 0.6$ s for an average light intensity $I_0 \simeq 0.5~W/cm^2$. Note that the anomalous selective properties of anisotropic (intermode) diffraction from a photorefractive grating in $LiNbO_3$ in this geometry were investigated earlier in [9.114].

High holographic sensitivity is exhibited by cubic $Bi_{12}SiO_{20}$. For these crystals, another method of increasing the bandwidth of the photorefractive deflector was proposed in [9.108-110]. The increase is achieved by an additional tilt of the volume phase grating with a variable spacing. For this purpose, a high-efficiency holographic grating, which produces a required change in the incidence angles of the recording beams when their wavelength changes, is placed immediately in front of the PRC (Fig.9.18b).

A photorefractive deflector with up to 200 resolvable positions was realized using this method in [9.110]. The experiment used a 2.7 mm thick $Bi_{12}SiO_{20}$ crystal, which provided a diffraction efficiency of $\eta \simeq 1\%$ for the diffusion recording mechanism ($\Lambda \simeq 1~\mu m$). Gratings with a variable spacing were recorded by a dye laser ($\lambda \simeq 530 \div 570nm$) and readout was performed by a semiconductor laser ($\lambda \simeq 840nm$). The experimental switching time of the deflector was $\tau_{sc} \simeq 0.1$ s for the recording intensity of $\simeq 10~mW/cm^2$ ($\lambda = 550nm$).

An alternative method to increase the bandwidth of a photorefractive deflector using a cubic PRC and combinational hologram recording [9.93-95, 115] was suggested in [9.110]. It involves a simultaneous recording of a discrete set of combinational holograms with the wave vectors

$$K_{nm} = nK' \pm mK'' , \qquad (9.25)$$

where n and m are any integers. To this end the sample is exposed simultaneously to two interference patterns with the wave vectors K' and K'' and nonlinear regime of holographic recording is used. We do not discuss in detail here the results obtained in [9.110,115], but note that the maximum bandwidth for the photorefractive deflector involved is equal to that given by (9.23). The deflected beam diffracts from the grating with wave vector $K_{1-1} = K' - K''$ (Fig.9.19). Here one of the grating vectors, for instance K'',

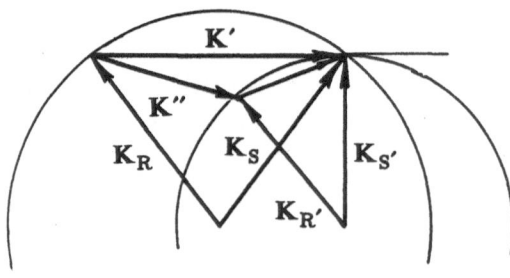

Fig.9.19. Vector diagram explaining the increase of the deviation angle in a photorefractive deflector using recording of combinational holograms ($K_{R,S}$: wave vectors of recording beams with a variable wavelength, K'': a constant auxiliary grating, $K_{R'}$ and $K_{S'}$: wave vectors of readout and deflected beams, respectively)

is assumed to be constant, and the second one, $\mathbf{K}'$, changes its length as the wavelength of the beams recording this grating varies. The configuration is chosen so that the wave vector $\mathbf{K}_{1-1}$ slips along the tangent to the corresponding wave vectors surface of the readout light similar the classical geometry of wideband anistropic diffraction (Fig.5.12a).

9.7.3 Enhanced Reflection from Spatial Light Modulators

Another scanning method [9.116] relies on enhancement of the light reflected from one or another region of the SLM's surface (Fig.9.20) through two- or four-wave mixing in a PRC. A nonlinear regime of this process is desirable, which at large Γd product affords an almost total transfer of the pump beam intensity into the signal light beam of a required direction.

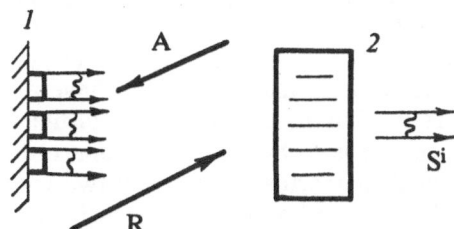

Fig.9.20. Switching of a light beam through two-wave energy exchange in PRC [9.116] (1: array of piezo mirrors, 2: PRC)

An experimental implementation of such a deflector using BaTiO$_3$ was also demonstrated in [9.116]. An array of piezomirrors (4×4) served as an SLM. All the piezomirrors except the one that corresponded to the chosen direction of deflection were excited by a ramp generator, which suppressed efficiently the two-wave amplification of the reflected light beam. As a consequence, only the beam reflected from the nonexcited mirror was amplified. The deflected beam reached about 10% of the pump intensity. The typical switching time was $\simeq 2.5$ s for $I_0 \simeq 0.5$ W/cm^2 and $\lambda = 514$ nm.

9.8 Spectrum and Correlation Analysis of Images

This section is concerned with the use of PRCs in coherent optical spectrum analyzers and correlators. The fundamental principles of spectrum analysis in coherent optical systems and the basic requirements to the SLMs as input devices for such systems were discussed in Sect.2.3.

9.8.1 Coherent Optical Correlators

Let us recall briefly how the coherent optical system can perform the correlation analysis. If we have two images - i.e., $T_{in}(x,y)$ and $T_r(x,y)$ - their correlation function is defined as

$$A(x'',y'') = \iint T_{in}(x,y)T_r^*(x+x'',y+y'')dxdy . \tag{9.26}$$

To show how transformation (9.26) can be performed by the optical system, let us use the properties of the Fourier transform and rewrite this equation in a more convenient form:

$$A(x'', y'') = \mathcal{F}^{-1}\left[\tilde{T}_{in}(x', y')\tilde{T}^*(x', y')\right], \tag{9.27}$$

where $\mathcal{F}^{-1}$ denotes the inverse Fourier transformation, $\tilde{T}_{in}(x', y')$ is the Fourier transform of $T_{in}(x, y)$, and $\tilde{T}_r^*(x', y')$ is the complex-conjugate Fourier transform of the image $T_r(x, y)$.

Figure 9.21 shows the optical system with the output light amplitude $A_{out}(x'', y'') \propto A(x'', y'')$. The system consists of two units that perform Fourier transformation. The image $T_{in}(x, y)$ is placed at the input of the first unit (with lens L_1) in the P_1 plane. A spatial filter with amplitude transmittance $\propto \tilde{T}_r^*(x', y')$ matched with the image T_r is put in the P_2 plane. Since the filter is illuminated by the light with amplitude $\propto \tilde{T}_{in}(x', y')$, the light amplitude behind it is proportional to the product $\tilde{T}_{in}(x', y')\tilde{T}_r^*(x', y')$. The lens L_2 performes the second Fourier transformation, and we have the correlation function in the output plane P_3, but in the inverted coordinate system. This is usually of no significance and is a characteristic property of the optical system shown in Fig.9.21 which performs successively two forward Fourier transformations, but not the forward and inverse transformations, as required by (9.27).

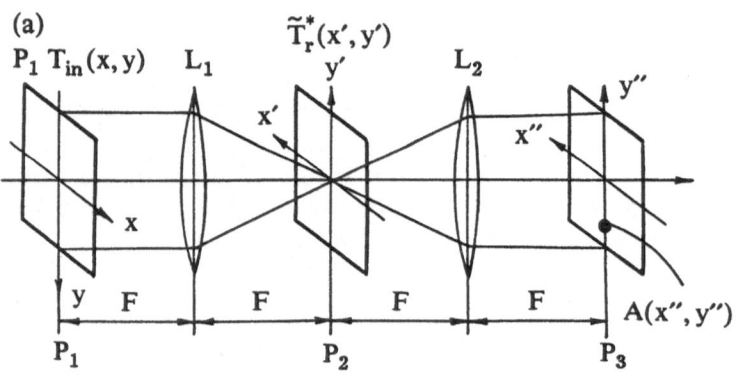

(a)

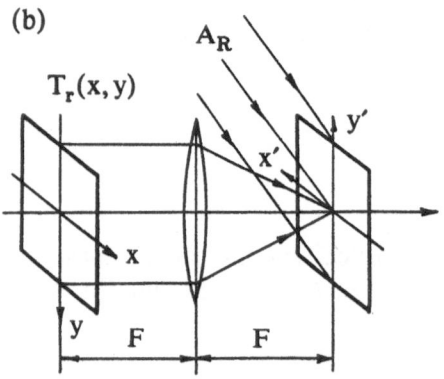

(b)

Fig.9.21. (a) Coherent optical correlator ($T_{in}(x, y)$: amplitude transmittance of a transparency in the input plane, $T_r^*(x', y')$: amplitude transmittance of a matched filter, $A(x'', y'')$: light amplitude in the output plane). (b) Recording of a matched filter (A_R: reference wave)

The matched filter with complex transmittance $\tilde{T}_r^*(x', y')$ is usually produced by means of the holographic technique [9.117]. For this purpose the Fourier hologram of the image $T_r(x, y)$ is to be formed. If we use the same scheme for recording which is shown in Fig.9.21, we have to place the image $T_r(x, y)$ in the input plane P_1 and illuminate the photographic plate in the plane P_2 by a reference beam $A_r(y', z) = A_r \exp[-i(K_y y' - K_z z)]$. A derivation similar to that given in Sect.1.2 yields the final transmittance of the developed plate:

$$T(x', y') = T_0 + T_1 f \left[A_r \tilde{T}_r(x', y') \right] + T_1 \tilde{T}_r(x', y') A_r^* e^{iK_y y'}$$

$$+ T_1 \tilde{T}_r^*(x', y') A_r e^{-iK_y y'} , \tag{9.28}$$

where $f[A_r \tilde{T}_r(x', y')]$ describes the transmittance dependence $T(x',y')$ at low spatial frequencies unimportant for our analysis. The last term in (9.28) is proportional to the desired transmittance distribution $\tilde{T}_r^*(x', y')$.

If this plate is inserted into the frequency plane P_2 of the correlator (Fig.9.21) and the input transparency $T_{in}(x, y)$ is illuminated by a plane wave A_{in}, the field amplitude in the output plane will be given by

$$A_{out}(x'', y'') = A_{in} C_1 T_{in}(x'', y'') + A_{in} C_2 f(x'', y'')$$

$$+ A_{in} C_3 \iint T_{in}(x, y) T_r(x''-x, y''-y - F\sin\theta) dx dy$$

$$+ A_{in} C_4 \iint T_{in}(x, y) T_r^*(x-x'', y-y''-F\sin\theta) dx dy, \tag{9.29}$$

where $\sin\theta = K_y \lambda / 2\pi$ and C_1, C_2, C_3, and C_4 are the numerical coefficients. The first two terms in (9.29) describe the light field distribution near the origin of the output coordinate system and are of no use here. The third and fourth terms represent the convolution and correlation between 2D functions $T_r(x, y)$ and $T_{in}(x, y)$, respectively. If $T_{in}(x, y) = T_r(x, y)$, these terms are autocorrelation and autoconvolution.

The autocorrelation function is known to have the maximum at its center. For a complicated image $T_{in}(x, y)$ with a rich spectrum of spatial frequencies, the autocorrelation function occupies rather a small area in the output plane, so the light is focused into a relatively small but bright spot with coordinates $x'' = 0$, $y'' = -F\sin\theta$. The intensity of the spot exceeding a fixed level can indicate that the reference image $T_r(x, y)$ and the image to be recognized $T_{in}(x, y)$ are identical. This is how the pattern recognition proceeds. If the coordinates of the bright correlation spot differ from those given above, this means that the input image is shifted from the origin of the input coordinate system, and the magnitude of the displacement corresponds directly to the displacement of the correlation spot.

As pointed out in Sect.2.3, PRCs in coherent optical correlators and spectrum analyzers are mainly used in Spatial Light Modulators (SLMs) in the input plane P_1. PRCs can also be used as reversible media for holographic matched filters. The reversible holographic filters must be linear with respect to the readout light amplitude, allow recycling, and have a low noise level. These and a number of other requirements coincide with those placed on the optically addressed SLMs. There are, however, two significant distinctions. The first is the spatial resolution of the medium used for recording the filter. It must be at least four times as high as that used for the SLM, to warrant a sufficient angular separation between the correlation and noninformative light beams (i.e., terms 1 and 2 in (9.29)). To this end, the incidence angle of the reference beam θ during the filter recording should be

$$\nu_c = \frac{\sin\theta}{\lambda} > 3\nu_{max} \, . \qquad\qquad (9.30)$$

Here ν_{max} is the maximum spatial frequency in the image and ν_c is called the "carrier" frequency. The output correlation function will then be formed in the frequency range $\nu_c \pm \nu_{max}$. This means that the maximum spatial frequency of the interference fringes in the filter is approximately equal to $4\nu_{max}$.

The second difference is the dynamic range requirements. If we assume that any image with $N \times N = 10^6$ pixels and dynamic range of 40 dB can be placed in the input plane, then the variation in the light intensity in the Fourier plane can reach 160 dB. Clearly such a photosensitive medium is unrealizable. Being, of course, an extreme case, this example shows, however, that the dynamic range of the medium for the matched filter should exceed remarkably that of the input device. Moreover, it is evident that the precise matched filter cannot be recorded for all images, because of a limited dynamic range of the medium. It should also be noted that the practical recognition problems do not always require recording of the filter that exactly corresponds to the matched one. However, if the recording medium has an insufficient dynamic range, special measures should be taken to provide the maximum signal-to-noise ratio at the output of the recognition system. In addition to the dynamic range of the medium, one should take into account here the type of image to be recognized and expected noise in the input signal and optical system.

9.8.2 SLMs in Input Systems

The input device using the SLM is to meet a number of specific requirements briefly discussed in this section. According to the approximate estimates given in Sect.2.3, for a coherent optical processor to be competitive, the input device must ensure recording of images of about 1000×1000 pixels at a rate of ~30 Hz. In addition, the required linearity is to be provided, and geometrical and phase distortions must be low.

Recording systems. Almost all photorefractive SLMs are optically addressed devices. This means that the image to be recorded must be imaged onto the modulator plane. The sensitivity of the photorefractive SLM (10^{-4} to 10^{-6} J/cm^2 in the blue-green region) does not, however, allow direct recording of objects in most cases, for instance, in daylight, and requires a special recording system. If the input data are in the form of electrical signals, the recording system must convert them into the images to be recorded on such SLMs. Recording systems involving image brightness amplifiers [9.118,119], CRTs [9.120], and laser scanning devices [9.121] were suggested for applications of protorefractive SLMs in real time.

Use of an image brightness amplifier together with the PROM SLM was proposed in [9.118] and was experimentally investigated with PRIZ SLM in [9.119]. The amplifier not only ensures the required recording light intensity but also converts its spectrum to the region of maximum SLM sensitivity. Figure 9.22 shows the recording device with the brightness amplifier and PROM SLM. The system is in fact an optically addressed SLM with enhanced sensitivity. The PROM is operating here with the readout in the reflection mode, which was achieved by incorporating a multilayered dielectric mirror into the modulator structure. The mirror reflects the readout light (λ = 633nm) almost completely and transmits the recording light from the brightness amplifier. The PROM modulator is in immediate contact with the output fiber-optic plate of the amplifier. The experimentally achieved sensitivity of PRIZ SLM/image amplifier device was 10^{-9} J/cm^2 [9.119]. The image amplifier here had a gain of 10^4.

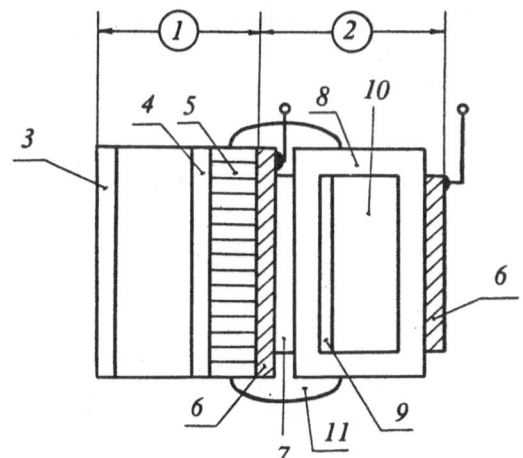

Fig.9.22. Illustrating the PROM/ image amplifier (1: image amplifier, 2: PROM, 3: photocathode, 4: phosphor screen, 5: fiber-optic plate, 6: transparent electrode, 7: interface oil, 8: parylene insulating layer, 9: dichroic reflector, 10: $Bi_{12}SiO_{20}$ wafer, 11: silicon rubber)

Recording devices involving CRTs can be fabricated in a way similar to that shown in Fig.9.22. The SLM is connected to the CRT through a fiber-optic plate [9.119]. In another version, the image from the CRT may be imaged onto the SLM by a lens [9.120]. To be used with the photorefractive SLM, the CRT must have an enhanced brightness in the blue region. If the device is used to record images, the TV system ensures the brightness amplification. Moreover, the CRT enables recording of electrical signals onto optically addressed SLMs.

One important problem encountered in using light amplifiers and CRTs in input devices is the elimination of geometrical distortions arising in the recorded image. The geometrical distortions should not exceed 100/N % for images containing N×N pixels. Otherwise they reduce the resolution of the spectrum analyzer, and hence the information capacity of the image. This means that any element of the image should not change its size by more than 100/N % after recording onto a SLM. If complicated correcting systems are not used, the CRT usually gives geometrical distortions from 1 to 10%. The distortions lower than 0.1% which are required for the processing of images containing 1000×1000 pixels can be achieved only by using complicated and expensive digital correcting devices.

Figure 9.23 presents a recording system using a laser scanning device [9.121]. It incorporates a one-channel acousto-optic modulator and a mechanical deflector. Such a system has been used to record electrical signals onto the PROM SLM. The electrical signal is applied continuously to the acousto-optic modulator, where it is first converted to a running acoustic wave and then read out by a laser pulse. The pulse duration should be short, so that the acoustic wave displacement during the readout period can be neglected. Thus one pulse is needed to read out a portion of the electrical signal of duration t (the propagation time of the acoustic wave through the aperture of the modulator) from the acousto-optic modulator and to record it onto the SLM as a line. The next line is recorded in a discrete time interval t when the information fed into the modulator is entirly renewed. This line is recorded onto the SLM below the preceding one, the required displacement of the recording beam being provided by the mechanical deflector (a rotating mirror prism in [9.121]).

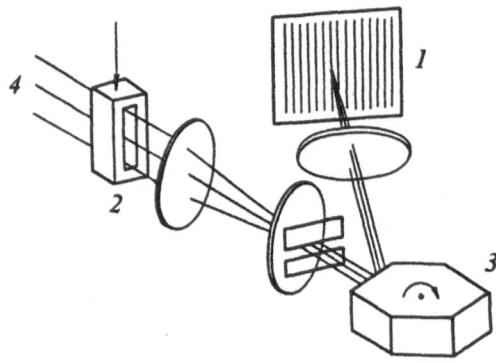

Fig.9.23. Recording of electric signals onto the PROM modulator (1: PROM SLM, 2: acoustooptic modulator, 3: rotating mirror, 4: light from a pulsed laser)

The requirement common to all the recording systems under discussion is linearity, i.e., the linear relation between the input signal and the modulation amplitude of the output readout light. Nonlinearities due to both the recording systems and the SLMs can lead to ambiguities in the results of signal (image) processing by the optical processor. It is impossible, however, to specify the level of allowable nonlinear distortions for a general case. It is clearly determined by the type of signal to be processed and the problem to be solved by the whole system.

The readout system should involve a source of coherent radiation (a CW or pulsed laser) and a collimating system providing illumination of the SLM by a light beam with a desired wave front. The readout system should ensure absence of phase distortions and a minimum noise level at the output of the optical processor.

As noted in Sect.2.3, the phase distortions in the readout light wave behind the SLM should not be in excess of $\lambda/4$, with due account for distortions introduced by the SLM itself. A higher level of the phase distortions, like the geometrical distortions of the images during recording, leads to a reduction of the equivalent information capacity of the processed image. The phase distortions arising from the SLMs can, in principle, be compensated for by complicating the readout system. To this end, the SLM should be illuminated by the light with the wave front that is a complex conjugate of the phase distortions of the SLM. This can be achieved in particular by using holographic elements in the readout system as was employed, for instance, to compensate for phase distortions of the PROM SLM [9.122]. Such a complicated readout system can, however, give rise to considerable losses in the readout light.

Since the results of information processing are typically converted into an electrical signal at the output of the optical processor, the inherent noises of the whole processor are determined by the noises of optical elements and the output photodetector. An increase of the readout-light intensity leads to an increase of both the output optical signal and optical noise, the signal-to-noise ratio remaining the same. Hence, the readout light intensity has to be high enough for the photodetector at the output to detect the optical noise.

To illustrate the results of application of photorefractive SLM in the input systems of the coherent optical processor, Fig.9.24a shows the spectrum of a harmonic electric signal obtained using the PRIZ SLM. For such a narrow-band signal, the signal-to-noise ratio at the output of the optical spectrum analyzer reaches 10^6, and the optical noise of the processor cannot be detected simultaneously with the signal because of the limited dynamic range of the photodetector (photographic film in this case). Figure 9.24b depicts the spectrum of a wideband 2D signal obtained by using the PRIZ device. The input image was generated by recording a Fourier hologram of a diffuse scatterer onto the SLM. The density of information recorded on the SLM was $\simeq 100$ Kbit/cm^2, the signal-to-noise ratio at the output of the processor was $\simeq 300$.

Figure 9.24c presents the spectrum of a noiselike image with texture (sea surface, Fig.9.24d). It was obtained in [9.123] to determine the texture direction. The image was recorded onto the PRIZ from the photographic film or from the CRT screen. In fact there is no necessity to re-record images from the photographic film on the SLM, since the film itself can be used as an input transparency. *Boyarchuk* et al. claimed, however, that this re-recording allows a more precise analysis, because the modulator PRIZ efficiently reduces the noises of the optical system.

An example of an electro-optic SLM as an input device in a coherent optical correlator was given in [9.124]. The Phototitus SLM is used in the Vander Lught geometry. The reference image is read from the same SLM during recording of a matched filter. Since a crossed analyzer is placed be-

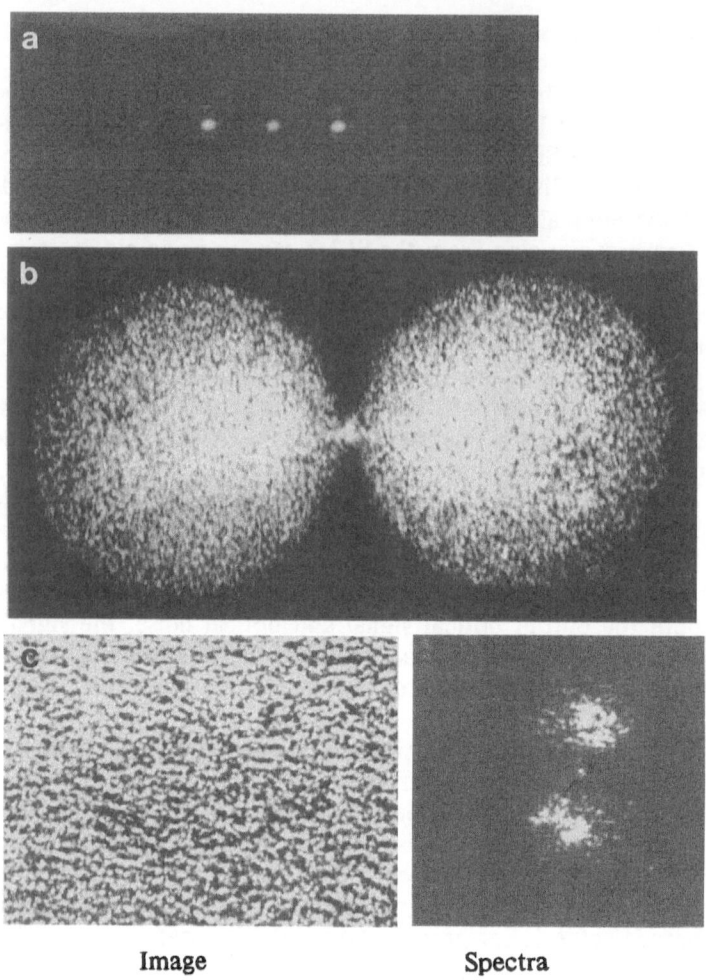

Image   Spectra

Fig.9.24. 2D spectra of images obtained in [9.123] using the PRIZ SLM: (a) Harmonic signal, (b) wideband signal, (c) sea surface image (a noise-like image with texture)

hind the SLM, the light polarization in the holographic filter plane is orthogonal to the initial one. Therefore, it is also necessary to change the reference-beam polarization by 90° to record the filter. This is accomplished by means of a $\lambda/4$ plate and polarizer. The correlation analysis of aerial images and printed text, which were recorded onto the Phototitus using incoherent light, was demonstrated. The intensities of correlation peaks were close to the theoretical values, and the output signal-to-noise ratio allows the image fragments to be recognized and located. In particular, for aerial images, the 6×6 mm^2 fragment used as a reference object was located in the 25×25 mm^2 input image.

9.8.3 Holographic Recording of Matched Filters in PRC

Volume samples of PRC can also be used as a photosensitive medium for holographic recording of the Fourier-transformed reference images in coherent optical correlators. In contrast to conventional hologram (recordered e.g., using a photoplate), a dynamic hologram can be read out immediately in the process of recording. To separate the reconstructed wave from the recording one, a counterpropagating readout wave (and hence the reconstructed one) is used (Fig.9.25). As a result, all four waves are simultaneously present in the volume of the medium, namely two Fourier-transformed signal waves $T_{in}(x, y)$ and $T_r(x, y)$, an auxiliary plane wave A_r, and the output wave corresponding to convolution or correlation. This indicates that the correlation analysis in the dynamic holographic medium proceeds via four-wave mixing [9.125].

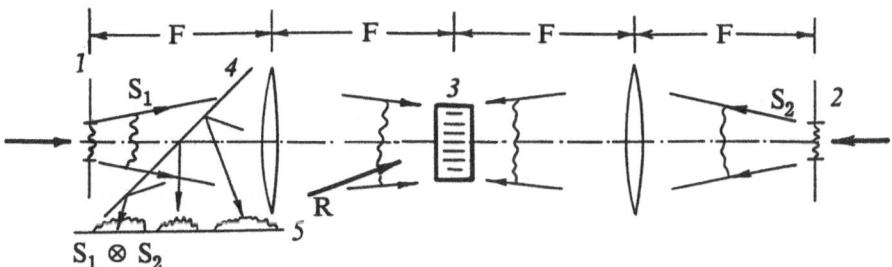

Fig.9.25. Correlation analysis of 2D images through four-wave mixing in PRC (*1,2*: processed images' planes, *3*: PRC, *4*: beam splitter, *5*: output plane)

The experimental results on the correlation analysis of 2D images through four-wave mixing via volume dynamic holograms in PRCs were presented in [9.126] (LiTaO$_3$, λ=442nm) and in [9.127-129] (Bi$_{12}$SiO$_{20}$, λ= 514nm). A more traditional approach, based on a joint Fourier transform geometry, or recording of a conventional Vander Lught matched filter, was employed in [9.130-132]. High-sensitivity cubic photorefractive crystals Bi$_{12}$SiO$_{20}$ [9.130, 131] and GaAs ($\lambda = 1.06\mu$m) [9.132] were used in these experiments as dynamic holographic media.

Note that the images to be processed were usually the simplest binary transparencies with rather a small number of pixels in the incidence plane (along the x axis). This is due to angular selectivity of the volume hologram formed in PRCs. The ultimately allowable linear size of the images to be processed and that of the resulting one along this coordinate axis is limited by the value

$$D_x \simeq F2n\Delta\theta' \simeq F \frac{n\lambda}{\theta d} . \tag{9.31}$$

In turn, the transverse size of the photorefractive sample L_x determines the diffraction-limited minimum element in these images by

$$\Delta_x \simeq \lambda \frac{F}{L_x} \, . \tag{9.32}$$

The total number of pixels along the x coordinate proves to be

$$N_x = \frac{D_x}{\Delta_x} \simeq \frac{nL_x}{\theta' d} \, . \tag{9.33}$$

For the typical values of $n = 2.5$, $d \simeq L_x \simeq 3$ mm, $\theta' \simeq 0$, 1, this yields a fairly small value of $N_x \simeq 25$.

The angular selectivity of the hologram along the orthogonal direction y is much lower (Sect.5.6) and the minimum number of pixels

$$N_y = \frac{D_y}{\Delta_y} \simeq L_y \sqrt{\frac{n}{\lambda d}} \, . \tag{9.34}$$

For the values of the parameters given above, it is much higher ($N_y \simeq 10^3$ for $\lambda = 0.5\,\mu$m).

This severe limitation on N_x can be removed to a great extent by using the six-wave mixing geometry [9.93-95]. An alternative way of overcoming an undesirably high angular selectivity of volume Vander Lugt filters was proposed in [9.133,134]. To this end, the 2D transparency to be processed is illuminated on reconstruction by a polychromatic plane wave.

9.9 Nonlinear Processing in the Image Plane

The literature reports at least several ways to use PRCs for nonlinear processing of coherent images in the plane where they are formed. They include edge enhancement, contrast inversion, division, and also incoherent-to-coherent image conversion (Sect.9.8.2).

9.9.1 Image Contouring

Edge enhancement of simple binary images by means of volume BSO samples was first demonstrated in [9.135]. In this experiment (Fig.9.26a) a slightly defocused binary image was holographically recorded in a usual geometry with the beam ratio differing from the traditional one ($I_{R_1}/I_{S_1} \simeq 0.1$). Simultaneously with recording, reconstruction of the recorded image by a counterpropagating plane readout beam R_2 ($I_{R_2}/I_{S_1} \ll 1$) was accomplished, to yield an image with enhanced edges (Fig.9.26b). *Huignard* and

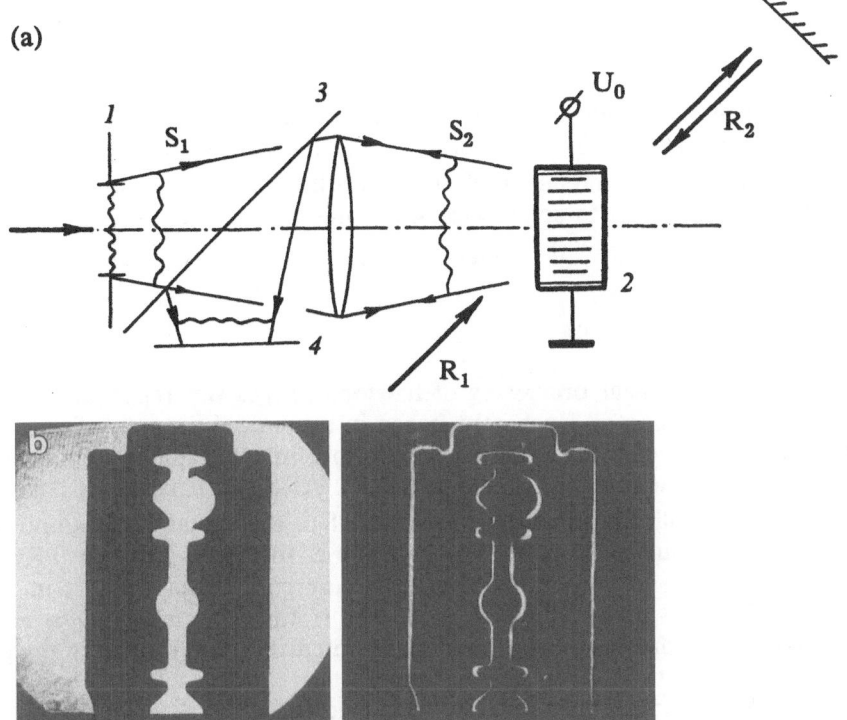

Fig.9.26. (a) Holographic experiments on edge enhancement of binary images [9.135] (*1*: image plane, *2*: PRC, *3*: beam splitter, *4*: output plane). (b) Typical example of the image with enhanced edges obtained in [9.135] with BSO crystal

Herriau attributed the effect to a remarkable reduction of a DC electric field applied to the sample in the bright regions of the recorded pattern $(E_0(x,y) \propto (I_{S_1}+I_{R_1}+I_{R_2})^{-1} \simeq I_{S_1}^{-1}(x,y))$. As a consequence, the efficiency of the dominating drift mechanism of recording and hence the intensity of the corresponding fragments of the reconstructed image become markedly reduced.

Later, similar experiments on edge enhancement were carried out with $BaTiO_3$ [9.136, 137]. Holographic recording was here performed, however, in the absence of an external electric field, i.e., through the diffusion mechanism. To explain the effect of edge enhancement, *Feinberg* invoked another mechanism of nonlinearity fairly common in most presently known PRCs used for holographic recording. Namely, they used the fact that the diffraction efficiency of the hologram in PRC does not depend usually on the total intensity of the recording beams. In fact, as shown in Sect.4.2.2, the stationary amplitude of the electric field grating proves to be proportional to the recording interference pattern contrast, i.e.,

$$|E_{sc}| \propto |m| = 2\frac{(I_{R_1}I_{S_1})^{1/2}}{I_{R_1}+I_{S_1}}. \tag{9.35}$$

The signal beam intensity for this regime of holographic recording is chosen to be much higher than the reference beam intensity ($I_{S_1} \gg I_{R_1}$). As predicted by (9.35), this provides a nonlinear regime of the hologram recording when its amplitude is inversely proportional to the signal wave amplitude ($|E_{sc}| \propto I_{S_1}^{-1/2}$). So the parts of the reconstructed image corresponding to the bright fragments of the recording beam prove to be dark like those corresponding to actually dark fragments where $I_{S_1} \ll I_{R_1}$, thus giving rise to contouring of the image.

9.9.2 Brightness Inversion

A more precise nonlinear processing of halftone images was reported in [9.138]. Here an inversion of a coherent image in a standard geometry for recording a hologram of a focused image by a low-intensity plane reference beam ($I_{R_1} \ll I_{S_1}$) was demonstrated for BGO. As a consequence, a hologram with the amplitude inversely proportional to the signal beam amplitude at a given point of a processed image was formed within the PRC volume. Its readout by a counterpropagating reference beam resulted in a reconstructed image with a contrast inversion. A fairly accurate inversion was demonstrated for the signal beams whose intensity varied by more than two orders of magnitude.

Similar BGO experiments were also reported in [9.139].

9.9.3 Optical Logic Elements

One more example of nonlinear processing in the image plane is the implementation of optical digital logic. Originally the PRC use for this purpose was based on the coherent image subtraction known earlier [9.140]. Holograms of two images were recorded in succession in the same region of a photosensitive medium, with an additional 180° phase shift in the reference beam during the second exposure. These experiments on coherent image subtraction and optical logic performed in PRCs [9.141-145] used a linear regime of holographic recording. They had no further development, partly because of high sensitivity to occasional phase shifts in the shoulders of the recording scheme. There were also severe requirements on the quality of optical elements and vibration isolation of the setup used. Note that this drawback can be overcome to a certain extent by using special compensators of phase distortions based on photorefractive PCMs [9.146].

The growing interest in photorefractive optical logic elements observed recently relies on the use of nonlinear energy coupling via dynamic gratings. Subtraction of binary images by nonlinear two-wave mixing in photorefractive $LiNbO_3$:Fe was first demonstrated in [9.147]. The experiment involved nonstationary two-wave energy coupling in $LiNbO_3$:Fe, but the steady-state energy coupling via shifted phase gratings can also be used to perform this operation.

Indeed, as shown in Sect.6.2, the intensity ratio between two interacting light waves at the output of the crystal changes proportionally to $\exp(\Gamma d)$ (Fig.9.27a). This means that if the Γd product is fairly high, the output energy of one (more powerful) beam R_1 can be nearly completely transferred to the output energy of the other one [9.148-150]. Thus, if two coherent light patterns are simultaneously imaged into the crystal volume, the intensity of one transmitted beam (the donor) will be suppressed. On the other hand, the respective portions of the other transmitted beam (the acceptor) will be amplified. Evidently such changes in the transmitted beams occur only in those areas where their intensities are not equal to zero.

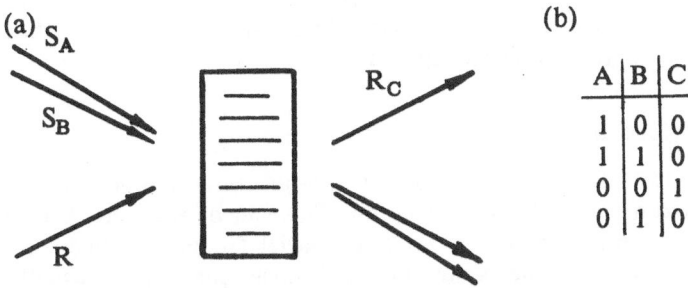

Fig.9.27. (a) Holographic experiment for implementation of logic operation "NOR" via two-wave coupling in PRC. (b) Truth table for logic operation "NOR"

This two-wave coupling geometry offers the possibility of implementing the logic operation "NOR", the truth table of which is also given in Fig.9.27b. The presence of a certain input intensity level at a given point of crossing light beams S_1 and R_1 should be taken there as logic 1 and its absence as logic 0. The output signal is the intensity of light beam S_1 at the PRC output.

Nonlinear optical configurations that can perform a wide variety of logic operations were described from the general point of view in [9.129], and similar geometries employing PRCs were considered in [9.151]. Limitations on the minimum size of the pixel in the processed images (Λ^{min}) were also discussed there. It proves to be determined by the sample thickness d and the crossing angle of the interfering light beams $2\theta'$:

$$\Lambda^{min} \simeq d \frac{2\theta'}{n} . \tag{9.36}$$

In particular, for the typical values $d \simeq 1$ mm and $\theta' \simeq 0.1$, $\Lambda^{min} \simeq 100 \ \mu m$. Some reduction in the spatial carrier frequency can increase spatial resolution to a certain degree. This is not, however, always desirable in crystals with the diffusion mechanism of recording of the shifted holograms (e.g., in $BaTiO_3$), where $\Gamma \propto K \propto 1/2\theta'$.

The highest sensitivity of presently known PRCs at such spatial frequencies ($S^{-1} \simeq 10^{-5} J/cm^2$) yields the upper estimate for the energy for switching of such a photorefractive logic element of $W \sim 10^{-9}$ J. Note, however, that the question of applicability of such devices is not yet solved.

9.10 Nonlinear Processing in the Spatial Frequency Plane

This section should also include the correlation analysis of two-dimensional images. However, because of the particular practical significance of this type of processing, we have considered it separately in Sect.9.8. Here we give other examples of processing in the spatial frequency plane that exploit the nonlinearity of holographic recording in PRCs.

9.10.1 Image Contouring

It is well known [9.152] that for the 2D pictures, edge enhancement is equivalent to their two-dimensional differentiation. This can be reached also by a suppression of the zero components in their spatial frequency spectrum. These components located in the center of the Fourier plane have usually the maximum amplitude. As shown in the preceding section, such a maximum in the spatial light distribution can be suppressed rather easily as a result of a nonlinear recording of the Fourier hologram in PRCs. The basic condition to be fulfilled is that the signal-beam intensity at the center of the Fourier plane be well above the average level of the reference beam. As a consequence, the steady-state hologram amplitude in the region corresponding to the zero spatial frequency of the image will be markedly reduced. Its reconstruction yields the desired image with the enhanced edges.

This technique was reported in [9.136] where holographic recording in the BaTiO$_3$ crystal at the argon-ion laser line ($\lambda = 514$nm) was used. The hologram was continuously read out by a counterpropagating plane reference beam reflected from the mirror placed behind the crystal; i.e., a standard four-wave mixing geometry was employed. Note that in contrast to similar experiments on edge enhancement through nonlinear recording of a focused image hologram (Sect.9.9.1), this method gives rise to intensity doublets at the edges, as should be expected.

Edge enhancement can be performed in a similar fashion, i.e., by suppressing the zero components of the Fourier spectrum of the image by the PROM SLM. As shown in Sect.8.1.5, this device can produce the negative of the recorded image. The brightest parts of the image will correspond in this case to the SLM parts with the lowest transmittance. In [9.153] the PROM was placed in the Fourier plane of the coherent optical processor, and the negative of the spectrum was recorded on it. When it was illuminated by the Fourier-transformed original image, the most intense low-frequency components of the image spectrum at the output of the SLM were thus attenuated to give rise to the edge enhancement.

9.10.2 Inspection of Periodic Structures

Detection of defects in integrated-circuit masks [9.154] can be considered as an interesting extension of the edge-enhancement technique. The technique is based on the periodicity of an ideal mask along both coordinates. Its spatial spectrum is an array of bright localized spikes. The spatial spectrum of a defect, because of its irregularity, is less intense and is more or less uniformly "spread" over the Fourier plane. If proper conditions are fulfilled during the nonlinear recording of the Fourier hologram of the mask under test, the hologram parts corresponding to the bright spikes of its periodic structure can be efficiently suppressed. The readout will result in a relative enhancement of the irregular defect image. Experiments were carried out at the argon-ion laser wavelength ($\lambda = 514$ nm) using BSO crystal under external DC field.

9.11 Additional Remarks

We have discussed the most common possible applications of PRCs and multilayered structures using the PRCs. Several recently suggested applications should also be noted:

a) Improvement of the quality of a wave front (cleaning up of the beam) through two-wave mixing of the pump beam having a complicated spatial structure with the zero-order transverse mode of the resonator [9.155, 156].

b) Compensation of the unwanted deflection of the modulated light beam in the acousto-optic light modulator after a return passage of the phase conjugate light wave through the modulator [9.157].

c) Limitation of the intensity of the coherent light wave transmitted in a certain range of propagation angles or suppression of the most intense spectral lines in this wave [9.158].

d) Optical bistable elements using self-pumped phase-conjugate geometries [9.159, 160].

e) Novelty filters (dynamic selection of images) [9.161-166].

f) Time-integrating optical correlators [9.167].

g) Interconnection systems [9.168].

Appendix: Photorefractive Crystals

This appendix is essentially a reference table that gives information on salient physical and holographic properties of the major, presently known photorefractive crystals. The data on most crystals are arranged according to the following scheme:[1]

1. Physical properties
    - 1.1   Basic growth technique
    - 1.2   Density, hardness
    - 1.3   Symmetry and lattice constants
    - 1.4   Ferroelectric properties
    - 1.5   Dielectric permeability
    - 1.6   Optical absorption
    - 1.7   Index of refraction, birefringence, optical activity
    - 1.8   Linear electro-optic effect
    - 1.9   Photoconductivity
    - 1.10 Bulk photovoltaic effect
    - 1.11 Photorefractive effect

2. Holographic Properties
    - 2.1   First experiments on holographic recording
    - 2.2   Recording geometries
    - 2.3   Basic microscopic mechanisms of holographic recording
    - 2.4   Sensitivity
    - 2.5   Dark relaxation time
    - 2.6   Optical erasure
    - 2.7   Techniques of nondestructive hologram readout
    - 2.8   Two-wave mixing
    - 2.9   Photoinduced noise
    - 2.10 Four-wave mixing
    - 2.11 Passive phase conjugate geometries
    - 2.12 Pulsed holographic recording
    - 2.13 Holographic recording in planar waveguides

---

[1] Some subjects are omitted below since either they are not relevant in the present context or are unknown to the authors.

A1 Lithium Niobate (Metaniobate) LiNbO₃

A1.1 Physical Properties

A1.1.1 High-quality crystals are grown using the Czochralski method at $T_m \gtrsim 1200°$ C from a congruent melt with a Li_2O/Nb_2O_5 ratio of $\simeq 0.946$ [A.1-3].

A1.1.2 Density: 4.628 g/cm³ [A.4], Mohs' hardness: 5 [A.3].

A1.1.3 Space group at room temperature is R3c (point group 3m). The hexagonal unit cell contains six $LiNbO_3$ formulas: a = 0.514829±2·10⁻⁶ nm, c = 1.38631±4·10⁻⁵ nm [A4].

A1.1.4 It is a ferroelectric with the spontaneous-polarization axis along the "c" axis. Remains single domain on heating nearly to the melting point. Is poled during cooling after growth under a DC field along the "c" axis [A.3].

A1.1.5 $\epsilon_a^T = 84$, $\epsilon_c^T = 30$, $\epsilon_a^S = 44$, $\epsilon_c^S = 29$ [A.5].

A1.1.6 Intentionally undoped crystals are transparent, pale yellow, $\Delta W = 3.72$ eV [A.6]. The optical absorption curves for undoped and Fe-doped $LiNbO_3$ are plotted in Fig.A.1 [A.7, 8].

A1.1.7 The refractive-index dispersion was studied in [A.9] (Table A.1).

A1.1.8 The tensor $\hat{r}$ matrix is given in Table A.2: $r_{33} = 30.8$, $r_{13} = 8.6$, $r_{42} = 28.0$, $r_{22} = 3.4$, ×10⁻¹⁰ cm/V (for a strain-free sample, $\lambda = 633$ nm) [A.10]. $LiNbO_3$ is one of the crystals most commonly used in electro-optic modulators [A.11-13].

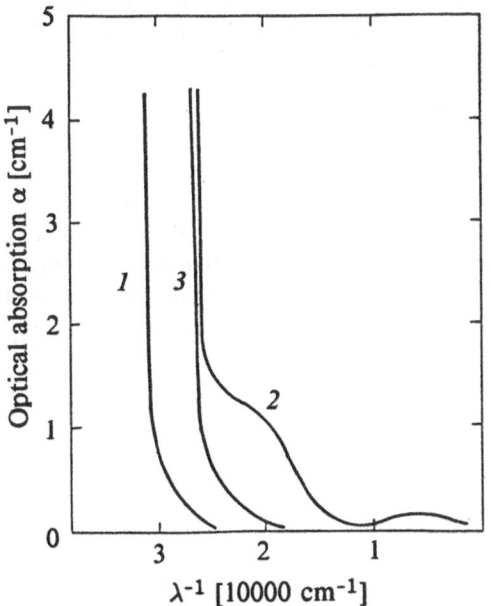

Fig.A.1. Typical optical absorption curves for undoped LiNbO₃ (*1*), LiNbO₃:Fe (*2*), and strongly oxidized LiNbO₃: Fe (*3*) [A.7]

Table A.1. Wavelength Dependence of the Refractive Indices of LiNbO$_3$ at T = 25° C [A.9]

λ [μm]	n$_e$	n$_o$
0.42	2.3038	2.4144
0.45	2.2765	2.3814
0.50	2.2446	2.3444
0.55	2.2241	2.3188
0.60	2.2083	2.3002
0.65	2.1964	2.2862
0.70	2.1874	2.2756
0.80	2.1741	2.2598
0.90	2.1647	2.2487
1.00	2.1580	2.2407
1.20	2.1481	2.2291

A1.1.9 The photoconductivity of nominally pure and doped LiNbO$_3$ crystals was investigated in [A.14-18].

A1.1.10 The bulk photovoltaic effect was first studied by *Glass* et al. [A.19]. The spectral dependence of the effect was thoroughly analysed in [A.16, 20]. Typical E$_G$ is = 50÷70 kV/cm for heavily doped LiNbO$_3$:Fe, it decreases to 10÷20 kV/cm for slightly doped samples [A.18], and is even lower for undoped reduced samples [A.21].

A1.1.11 The photorefractive effect in LiNbO$_3$ was first observed as "optical damage" by *Ashkin* et al. [A.22] and then studied in detail in [A.23]. The spectral dependence was investigated in [A.24, 25]. In accordance with [A.26, 27], doping with Mg results in a drastic reduction of "optical damage" in LiNbO$_3$.

A1.2 Holographic Properties

A1.2.1 First holographic recording experiments were carried out using pure LiNbO$_3$ samples by *Chen* et al. [A.28] and using crystals doped with transition elements (Fe, Cu, Mn) by *Amodei* et al. [A.29-31]. Samples doped with rhodium [A.32], uranium [A.33], and also γ-irradiated LiNbO$_3$ [A.30] have also been used for hologram recording.

Table A.2. Point group 3m

$$\hat{\epsilon}: \begin{vmatrix} \epsilon_a & 0 & 0 \\ 0 & \epsilon_a & 0 \\ 0 & 0 & \epsilon_c \end{vmatrix} \; ; \quad \hat{r}: \begin{vmatrix} 0 & -r_{22} & r_{13} \\ 0 & r_{22} & r_{13} \\ 0 & 0 & r_{33} \\ 0 & r_{51} & 0 \\ r_{51} & 0 & 0 \\ -r_{22} & 0 & 0 \end{vmatrix}$$

A1.2.2 A conventional geometry for recording and readout of transmission holograms [A.28] involves the "c" axis in the incidence plane along **K**, ordinarily or extraordinarily polarized recording beams, and the extraordinarily polarized readout and reconstructed beams (Fig.5.7). Denisyuk's holograms in $LiNbO_3$ were recorded in [A.34, 35].

A fairly efficient anisotropic (intermode) diffraction has been observed [A.36-39]. Polarization and orientation dependences of different types of diffraction in $LiNbO_3$ were studied in [A.40-42].

A1.2.3 The diffusion mechanism of holographic recording is, in general, most common for nominally pure $LiNbO_3$ samples [A.43-45] and allows recording of shifted holograms with nearly 100% diffraction efficiency [A.44]. It is most pronounced in reduced $LiNbO_3$ samples [A.21, 46].

In crystals doped with the transition elements (Fe,Cu,Mn), a dominant mechanism is due to the photovoltaic effect [A.16, 20, 47, 48] that produces unshifted holograms with nearly 100% diffraction efficiency [A.49]. A similar mechanism has also been observed in nominally pure (but not reduced) $LiNbO_3$ samples, this fact being attributable to the presence of uncontrollable Fe impurities [A.50]. An applied electric field E_0 gives rise to an additional mechanism of recording an unshifted phase hologram [A.45, 51, 52].

Hologram recording in $LiNbO_3$:Fe via the transverse diagonal components of the photovoltaic tensor and via the circular photovoltaic effect was studied in [A.53.54].

A1.2.4 The highest sensitivity for the nominally pure and doped $LiNbO_3$ samples is observed in the blue-green spectral region. The typical sensitivity of undoped $LiNbO_3$ crystals $S^{-1} \simeq 10 \div 100$ J/cm^2 [A.44]. In $LiNbO_3$:Fe, $S^{-1} \simeq 1 \div 0.1$ J/cm^2 [A.55]. It is strongly dependent on the degree of reduction [A.8] and nearly independent on spatial frequency up to $\Lambda^{-1} \simeq 10^4$ mm^{-1} [A.47].

A1.2.5 As shown in [A.56-58], the dominant mechanism of dark conductivity in $LiNbO_3$ is the ionic conductivity with the activation energy $W_a \simeq 1.1$ eV and typical dielectric relaxation time $\tau_M^d \sim 10^6$ s at room temperature and $\tau_M^d \sim 10^{-1}$ s at T = 240° C. The latter is likely to be associated with mobile ionic OH^{1-} complexes whose average concentration in $LiNbO_3$ independent of the doping level is $\sim 10^{18}$ cm^{-3} [A.59, 60].

$LiNbO_3$:Fe crystals are also characterized by electron excitation from the donor Fe^{2+} level with an activation energy W_a of $\simeq 1.3$ eV and a typical dielectric relaxation time (for moderately doped $LiNbO_3$ samples containing ~ 0.1 mol.% Fe) $\tau_M^d \sim 10^{10}$ s at room temperature [A.56-58, 61]. The hologram relaxation processes in the presence of dark conductivity and under illumination were considered in detail in [A.18, 60]. Previously recorded holograms in $LiNbO_3$ and $LiNbO_3$:Fe samples are erased by heating the samples up to T $\simeq 300 - 350°$ C [A.56].

A1.2.6 Optical erasure of holograms in $LiNbO_3$ was studied in [A.55, 62].

A1.2.7 Practically nondestructive readout of elementary sinusoidal gratings is typically accomplished by a HeNe laser (λ = 633 nm). Several techniques for nondestructive readout of a hologram of a complicated object have been proposed. They include thermal fixing [A.18, 30, 56-58, 60, 63, 64], hologram recording via two-photon and two-step absorption [A.47, 65, 66] for reconstruction at the recording wavelength, and also hologram readout using the anisotropic diffraction [A.67, 68], and also nonlinear holographic recording of combinational holograms [A.69] for reconstruction at different wavelength.

A1.2.8 Energy exchange via shifted phase gratings in $LiNbO_3$ was first observed by *Staebler* et al. [A.45] and *Gaylord* et al. [A.70]. The gain factor for two-wave mixing via diffusion gratings in nominally-pure reduced $LiNbO_3$ $\Gamma \simeq 10$ cm^{-1} ($\lambda = 633$nm, $\Lambda = 0.3\mu$m) [A.71].

The gain factors obtained for the transient energy exchange via unshifted gratings in $LiNbO_3$ under an externally applied field $E_0 = 5 \div 10$ kV/cm [A.72, 73] were also about several tens of cm^{-1}.

A1.2.9 For the conventional holographic arrangement (the "c" axis in the incidence plane, extraordinary polarization), the photoinduced noise in $LiNbO_3$ has a shape of two lobes lying along the "c" axis [A.74-76]. These lobes are rotated by a small angle in the crystal plane when the ordinary light polarization is used [A.77]. At the initial stage of hologram recording, the noise intensity grows as the fourth power of exposure [A.34]. Complicated noise structures arising upon a simultaneous illumination of $LiNbO_3$ with several beams were studied in [A.57]. The amplified backward scattering in $LiNbO_3$:Fe was reported in [A.78].

A1.2.10 Degenerate four-wave mixing via shifted gratings in nominally pure $LiNbO_3$ was first studied by *Kukhtarev* et al. [A.79-81]. Coherent oscillation through four- and six-wave mixing with positive feedback in $LiNbO_3$:Cu without external cavity was studied in [A.82].

A1.2.11 Coherent oscillations through four-wave mixing for different recording mechanisms in a linear cavity were studied in [A.83, 84].

A1.2.12 Holographic recording in pure and doped samples of $LiNbO_3$ by nanosecond light pulses was investigated in [A.85-89]. Pulsed recording in $LiNbO_3$ via two-photon absorption and via two-step absorption in $LiNbO_3$:Cr was also studied in [A.47, 65, 66].

A1.2.13 Hologram recording in $LiNbO_3$ waveguide layers was analysed in [A.90-94].

A2 Lithium Tantalate LiTaO₃

A2.1 Physical Properties

A2.1.1 Crystals are grown by the Czochralski method at $T_m \simeq$ 1650°C [A.1, 3].

A2.1.2 Density: 7.454 g/cm³, Mohs' hardness: 5 [A.3].

A2.1.3 Isomorphous to $LiNbO_3$, a = 0.5143 nm, c = 1.3756 nm [A.3].

A2.1.4 Similar to $LiNbO_3$, $T_c = 665°$ C [A.95].

A2.1.5 $\epsilon_a^T = 51$, $\epsilon_c^T = 45$, $\epsilon_a^S = 41$, $\epsilon_c^S = 43$ [A.3, 5].

A2.1.6 Crystals are transparent, optical absorption spectra were measured in [A.7, 14].

A2.1.7 $n_o = 2.175$, $n_e = 2.180$ at $\lambda = 633$ nm [A.96]; $n_o = 2.2160$, $n_e = 2.2205$ at $\lambda = 0.5$ μm [A.97].

A2.1.8 For tensor $\hat{r}$ matrix see Table A.2; $r_{33} = 30.3$, $r_{13} = 7$ [10^{-10} cm/V] ($\lambda = 633$ nm, a strain-free sample) [A.98].

A2.1.9 Photoconductivity of $LiTaO_3$ was studied in [A.14].

A2.1.10 The bulk photovoltaic effect in $LiTaO_3$ has been observed by *Glass* et al. [A.19]. The photovoltaic field is $E_G = 3 \div 5$ kV/cm [A.99] and can be reduced to $E_G \simeq 0.5$ kV/cm by annealing in hydrogen [A.81]. From estimates of [A.100] E_G in $LiTaO_3$:Fe can, however, be as high as $\simeq 300$ kV/cm.

A2.1.11 The photorefractive effect as "optical damage" was discovered in $LiTaO_3$ by *Ashkin* et al. [A.22] (see also [A.23]).

A2.2 Holographic Properties

A2.2.1 As far as we know, the first holographic recording experiments in $LiTaO_3$ were reported by *Spinhirne* et al. [A.101].

A2.2.2 Similar to $LiNbO_3$.

A2.2.3 Basic mechanisms of holographic recording (diffusion, drift, photovoltaic) are similar to those observed in $LiNbO_3$ [A.99, 102, 103].

A2.2.4 Similar to $LiNbO_3$. Sensitivity of pure samples of $LiTaO_3$ in the blue region of the spectrum is $S^{-1} \simeq 1 \div 0.1$ J/cm^2 [A.101]. Doping with Fe, Cu, Mn can noticeably enhance the photorefractive effect in the blue-green region [A.103, 104].

A2.2.5 A typical hologram erase time in pure $LiTaO_3$ upon a continuous readout by a HeNe laser is $> 10^7$ s [A.101].

A2.2.7 A nondestructive hologram readout through two-step recording in $LiTaO_3$:Fe was studied in [A.105].

A2.2.9 Polarization-dependent photoinduced noise in $LiTaO_3$:Cu was studied in [A.106-108].

A2.2.10 Four-wave mixing by shifted phase gratings in reduced samples of $LiTaO_3$ was investigated in [A.79.81]. For four-wave mixing in an externally applied DC field ($E_0 = 10$ kV/cm) phase-conjugate reflectivity R $\simeq 4 \div 5$ was observed in [A.109, 110].

A3 Barium Titanate BaTiO3

A3.1 Physical Properties

A3.1.1 Crystals in the tetragonal phase, stable at room temperature, are grown by the TSSG technique [A.111] at about $1400°$ C from a melt of $BaTiO_3 + TiO_2$ with a stoichiometric excess of TiO_2 ($\simeq 65\%$).

A3.1.2 Density: 6.020 g/cm^3 [A.112].

A3.1.3 Space group at room temperature is P4mm (point group 4mm), a = 0.3992 nm, c = 0.4036 nm [A.113].

A3.1.4 At room temperature (in the tetragonal phase) it is a ferroelectric with a spontaneous polarization axis along the "c" axis. On cooling from the melting point three phase transitions occur [A.113]: a cubic modification (m3m) at T ⪆ 120° C; a tetragonal modification (4mm) at 120° C ⪆ T ⪆ 5° C; an orthorhombic modification (mm2) at +5° C ⪆ T ⪆ -90° C; a rhombohedral modification (3m) at T ⪅ -90° C. According to [A.114] the Curie temperature of the samples grown by TSSG technique T_C = 130± 2° C; it can vary over a wide range (120÷130° C) after an oxidation–reduction treatment [A.115].

The procedure of poling that is extremely important for this PRC was described in detail in [A.115]. It involves a slow cooling of the sample from the temperature above T_C to room temperature in external DC field E_0 = 0.5÷2 kV/cm. The degree of poling can be improved by cycling of this procedure with an additional polishing of the sample (e.g., [A.116]).

A3.1.5 Dielectric constants of $BaTiO_3$ given in the literature differ remarkably, in particular, as given in [A.117] ϵ_c^S = 106, ϵ_a^S = 4300.

A3.1.6 Crystals are transparent, typically pale yellow. The absorption edge is near 410 nm ($\Delta W \simeq 3eV$) [A.115]. A typical optical absorption coefficient in intentionally undoped as-grown crystals $\alpha \simeq 0.3$ cm^{-1} is nearly constant in a wide spectral region 450÷1000 nm and can be markedly increased by oxidation or reduction [A.115]. As it was reported in [A.118], the optical absorption coefficients for different $BaTiO_3$ samples obtained even from one supplier can differ by nearly an order of magnitude.

A3.1.7 Crystal is optically uniaxial, n_o = 2.458, n_e = 2.399 at λ = 550 nm [A.119].

A3.1.8 The linear electro-optic tensor $\hat{r}$ matrix is given in Table A.3. The values of the coefficients given in literature also differ greatly (e.g., [A.114]); as approximate values r_{13} = 8, r_{33} = 28, r_{42} = 820 [10^{-10}cm/V] [A.13] can be given.

A3.1.9 As shown in [A.120], the hole photoconductivity dominates in most $BaTiO_3$ samples studied. There are, however, crystals where photoelectrons are dominant. Along with this, annealing in reduction atmosphere can transform the hole photoconductivity into the electron one [A.115, 118].

At the present time, the basic microscopic model for undoped photorefractive $BaTiO_3$ is that where impurity centers associated with uncontrollable incorpration of Fe atoms serve as acceptors and donors [A.114, 118]. Like in $LiNbO_3$ Fe^{2+} centers can be considered as donors for photoelectrons (and traps for holes, correspondingly) and Fe^{3+} centers as acceptors (correspondingly, donors for photoinduced holes). The average concentration of Fe^{3+} centers in intentionally undoped samples measured by EPR was $(2÷8)\cdot10^{18}$ cm^{-3}. In as grown samples without oxidation or reduction treatment $N^{Fe^{2+}}$ = $(2÷9)\cdot10^{16}$ cm^{-3}.

Table A.3. Point group 4mm

$$\hat{\epsilon}: \begin{vmatrix} \epsilon_a & 0 & 0 \\ 0 & \epsilon_a & 0 \\ 0 & 0 & \epsilon_c \end{vmatrix}; \quad \hat{r}: \begin{vmatrix} 0 & 0 & r_{13} \\ 0 & 0 & r_{13} \\ 0 & 0 & r_{33} \\ 0 & r_{51} & 0 \\ r_{51} & 0 & 0 \\ 0 & 0 & 0 \end{vmatrix}$$

A detailed analysis of the impurity structure related to the photorefractive effect in BaTiO$_3$ is given in [A.121].

A3.1.11 For the first time the photorefractive effect in BaTiO$_3$ as an "optical damage" was also observed by *Ashkin* et al. [A.22].

A3.2 Holographic Properties

A3.2.1 *Townsend* et al. [A.122] were the first to perform holographic recording in BaTiO$_3$. However, the wide use of the crystals began only with the experiments of *Feinberg* et al. [A.117, 123].

A3.2.2 The first experiments [A.117, 122, 124] involved the use of a symmetric geometry (Fig.5.7). An asymmetric configuration [A.123] allowing the use of extremely high nondiagonal electro-optic coefficient r_{51} proved to be more efficient (Fig.5.9). To achieve an optimum angle of $\simeq 55°$ between K and the "c" axis, sample immersion is used in this scheme [A.123, 125, 126]. An efficient anisotropic (intermode) diffraction accompanied by a change of the polarization type can also be observed in this PRC [A.116, 127].

A3.2.3 In most experiments using BaTiO$_3$ efficient holographic recording is performed in the absence of an applied field through the diffusion mechanism with a typical dependence $\eta^{st} \propto K^2$ [A.117, 122]. With the external DC field $E_0 \gtrsim E_D$ applied, recording is dominated by the drift mechanism [A.117].

A3.2.4 The sensity region of BaTiO$_3$ covers almost entirely the visible spectral region with a typical value $S^{-1} \simeq 10^{-2}$ J/cm^2 ($\Lambda = 3\mu m$) which falls sharply at $\lambda \gtrsim 1$ μm [A.122]. The characteristic times of a passive phase-conjugate mirror formation in BaTiO$_3$ of $\simeq 5$ sW/cm^2 at $\lambda = 514$, 815 nm, and $\simeq 5 \cdot 10^2$ s·W/cm^2 at $\lambda = 1090$ nm were reported in [A.126].

A3.2.5 A typical dark relaxation time of holograms at room temperature $\tau_{sc}{}^d = 0.1 \div 1$ s decreases markedly with increasing spatial frequency K and samples temperature [A.128]. Recording of low-efficiency holograms with dark relaxation time of the order several days was also reported in [A.122].

A3.2.6 Incoherent optical erasure of holograms was studied in [A.124, 128].

A3.2.7 *Micheron* et al. [A.124] investigated electric fixing of the holograms in BaTiO$_3$.

A3.2.8 Two-wave mixing in BaTiO$_3$ was studied in [A.117, 120, 125, 129-131]. In experiments using an asymmetric recording geometry (Fig.5.9) a 4000-fold amplification of a weak signal beam, which corresponds to $\Gamma \simeq 30$ cm^{-1}, was observed [A.125]. Effects of the pump beam "depletion" on amplification of a strong signal beam were studied in [A.132]. Oscillation in a ring resonator using two-wave mixing in BaTiO$_3$ was observed in [A.125, 133, 134], the frequency shift of the wave generated in this geometry was studied in [A.135].

A3.2.9 Photoinduced noise in BaTiO$_3$ was studied in [A.136].

A3.2.10 Four-wave mixing in $BaTiO_3$ in an active phase-conjugate geometry (with external pump beams) was studied in [A.117, 123, 125, 130, 133]. The experimental comparison of the four-wave mixing geometries with similarly or orthogonally polarized pump waves was performed for $BaTiO_3$ in [A.137]. The highest reflectivity for a weak signal beam $R \simeq 10^2$ was obtained in [A.123, 138]. Effects of pump-beam depletion in a case of a "strong" signal beam were investigated in [A.132]. Oscillation in a resonator formed by two active phase conjugate mirrors using $BaTiO_3$ was observed in [A.139].

A3.2.11 Passive phase-conjugate geometries (without external pump beams) in $BaTiO_3$ were studied experimentally by a number of researchers: linear and semilinear geometries [A.133, 140-142] and, in particular, conjugation of a multicolor wavefront [A.142]; geometries using internal reflections [A.143-145]; and ring geometries [A.126, 146-150]. Double phase-conjugate mirror geometry using $BaTiO_3$ was investigated in [A.151-153], more complicated arrangements for interactions of mutually noncoherent beams in this PRC in [A.154-157].

A3.2.12 Pulsed nanosecond holographic recording in $BaTiO_3$ was studied in [A.158] where nearly 10-fold decrease of sensitivity as compared with recording by millisecond pulses was observed. Long-living photorefractive holograms were recorded in $BaTiO_3$ by 100-ps pulses [A.159].

A4 Strontium Barium Niobate $Sr_xBa_{1-x}Nb_2O_6$ (SBN)

A4.1 Physical Properties

A4.1.1 Crystals are grown from the melt by the Czochralski technique [A.160-162] at $T_m = 1500 \pm 10°$ C (for $x = 0.75$).

A4.1.2 Density: 5.4 g/cm^3 at $x = 0.75$ [A.160].

A4.1.3 At room temperature SBN has a structure of potassium-tang-sten bronze, space group P4bm (point group 4mm); a = $1.243024 \pm 2 \cdot 10^{-6}$ nm, c = $0.3941341 \pm 1 \cdot 10^{-6}$ nm at $x = 0.75$ [A.160-162].

A4.1.4 At room temperature it is a ferroelectric with a diffuse phase transition at $T_C \simeq 60°$ C (for $x = 0.75$). Above T_C SBN transforms into a tetragonal nonpolar modification (point group $\bar{4}$2m). Samples are poled by cooling from $65 \div 150°$ to room temperature in a DC external field $E_0 = 10$ kV/cm [A.162].

A4.1.5 $\epsilon_c^S \simeq 3400$ [A.163]; $\epsilon_a \simeq 500$, $\epsilon_c \simeq 3000$ (for $x = 0.75$) [A.162].

A4.1.6 Intentionally undoped samples are transparent in the visible spectral region [A.160-162, 164].

A4.1.7 SBN exhibits noticeable birefringence ($n_o = 2.312$, $n_e = 2.299$ at $\lambda = 633$ nm, $x = 0.75$) [A.165].

A4.1.8 Tensor $\hat{r}$ matrix is given in Table A.3. $|r_{33}| = 134$, $|r_{13}| = 6.6$, $|r_{51}| = 4.2$ [$\times 10^{-9}$ cm/V] [A.163], see also [A.161].

A4.1.9 Photoconductivity spectra of nominally pure crystals were studied in [A.164].

A4.1.11 Spectral and temperature dependences of the photorefractive effect in SBN were studied in [A.166].

A4.2 Holographic Properties

A4.2.1 First holographic recording experiments in undoped SBN samples were carried out by *Thaxter* et al. [A.167, 168]. Efficient holographic recording can also be performed in SBN crystals doped with cerium [A.169-172].

A4.2.2 As a rule, a symmetric recording-readout geometry similar to that used for LiNbO$_3$ was utilized (Fig. 5.7).

A4.2.3 Highly efficient phase holograms can be recorded both in the external DC field [A.167, 168, 173, 174] and through the diffusion recording mechanism [A.170].

A4.2.4 Sensitive in the blue-green region. A typical sensitivity of undoped SBN is $S^{-1} \simeq 3 \cdot 10^{-3}$ J/cm^2 at $\Lambda = 15$ μm, $E_0 = 3$ kV/cm [A.168]. For SBN:Ce $S^{-1} \simeq 5 \cdot 10^{-3}$ J/cm^2 at $\Lambda = 1$ μm, $E_0 = 0$ [A.170].

A4.2.5 Dark storage time $\tau_{sc}^{d} \gtrsim 5 \cdot 10^5$ s [A.168]; $\tau_{sc}^{d} \gtrsim 2 \cdot 10^6$ s for SBN:Ce [A.170].

A4.2.6 SBN allows multiple optical erasure of previously recorded holograms [A.168].

A4.2.7 Electrical fixing of holograms by applying a pulsed electric field to the sample was reported in [A.175].

A4.2.8 Two-wave mixing in SBN:Ce was studied in [A.176] in a linear resonator ($\Gamma \gtrsim 50$ cm^{-1} at $\lambda = 488$ nm, $\Lambda = 0.5 \mu$m), and also in [A.177, 178]. Dynamic self-diffraction of phase-modulated light beams in SBN was reported in [A.179, 180]. Wavelength (at $\lambda = 514, 840, 1090$ nm) and temperature ($-30°$ C $< T < +40°$ C) dependences of the gain factor Γ and characteristic time for hologram formation τ_{sc} were obtained in [A.181] for cerium- and calcium-doped SBN crystals.

A4.2.9 Photoinduced noise in SBN:Ce was studied in [A.182].

A4.2.10 Four-wave mixing in SBN was investigated in [A.177] ($\lambda = 514$ nm, $R \lesssim 0.5$), and in [A.183] ($R > 1$).

A4.2.11 Passive phase conjugation in a ring geometry using SBN:Ce was investigated in [A.183, 184] where a reflectivity $R = 0.2 \div 0.3$ was observed for $\lambda = 442$ nm. For the same wavelength, the reflectivities $R = 0.6$ and $R = 0.2$ for nominally pure and Ce-doped SBN:60 (Sr$_{0.6}$Ba$_{0.4}$Nb$_2$O$_6$), respectively, were obtained in [A.185] in the geometry with a corner cube reflector.

Reflectivity $R = 6\%$ and characteristic time for hologram formation $\simeq 8$ s (for $I_0 \simeq 200$ mW/cm^2) were reported in [A.186] for Ce-doped SBN:75 (Sr$_{0.75}$Ba$_{0.25}$Nb$_2$O$_6$) under similar conditions. For the red region of the spectrum, reflectivity $4 \div 7\%$ and the characteristic recording time of about several minutes (for $I_0 \simeq 1$ W/cm^2) were also observed for SBN:60 in [A.187].

A4.2.13 Holographic recording in SBN single-crystal optical fibers was experimentally demonstrated in [A.188].

A5 Sodium Barium Niobate Ba$_2$NaNb$_5$O$_{15}$ (NBN)

The crystals are ferroelectric and optically biaxial at room temperature (point group mm2). Dielectric permeability (Table A.4) is $\epsilon_a = 238 \pm 5$, $\epsilon_b = 228 \pm 5$, $\epsilon_c = 43 \pm 2$, electro-optic coefficients are $r_{13} = 15 \pm 1$, $r_{23} = 13 \pm 1$, $r_{33} = 48 \pm 2$, $r_{42} = 92 \pm 4$, $r_{51} = 90 \pm 4$ [$\times 10^{-10}$ cm/V] ($\lambda = 633$ nm, frequency f = 0) [A.162, 189].

Holographic recording in NBN:Fe was first performed by *Amodei* et al. [A.190]. Thermal fixing of holograms, like in LiNbO$_3$, was reported in [A.56]. Two-wave mixing was studied in nominally pure NBN samples [A.191] ($\Gamma_{max} \simeq 8$ cm^{-1}, $\lambda = 488$ nm, $\Lambda \stackrel{\sim}{<} 1\mu$m, $E_0 = 0$). The use of two-wave mixing in NBN for low-noise preamplification in a copper-vapor coherent amplifier was investigated in [A.192].

In [A.191] oscillation via two-wave mixing in a ring and linear resonators was reported. Fidelity of phase conjugation in a NBN-based four-wave mixing arrangement with independent pump waves (R = 5%) and in a linear passive phase-conjugate geometry (R = 1.5%) was studied in [A.193].

A6 Potassium Niobate KNbO$_3$

KNbO$_3$ (point group mm2, Table A.4) is a ferroelectric with dielectric constants $\epsilon_a = 140$, $\epsilon_b = 1200$, $\epsilon_c = 40$ (at f = 100 kHz) at room temperature [A.194]. It is optically biaxial ($n_a = 2.280$, $n_b = 2.329$, $n_c = 2.169$) and has fairly high electro-optic coefficients: $r_{13} = 28 \pm 2$, $r_{23} = 1.3 \pm 0.5$, $r_{33} = 64 \pm 5$, $r_{42} = 380 \pm 50$, $r_{51} = 105 \pm 13$ [$\times 10^{-10}$ cm/V] ($\lambda = 633$ nm) [A.195]. Optical and optoelectrical properties of nominally pure and Fe-doped KNbO$_3$ were investigated in [A.196].

Holographic recording was studied first by *Günter* et al. both in doped KNbO$_3$:Fe [A.197, 198] and nominally pure reduced KNbO$_3$ [A.199, 200]. The typical holographic orientation of KNbO$_3$ coincides with that used for LiNbO$_3$: the "c" axis in the sample and incidence planes, polarization of the light beams - extraordinary (H). The highest sensitivity $S^{-1} \sim 10^{-4}$ J/cm^2 ($\lambda = 488$ nm, $\Lambda = 2\mu$m) was observed in reduced samples of KNbO$_3$ [A.201]. Two-wave mixing experiments using KNbO$_3$:Fe [A.196, 202-204] succeeded in achieving a gain factor $\Gamma = 11.5$ cm^{-1} ($\lambda = 488$ nm, $\Lambda \simeq 7\mu$m, $E_0 = 16$ kV/cm) which allows phase conjugation with R = 0.1 in four-wave mixing geometry [A.203, 204]. Efficient intramode (anisotropic) diffraction

Table A.4. Point group mm2

$$\hat{\epsilon}: \begin{vmatrix} \epsilon_a & 0 & 0 \\ 0 & \epsilon_b & 0 \\ 0 & 0 & \epsilon_c \end{vmatrix} ; \quad \hat{r}: \begin{vmatrix} 0 & 0 & r_{13} \\ 0 & 0 & r_{23} \\ 0 & 0 & r_{33} \\ 0 & r_{42} & 0 \\ r_{51} & 0 & 0 \\ 0 & 0 & 0 \end{vmatrix}$$

[A.205, 206] and anisotropic self-diffraction [A.207] can also be observed in KNbO$_3$.

Holographic currents in KNbO$_3$:Fe were investigated in [A.208].

A7 Bismuth Silicon Oxide Bi$_{12}$SiO$_{20}$ (BSO)

A7.1 Physical Properties

A7.1.1 Crystals are typically grown using the Czochralski technique from the melt Bi$_2$O$_3$+SiO$_2$ at T$_m$ $\simeq$ 900° C [A.209, 210].

A7.1.2 Density: 9.14÷9.22 g/cm^3, Mohs' hardness: $\simeq$6 [A.112].

A7.1.3 Crystals belong to space group I23 (point group 23) with a body-centered cubic cell at room temperature, a = 1.010433 (5) nm [A.211].

A7.1.4 Being cubic the crystal does not possess spontaneous polarization.

A7.1.5 ϵ = 56 [A.212].

A7.1.6 Nominally pure crystals are transparent, yellow in color; bandgap ΔW = 3.15÷3.25 eV; a typical optical absorption curve is plotted in Fig.A.2 [A.212, 213]. It is customary to assume [A.213, 214] that the shoulder in the absorption spectrum with photon energy $\hbar\omega \lesssim 3$ eV is due to localized energy states near the conduction band bottom with concentration N$_D$ ~ 10^{19} cm^{-3} associated with oxygen vacancies created by Si deficiency [A.211].

A7.1.7 n = 2.54 for λ = 633 nm, the dispersion of index of refraction in BSO was studied in [A.212]. The crystal is optically active [A.215, 216]

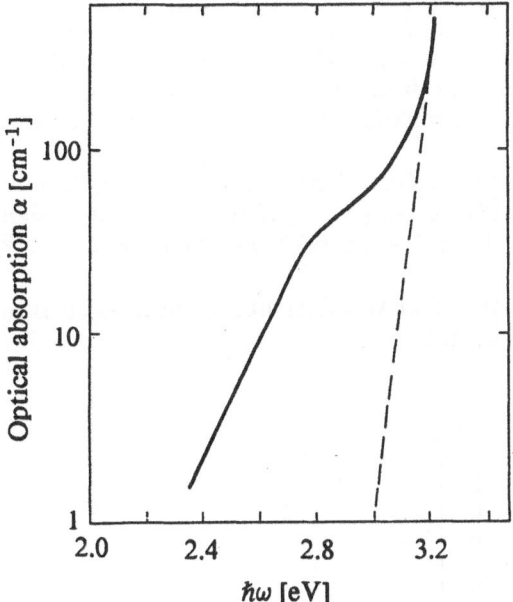

Fig.A.2. Typical optical absorption curve for undoped BSO crystal [A.213]

Table A.5. Optical activity of BSO and BGO [A.216]

λ [μm]	ρ [deg/mm]	
	BSO	BGO
0.45	60.2	60.9
0.50	42.2	41.5
0.55	31.8	30.8
0.60	25.2	24.1
0.65	20.5	19.5

(Table A.5), both the right and left rotating modifications exist. Faraday effect and electrogiration of BSO were studied in [A.215].

A7.1.8 Tensor matrix $\hat{r}$ is given in Table A.6. According to [A.212] $r_{41} = 5 \cdot 10^{-10}$ cm/V for $\lambda = 633$ nm, the corresponding half–wave voltage is $U_{\lambda/2} = \lambda/2n^3 r_{41} = 3.9$ kV. However, the literature reports also somewhat higher halfwave voltages in BSO of $\simeq 4 \div 4.5$ kV (e.g., [A.217]).

A7.1.9 Undoped BSO is a good photoconductor of n–type [A.212, 213, 218] in the region $400 \div 550$ nm. Quantum efficiency of photoconductivity $\beta \simeq 1$ at $\lambda \gtrsim 500$ nm with further decay to $\simeq 0.1$ at 550 nm and to $\simeq 0.01$ at 650 nm [A.219, 220]. The typical value of $\mu\tau$ is in the range $10^{-7} \div 10^{-6}$ cm^2/V ($8.5 \cdot 10^{-7}$ cm^2/V according to [A.213], $1.4 \cdot 10^{-7}$ cm^2/V according to [A.221], and $5.2 \cdot 10^{-6}$ cm^2/V according to [A.220]).

BSO commonly has a complicated structure of shallow traps with concentration of $10^{14} \div 10^{15}$ cm^{-3} [A.213, 222, 223]. It results, in particular, in photoconductivity increase in the red spectrum region stimulated by a preliminary blue–green illumination [A.222, 224], and in a decrease of the photoconductivity after preliminary IR irradiation of the crystal [A.225], a slow decay of conductivity (after illumination is turned off) is also typical of some BSO samples.

A7.1.10 The bulk photovoltaic effect (both linear and circular) in BSO was studied in detail by *Grachev* et al. [A.226, 227]. Because of a high photoconductivity, photovoltaic fields E_G in BSO do not exceed $2 \cdot 10^{-2}$ V/cm ($\lambda = 488$ nm).

A7.1.11 Photorefraction as an effect was first studied in the structure of the PROM type by *Hou* et al. [A.228].

Table A.6. Point group 23, $\bar{4}$3m

$$\hat{\epsilon} : \begin{vmatrix} \epsilon & 0 & 0 \\ 0 & \epsilon & 0 \\ 0 & 0 & \epsilon \end{vmatrix} \; ; \quad \hat{r} : \begin{vmatrix} 0 & 0 & 0 \\ 0 & 0 & 0 \\ 0 & 0 & 0 \\ r_{41} & 0 & 0 \\ 0 & r_{41} & 0 \\ 0 & 0 & r_{41} \end{vmatrix}$$

A7.2 Holographic Properties

A7.2.1 First holographic recording experiments using BSO were carried out by *Huignard* et al. [A.221].

A7.2.2 Conventional geometries for recording and readout of transmission holograms [A.221] are the same as traditional orientations of cubic noncentersymmetric crystals in transverse electro-optic modulators [A.11-13]. Crystals are cut in the (110) crystallographic plane; either the [001] or [1$\bar{1}$0] axis is in the incidence plane; light beam polarizations eigenstates at which diffraction with polarization preservation is observed are shown in Fig.5.10. For $K \parallel [1\bar{1}0]$ the use of readout H- and E-polarizated beams results in diffraction with polarization-plane rotation [A.229-231] which can be used to increase the signal-to-noise ratio in the reconstructed image [A.229].

Polarization dependences of η^{st} in the geometries indicated were studied in [A.41, 232-235]. Bragg maximum splitting due to optical activity has been investigated in [A.236, 237]. As was shown in [A.238], the absolute maximum of η^{st} in cubic crystals is obtained in the (110) cut samples at $K \parallel [1\bar{1}1]$ [A.238] (Sect.5.5).

A7.2.3 The diffusion mechanism of recording in BSO with a typical $\eta^{st} \propto K^2$ dependence was analysed in [A.221, 239] (λ=514nm, d=10nm, Λ=1μm, $\eta^{st} \simeq$0.25%). In further experiments on four- [A.240] and two-wave [A.241] mixing via shifted diffusion gratings in BSO a peak and a subsequent decay in the $\eta^{st}(K)$ dependence was observed to yield an estimate for acceptor impurity concentration of $N_A \simeq 10^{15}$ cm^{-3} (for λ = 514 nm).

Drift mechanism of recording in BSO in an external DC field E_0 was studied in [A.221, 239] (d=10mm, E_0=10 kV/cm, η^{st}=5% being constant on K) and in [A.75, 232]. Saturation of the drift recording due to limited N_A was observed in [A.240]. A similar effect was also observed in experiments on simultaneous measurements of the diffraction efficiency of the drift hologram and the stationary holographic current in BSO at λ = 442 nm [A.242].

Nonstationary recording mechanism in an external DC field was studied by *Stepanov* et al. [A.243, 244] and later in [A.245]. It was shown that in an external DC field the hologram in BSO moves along E_0 with a characteristic velocity V_0 (K,E_0,I_0). The use of the recording interference pattern moving in a resonance manner allows at m<<1 a nearly 10-fold increase in η^{st} [A.243, 244].

The complicated structure of impurity centers in BSO gives rise to typical dynamic effects. In particular, a transient increase of the diffusion hologram efficiency after a preliminary IR irradiation of the sample [A.246] and also dynamic holographic recording of time-varying signal waves in the red light with a simultaneous incoherent green illumination [A.247] were reported.

A7.2.4 At present BSO is one of the most sensitive photorefractive crystals (S$^{-1} \simeq 2 \cdot 10^{-3}$ J/cm^2 at λ = 514nm, Λ = 1μm) [A.221]. Because of long drift lengths $L_0 = \mu \tau E_0 \gtrsim 10 ~\mu$m, an increase of BSO sensitivity up to $S^{-1} \simeq$ $2 ~10^{-5}$ J/cm^2 at $\Lambda = 100 ~\mu$m can be expected.

The highest sensitivity of BSO is in the blue-green. On transition to λ = 633 nm, sensitivity decreases by $1 \div 2$ orders of magnitude because of reduction in α and β, it can be increased, however, by a factor of 10-20 by heating up to 300° C [A.248]. Changes in the kinetics of hologram recording in BSO due to doping by Al, Mn, Ni were studied in [A.249].

A7.2.5 A low dark conductivity $\sigma^d \simeq 2 \cdot 10^{-14}$ $(\Omega \cdot$cm$)^{-1}$ [A.212]; $\simeq 1.6$ $\times 10^{-15}$ $(\Omega \cdot$cm$)^{-1}$ [A.221] must provide the dark hologram storage time of $\tau_{sc}^d \gtrsim 3 \cdot 10^3 \div 3 \cdot 10^4$ s. An experimental value $\tau_{sc}^d \simeq 10^5$ s was reported in [A.221].

A7.2.6 Optical erase time of diffusion hologram (equal to the recording time) in an external field can be increased by nearly an order of magnitude [A.239, 243, 244, 250].

A7.2.7 For nondestructive monitoring of the elementary sinusoidal grating in BSO an attenuated beam from a HeNe laser is typically used [A.221]. Nondestructive readout of a complicated object hologram at a different wavelength can be accomplished by recording nonlinear combinational holograms [A.231, 251, 252]. During hologram recording at λ = 633 nm in BSO, relatively long-living "latent" holograms which can be developed when the external field is turned on are observed [A.253]. An effect similar to thermal fixing in LiNBO$_3$ at room temperature was also observed in [A.254, 255].

A7.2.8 Stationary energy coupling via a shifted hologram in BSO was first observed in [A.241] (λ=514nm, Γ=0.4 cm^{-1} at Λ=1μm; Γ=1cm^{-1} at Λ= 4μm, E_0=19kV/cm). A noticeable increase of energy coupling efficiency in BSO in an external field was obtained for recording of moving interference patterns: Γ = 2.4 cm^{-1} for λ = 514 nm, Λ = $3.5 \div 9.5 ~\mu$m [A.256]; Γ = 7 cm^{-1} for λ = 568 nm, Λ = 22 μm [A.257]; and Γ = 12 cm^{-1} for λ = 568 nm, Λ = 23 μm in a thin sample of thickness d = 1.27 mm [A.258]. For nonstationary holographic recording under an external alternating electric field, the gain factor of BSO proved to be $\Gamma \simeq 5$ cm^{-1} (λ = 633nm, Λ = 25μm, $E_{\sim} \simeq 12$ kV/cm) [A.217].

A7.2.10 *Huignard* et al. [A.240, 259] were the first to study four-wave mixing in BSO (λ=514nm, d=3mm, r=1, R=10^{-4} for diffusion recording at $\Lambda \lesssim 1 \mu$m; R=$2 \cdot 10^{-3}$ for drift recording at E_0=6kV/cm). A much higher reflectivity R = 2.7 was obtained in [A.260] for nonstationary recording of a moving grating with asymmetrical pumps (r=0.2, λ=568nm, d=5mm, E_0= 10 kV/cm, Λ=30μm).

A7.2.11 Oscillation in a linear, semilinear and ring resonators using BSO was observed in [A.261-263].

A7.2.12 Pulsed holographic recording in BSO was studied in [A.264-268].

A8 Bismuth Germanium Oxide Bi$_{12}$GeO$_{20}$ (BGO)

A8.1 Physical Properties

A8.1.1 Crystals are grown using the Czochralski technique from a melt of Bi$_2$O$_3$ and GeO$_2$ at T$_m$ of about 930° C [A.209].

A8.1.2 Density: 9.22÷9.39 g/cm^3; Mohs' hardness: 4.5 [A.112].

A8.1.3 Isomorphous to BSO [A.269], a = 1.0143÷1.0145 nm [A.112].

A8.1.4 Similar to BSO.

A8.1.5 ϵ = 40 [A.209, 212].

A8.1.6 Crystals are transparent, lemon yellow, ΔW = 3.15÷3.25 eV. Optical absorption spectra in the visible were studied in [A.212, 270]. On doping with Al, no shoulder in the optical absorption at $\hbar\omega \lesssim 3$ eV is observed [A.214].

A8.1.7 Dependence n(λ) coincides with the curve for BSO within the experimental error [A.212]. Crystal is optically active [A.215, 216, 271] (Table A.5). The Faraday effect in BGO was observed in [A.215], the electrogiration in [A.271].

A8.1.8 Tensor matrix $\hat{r}$ is given in Table A.6, r_{41} = 3.2 10^{-10} cm/V for λ = 633 nm [A.212, 271].

A8.1.9 The conducting properties are similar to those of BSO. The photoconductivity is of the electron type [A.212, 218], however it is somewhat lower than in BSO [A.212, 270] in the entire visible region, $\mu\tau$ = 1.2·10^{-7} cm^2/V [A.221]. In accordance with [A.272] BGO can, however, demonstrate both the n-type and p-type photoconductivity with a characteristic diffusion length of photocarriers L$_D$ = 2.3÷8.0 μm.

A8.1.10 The bulk photovoltaic effect manifests itself like in BSO [A.227].

A8.2 Holographic Properties

A8.2.1 First holographic recording was performed by *Huignard* et al. [A.221].

A8.2.2 Similar to BSO.

A8.2.3 With the allowance made for a 1.5-fold reduction in the electrooptic effect, the main characteristics of diffusion and drift holograms are similar to those of BSO [A.221]. Diffraction efficiency η = 95% was observed experimentally in BGO under an external DC field E$_0$ = 14 kV/cm (λ=514nm, Λ=20μm, d=10mm) [A.273].

A8.2.4 Sensitive in the 400÷550 nm region (S^{-1}=8mJ/cm^2 at λ=514 nm, Λ=1μm [A.221]). Like BSO, exhibits a sensitivity increase in the red when heated up to 300° C [A.248]. Changes in the kinetics of hologram recording in BGO resulting from Al and Ca doping were reported in [A.249].

A8.2.5 Dark conductivity is $\sigma^d \simeq$ 1.2·10^{-11} (Ω·cm)$^{-1}$ [A.212] or $\simeq$3 ×10^{-14} (Ω·cm)$^{-1}$ [A.221], which should ensure the dark hologram storage time $\tau^d_{sc} \simeq$ 1÷10^3 s.

A8.2.8 Two-wave mixing in BGO was demonstrated by *Günter* [A.274]. For λ = 605 nm, d = 9.15 mm, and E$_0$ = 18.8 kV/cm, a steady-state amplification of a weak signal beam (T=4 at Λ=4μm), a more efficient nonstationary one (T=10 at Λ=10μm), and that for a resonantly moving grating (T=8, Γ=2.3cm^{-1} at Λ=10μm, E_0=11kB/cm) were observed.

A8.2.10 Four-wave mixing via transmission hologram was studied in [A.275, 204] (R=10^{-2} at λ=605nm, d=9.15mm, Λ=4μm, E_0=10÷15kV/cm). As shown by *Ja* [A.276] four-wave mixing by reflection gratings in BGO can also be observed.

A9 Bismuth Titanium Oxide Bi$_{12}$TiO$_{20}$ (BTO)

A9.1 Physical Properties

A9.1.1 Crystals are grown using the Czochralski method from the melt of Bi$_2$O$_3$ + TiO$_2$ at T$_m$ $\simeq$ 950° C [A.209, 277].

A9.1.2 Density: 9.1 g/cm^3 [A.278].

A9.1.3 Isomorphous to BSO; a = 1.0177 nm [A.278].

A9.1.4 Similar to BSO.

A9.1.5 ϵ = 47 at f = 1 kHz [A.279].

A9.1.6 Crystals are transparent amber; optical absorption spectra were studied in [A.280]; band gap is $\Delta W \simeq 3.47$ eV. Typical α at λ = 633 nm is 0.3÷0.5 cm^{-1}.

A9.1.7 n = 2.25, ρ = 6.3 degr/mm at λ = 633 nm [A.279]; the dispersion of optical activity was studied in [A.215].

A9.1.8 For tensor $\hat{r}$ matrix see Table A.6; r_{41} = 5.17·10^{-10} cm/V at λ = 633 nm [A.279]; $U_{\lambda/2}$ = $\lambda/2n^3 r$ = 3.3 kV [A.217, 281].

A9.1.9 The spectral dependence of photoconductivity was also studied in [A.280]. BTO has a dominant electron photoconductivity in the blue-green [A.218]; $\mu\tau$ = 2.4·10^{-8} cm^2/V (L$_D$ = 0.25μm) at λ = 633 nm [A.282].

A9.1.10 Bulk photovoltaic properties are similar to those of BSO [A. 227].

A9.2 Holographic Properties

A9.2.1 First holographic recording in BTO was carried out by *Pencheva* et al. [A.218].

A9.2.2 Similar to BSO. The maximum diffraction efficiency was observed in the (110)-cut samples with the [1$\bar{1}$1] axis oriented in the incidence plane for H-polarized light beams [A.238, 283]. According to [A.284], holographic recording via the electrogyratory effect is also possible in BTO.

A9.2.3 Diffusion recording in BTO was used in [A.218, 238, 285]. The diffraction efficiency of up to 30% was obtained in [A.286] for the diffusion grating recorded by the light beams entering the crystal through adjacent orthogonal faces. The most thorough investigation concerns the nonstationary recording mechanism in alternating field (see Item A9.2.8 below).

A9.2.4 Because of optical absorption increase in the blue-green, the operating spectrum region of BTO is yellow-red; S^{-1} = 10÷3·10^{-3} J/cm^2 for λ = 633 nm, Λ = 3 μm [A.281].

A9.2.8 Efficient two-wave mixing by shifted gratings recorded in an external alternating field was observed by *Stepanov* et al. [A.281,282]. The maximum $\Gamma = 10 \div 15$ cm^{-1} at $E_\sim = 15$ kV/cm, $\Lambda = 2.5 \div 10$ μm; amplification factor for weak signal beams $T \gtrsim 100$. Strong signal beam amplification was studied in [A.283,287].

A9.2.9 Photoconduced noise in BTO was observed in [A.217,288].

A9.2.10 Four-wave mixing in BTO was investigated [A.217,288,289]. The highest phase-conjugate reflectivity $R \simeq 30$ for a weak signal beam was achieved in an efficient four-wave geometry with orthogonal polarizations of pump waves.

A9.2.11 Passive phase-conjugation in the ring geometry using BTO ($R = 1.5\%$) was reported in [A.283,290]. A similar reflectivity was also observed in a linear passive phase-conjugate geometry [A.283,291]. A remarkably higher reflectivity ($R \approx 40\%$) was obtained for BTO in a double phase-conjugate mirror geometry [A.283,292]. Note that in all these experiments with BTO listed above, efficient nonstationary recording under an external AC field and HeNe wavelength $\lambda = 633$ nm were utilized.

A9.2.12 Pulsed holographic recording in BTO was investigated in [A.293].

A10 Semiconductor Photorefractive Crystals

Recent considerable interest in semiconductor photorefractive crystals is mainly due to the possibility of operation in the IR region and fast hologram formation rates. The salient electro-optic properties of noncentrosymmetric cubic semiconductors which are most promising from this point of view were summarized in [A.294,295] (Table A.7). Other important physical characteristics of these PRCs have been summarized in [A.296].

A10.2 Holographic characteristics

A10.2.1 Photorefractive holographic recording in semiconductor GaAs:Cr and InP:Fe was demonstrated for the first time by *Glass* et al. [A.294] using a CW Nd:YAG laser ($\lambda = 1.06 \mu$m).

Table A.7. Characteristics of photorefractive semiconductor crystals [A.295]

Crystal	Operating wavelength range [μm]	Dark relaxation time [s]	n^3r [pM/B]	n^3r/ϵ [pM/B]
InP:Fe	0.85 - 1.3	10^{-3}	52	4.1
GaAs:Cr	0.8 - 1.8	10^{-3}	43	3.3
CdTe:In	0.9 - 1.6	10^{-3}	152	16

A10.2.2 Basic holographic orientations are similar to those used in the experiments with cubic PRCs of the sillenite family. Polarization dependences of the diffraction efficiency observed for transmission holograms in (110)-cut GaAs were investigated in [A.297]. Similar measurements were also performed in [A.298] for CdTe. For (001)-cut GaAs, polarization dependences of the diffraction efficiency for the reflection holograms were obtained in [A.299]. Diffraction with rotation of the light polarization was observed in [A.300] for transmission gratings in GaAs.

A significant contribution of the amplitude component to the photorefractive grating diffraction efficiency was demonstrated in [A.297] and also in [A.301] for GaAs.

A10.2.3 Practically all the main recording mechanisms have been demonstrated in the semiconductor PRCs. Diffusion recording was utilized, in particular, for GaAs [A.297, 299, 300, 302-305], InP:Fe [A.303, 304], and for CdTe [A.298]. Note that this recording mechanism needs no deposition of the electrodes onto the sample for application of an external voltage. Thus it can be used for local nondestructive monitoring of the electrophysical parameters of semiinsulating GaAs and InP:Fe wafers [A.304].

Drift recording under external DC electric field was also used for GaAs [A.294, 296, 306] and for InP:Fe [A.294, 296, 307]. Nonstationary mechanisms were also reported for GaAs, and recording of a moving pattern under external DC field [A.308-310] and recording under external alternating field [A.311, 312], in particular.

A10.2.4 Because of long transport lengths of photoinduced carriers ($\mu\tau \simeq 1.5 \cdot 10^{-4}\,\mathrm{cm^2/V}$ in GaAs:Cr and InP:Fe [A.294]) and acceptable value of $n^3 r/\epsilon$ (Table A.7), semiconductor PRCs can be regarded as most promising for fast and high-sensitivity holographic recording. In fact, the characteristic recording time in InP:Fe proved to be $\tau_{sc} = 10^{-4}$ s for $I_0 = 50$ mW/cm^2 ($\lambda = 1.06\,\mu m$, $\Lambda = 5\,\mu m$, $E_0 = 2.7\mathrm{kV/cm}$, $\eta = 0.2\%$) [A.294]. This corresponds to the holographic sensitivity $S^{-1} \sim 10^{-5}$ J/cm^2.

For GaAs [A.302], the characteristic recording time $\tau_{sc} = 2 \cdot 10^{-5}$ s was experimentally observed for $I_0 = 4$ W/cm^2 ($\lambda = 1.06\,\mu m$, $\Lambda = 1\,\mu m$, $E_0 = 0$, $\eta = 1\%$). This allows the holographic sensitivity of this PRC to be estimated as $S^{-1} \sim 10^{-4}$ J/cm, which is also approaching the theoretical limit (4.18).

A10.2.5 Comparatively short dark relaxation times that are characteristic of the semiconductor PRCs ($r_{sc}^d = 10^{-4}$s in GaAs:Cr and InP:Fe [A.294]) are determined by their high dark conductivity. Note that this typical dark relaxation time of the space-charge gratings mentioned above for GaAs:Cr was also obtained in the experiments on non-steady-state photo-EMF [A.313], when the crystal was illuminated by a vibrating interference pattern and the measured parameter was an alternating short-circuit current through the sample.

GaAs crystals with remarkably longer dark relaxation times $\tau_{sc}^d \gtrsim 8$ s were reported however in [A.305].

A10.2.8 Energy transfer via two-wave mixing in GaAs has been investigated for all recording mechanisms resulting in formation of shifted phase gratings. The following gain factors were experimentally observed: Γ

= 0.3÷0.4 cm^{-1} for diffusion recording at λ = 1.06 μm and Λ = 0.6÷1.2 μm [A.302]; Γ = 2.6 cm^{-1} for drift recording under saturation of trapping centers at λ = 1.06 μm, Λ = 6÷8 μm, and E_0 = 15 kV/cm [A.306]; Γ = 2.4 cm^{-1} for nonstationary recording under an alternating electric field at λ = 1.06 μm, Λ = 16 μm, $E_\sim$ = 0.8 kV/cm, and f = 20 kHz [A.312]; and Γ = 6-7 cm^{-1} for recording of a resonancely moving pattern under an external DC electric field at λ = 1.06 μm, Λ = 20 μm, and E_0 = 8.5 kV/cm [A.309].

For InP:Fe, Γ = 4÷4.4 cm^{-1} was observed in [A.307] for recording of a fixed pattern with a spacing Λ = 5÷10 μm under a DC field E_0 = 8 kV/cm.

A10.2.10 In a conventional four-wave mixing geometry with a GaAs crystal, the phase-conjugate reflectivity R = 14% for recording under an AC field and R $\simeq$ 1 for recording of a moving pattern under a DC field were obtained in [A.312 and 310], respectively. Four-wave mixing with an amplified phase-conjugate wave (R$\approx$5) was also observed in GaAs for crossed polarizations of the pump beams [A.310].

Reflectivity R = 1% was obtained for a conventional four-wave mixing arrangement with InP:Fe [A.296] as well.

A10.2.12 Two- and four-wave interactions of picosecond (43 ps) light pulses via photorefractive gratings in GaAs were investigated in [A.314, 315]. Two-wave mixing of picosecond (35ps) pulses in GaAs was also observed in [A.316].

Two-wave interaction of nanosecond light pulses in InP (and GaAs as well) was also investigated in [A.317].

A11 Other Photorefractive Crystals and Media

Holographic recording through the photorefractive effect has also been studied in cubic KTN crystals [A.318-321] and isotropic PLZT ceramics under external DC electric field [A.322-330]. Recording of holograms in these media has, however, many specific features, so we refer interested readers to the original publications.

Recently, investigations of photorefractive holographic recording in a new photorefractive crystal $Ba_{2-x}Sr_xK_{1-y}Na_yNb_5O_{15}$ (BSKNN) [A.161] were also reported.

References

Chapter 1

1.1 A. Ashkin, G.D. Boyd, J.M. Dziedzic, R.G. Smith, A.A. Ballmann, H.J. Levin-stein, K. Nassau: Appl. Phys. Lett. **9**, 72-74 (1966)

1.2 M.P. Petrov, S.I. Stepanov, A.V. Khomenko: *Photosensitive Electro-optic Media in Holography and Optical Information Processing* (Nauka, Leningrad 1983)

1.3 A.M. Glass, D. von der Linde, D.H. Auston, T.J. Negran: J. Electron Mater. **4**, 915-943 (1975)

1.4 P. Günter, F. Micheron: Ferroelectrics **18**, 27-38 (1978)

1.5 M.B. Klein: Photorefractive properties of BaTiO$_3$, in *Photorefractive Materials and their Applications I*, ed. by P. Günter and J.-P. Huignard, Topics Appl. Phys., Vol.61 (Springer, Berlin, Heidelberg 1988) Chap.7

1.6 B. Dishler, A. Rauber: Sol. State Commun. **17**, 953-956 (1975)

1.7 S.L. Hou, R.B. Lauer, R.E. Aldrich: J. Appl. Phys. **44**, 2652 (1973)

1.8 V.I. Berezkin, M.V. Krasin'kova: Pizma v Zh. Tech. Fiz. **9**, 467-471 (1983) [Engl. transl.: Sov. Tech. Phys. Lett. (USA) **9**, 202-203 (1983)]

1.9 O.A. Gudaev, V.A. Detinenko, V.K. Malinovskii: Fiz. Tverd. Tela **23**, 195-201 (1981) [Engl. transl.: Sov. Phys.-Solid State **23**, 109-113 (1981)]

1.10 R. Obershmid: Phys. Status Solidi (a) **89**, 263-270 (1985)

1.11 R. Orlowski, E. Krätzig: Sol. St. Commun. **27**, 1351-1353 (1978)

1.12 M.B. Klein, G.C. Valley: J. Appl. Phys. **57**, 4901-4905 (1985)

1.13 S. Ducharme, J. Feinberg: J. Opt. Soc. Am. **B3**, 283-292 (1986)

1.14 S.I. Stepanov: Zh. Tech. Fiz. **52**, 2114-2116 (1982) [Engl. transl.: Sov. Phys.-Tech. Phys. **27**, 1300-1301 (1982)]

1.15 G. Pauliat, M. Allain, J. Launoy, G. Roosen: Opt. Commun. **61**, 321-324 (1987)

1.16 V.I. Belinicher, B.I. Sturman: Usp. Fiz. Nauk **130**, 415-458 (1980) [Engl. transl.: Sov. Phys.-Usp. (USA) **23**, 199-223 (1980)]

1.17 M.P. Petrov, A.I. Grachev: Pis'ma v Zh. Eksp. & Teor. Fiz. **30**, 18-21 (1979) [Engl. transl.: JETP Lett. **30**, 15-18 (1979)]

1.18 J.J. Amodei: Appl. Phys. Lett. **18**, 22-24 (1971)

1.19 J.J. Amodei: RCA Rev. **32**, 185-198 (1971)

1.20 N.V. Kuchtarev, V.B. Markov, S.G. Odulov, M.S. Soskin, V.L. Vinetskii: Ferro-electrics **22**, 949-960 (1979)

1.21 M.G. Moharam, T.K. Gaylord, R. Magnusson, L. Young: J. Appl. Phys. **50**, 5642-5651 (1971)

1.22 Ch. Kittel: *Introduction to Solid State Physics* (Wiley, New York 1956)

1.23 A. Yariv, P. Yeh: *Optical Waves in Crystals* (Wiley, New York 1983)

1.24 A.S. Sonin, A.S. Vasilevskaya: *Electro-optic Crystals* (Atomizdat, Moscow 1971)

1.25 V.V. Bryksin, L.I. Korovin, M.P. Petrov: Zh. Tekh. Fiz. **58**, 1641-1648 (1988) [Engl. transl.: Sov. Phys. - Tech. Phys. **33**, 995-999 (1988)]

Chapter 2

2.1   D. Gabor: Nature **161**, 777-778 (1948)
2.2   E.N. Leith, J. Upatniecs: J. Opt. Soc. Am. **52**, 1123-1130 (1962)
2.3   Yu.N. Denisyuk: Dokl. Akad. Nauk SSSR, **144**, 1275-1278 (1962) [Engl. transl.:
      Sov. Phys.-Dokl. 7, 543-545 (1962)]; Optika i Spektrosk. 15, 522-532 (1963)
      [Engl. transl.: Optics and Spectrosc. 15, 279-286 (1963)]
2.4   W.R. Klein: Proc. IEEE **54**, 803 (1966)
2.5   R.J. Collier, C.B. Burckhardt, L.H. Lin: *Optical Holography* (Academic, New
      York 1971)
2.6   H. Kogelnik: Bell Syst. Techn. J. **48**, 2909 (1969)
2.7   M.P. Petrov, S.I. Stepanov, A.V. Khomenko: *Photosensitive Electro-optic Media
      in Holography and Optical Information Processing* (Nauka, Leningrad 1983)
2.8   D. Cassasent (ed.) *Optical Data Processing Applications*, Topics Appl. Phys.,
      Vol.23 (Springer, Berlin, Heidelberg 1978)
2.9   J.W. Goodman: *Introduction to Fourier Optics* (McGraw-Hill, New York 1968)
2.10 J.D. Gaskill: *Linear Systems, Fourier Transforms and Optics* (Wiley, New York
      1978)
2.11 A. Vander Lugt: IEEE Trans. **IT-10**, 139-145 (1964)
2.12 G.I. Vasilenko: *Holographic Pattern Recognition* (Sov. Radio, Moscow 1977)
2.13 H.J. Nussbaumer: *Fast Fourier Transform and Convolution Algorithms*, 2nd edn.,
      Springer Ser. Inf. Sci., Vol.2 (Springer, Berlin, Heidelberg 1982)

Chapter 3

3.1   M.P. Petrov: Introduction to optical signal processing with photorefractive mat-
      erials. *Electro-optic and Photorefractive Materials*, ed. by P. Günter, Springer
      Proc. Phys. **18**, 284-290 (Springer, Berlin, Heidelberg 1987)
3.2   V.V. Bryksin, L.I. Korovin, M.P. Petrov, A.V. Khomenko: Zh. Tekh. Fiz. 57,
      1918-1924 (1987) [Engl. transl.: Sov. Phys. Tech. Phys. 32, 786-790 (1987)]
3.3   P.B. Fellgett, E.K. Linfoot: Phil. Trans. Roy. Soc. (London) A 247, 369-407
      (1955)
3.4   R.J. Collier, C.B. Burchardt, L.H. Lin: *Optical Holography* (Academic, New
      York 1971)
3.5   P. Günter, H.J. Eichler: Introduction to photorefractive materials. *Electro-optic
      and Photorefractive Materials*, ed. by P. Günter, Springer Proc. Phys. 18,
      206-228 (Springer, Berlin, Heidelberg 1987)
3.6   S.K. Wemple, M. DiDomenico Jr.: In *Applied Solid State Science*, ed. by R.
      Wolf (Academic, New York 1972) pp.263-383

Chapter 4

4.1   J.J. Amodei: Appl. Phys. Lett. **18**, 22-24 (1971); RCA Rev. **32**, 185-198 (1971)
4.2   M.F. Deigen, S.G. Odulov, M.S. Soskin, B.D. Shanina: Fiz. Tv. Tela 16,
      1895-1902 (1974) [Engl. transl.: Sov. Phys. - Sol. State 16, 1237-1241 (1975)]
4.3   G.A. Alphonse, R.C. Alig, D.L. Staebler, W. Phillips: RCA Rev. 36, 213-229
      (1975)
4.4   M. Peltier, F. Micheron: J. Appl. Phys. **48**, 3683-3690 (1977)
4.5   M.G. Moharam, T.K. Gaylord, R. Magnusson, L. Young: J. Appl. Phys. 50,
      5642-5651 (1979)
4.6   N.V. Kukhtarev, V.B. Markov, S.G. Odulov, M.S. Soskin, V.L. Vinetskii: Ferro-
      electrics 22, 949-960 (1979)

4.7   P. Günter: Phys. Reports **93**, 199-299 (1982)
4.8   Ph. Refregier, L. Solymar, H. Rajbenbach, J.P. Huignard: J. Appl. Phys. **58**, 45-57 (1985)
4.9   F.S. Chen: J. Appl. Phys. **40**, 3389-3396 (1969)
4.10 W.D. Cornish, M.G. Moharam, L. Young: J. Appl. Phys. **47**, 1479-1484 (1976)
4.11 A.A. Kamshilin, M.G. Miteva, M.P. Petrov: Pis'ma v Zh. Tech. Fiz. **7**, 251-255 (1981) [Engl. transl.: Sov. Tech. Phys. Lett. **7**, 108-109 (1981)]
4.12 M.P. Petrov, S.I. Stepanov, A.V. Khomenko: *Photosensitive Electro-optic Media in Holography and Optical Information Processing* (Nauka, Leningrad 1983)
4.13 R.H. Bube: *Photoconductivity of Solids* (Wiley, New York 1960)
4.14 S.M. Ryvkin: *Photoelectric Effects in Semiconductors* (Consultants Bureau, New York 1964)
4.15 J.I. Pankove: *Optical Processes in Semiconductors* (Prentice-Hall, Inc. Englewood Cliffs, New Jersey 1971)
4.16 A.M. Glass, D. von der Linde, T.J. Negran: Appl. Phys. Lett. **25**, 233-235 (1974)
      M.E. Lines, A.M. Glass: *Principles and Applications of Ferroelectrics and Related Materials* (Clarendon, Oxford 1977)
      V.I. Belinicher, B.I. Sturman: Usp. Fiz. Nauk **130**, 415-458 (1980) [Engl. transl.: Sov. Phys.-Usp. **23**, 199-223 (1980)]
4.17 S.I. Stepanov, G.S. Trofimov: Zh. Tech. Fiz. **55**, 559-566 (1985) [Engl. transl.: Sov. Phys.-Tech. Phys. **30**, 331-334 (1985)]
4.18 R.J. Collier, C.B. Burckhardt, L.H. Lin: *Optical Holography* (Academic, New York 1971)
4.19 A. Marrakchi, J.P. Huignard, P. Günter: Appl. Phys. **24**, 131-138 (1981)
4.20 S.H. Wemple, M. Di Domenico, Jr.: Electro-optic and Nonlinear Optical Properties of Crystals, in *Appl. Sol. St. Science*, ed. by R. Wolf (Academic, New York 1972) pp.264-383
      A.M. Glass: Holographic Storage, in *Photonics*, ed. by M. Balkanski, P. Lallemand (Gauthier-Villars, Paris 1973) pp.163-192
4.21 J.P. Huignard, F. Micheron: Appl. Phys. Lett. **29**, 591-593 (1976)
4.22 J.P. Huignard, J.P. Herriau, G. Rivet, P. Günter: Opt. Lett. **5**, 102-104 (1980)
4.23 R. Orlowski, E. Krätzig: Solid State Commun. **27**, 1351-1354 (1978)
4.24 V.L. Vinetskii, N.V. Kukhtarev: Pis'ma v Zh. Tech. Fiz **1**, 176-181 (1975) [Engl. transl.: Sov. Tech. Phys. Lett. **1**, 84-86 (1975)]
4.25 M.B. Klein, G.C. Valley: J. Appl. Phys. **57**, 4901-4905 (1985)
4.26 S. Ducharme, J. Feinberg: J. Opt. Soc. Am. **B3**, 283-292 (1986)
4.27 S.I. Stepanov: Zh. Tech. Fiz. **52**, 2114-2116 (1982) [Engl. transl.: Sov. Phys.-Tech. Phys. **27**, 1300-1301 (1982)]
4.28 G.C. Valley: J. Appl. Phys. **59**, 3363-3366 (1986)
4.29 C. Medrano, E. Voit, P. Amrhein, P. Günter: J. Appl. Phys. **64**, 4668-4673 (1988)
4.30 F.P. Strohkendle, P. Tayebati, R.W. Hellwarth: In Digest, Topical Meeting on Photorefractive Materials (Los Angeles, CA 1987), pp.32-34
4.31 S.I. Stepanov, V.V. Kulikov, M.P. Petrov: Pis'ma v Zh. Tech. Fiz. **8**, 527-531 (1982) [Engl. transl.: Sov. Tech. Phys. Lett. **8**, 229-230 (1982)]
      S.I. Stepanov, V.V. Kulikov, M.P. Petrov: Opt. Commun. **44**, 19-23 (1982)
4.32 R.F. Kazarinov, R.A. Suris, B.I. Fuks: Fiz. i Tech. Polupr. **7**, 149-158 (1973) Engl. transl.: Sov. Phys.-Semiconductors **7**, 102-107 (1973)
4.33 J.P. Huignard, A. Marrakchi: Opt. Commun. **38**, 249-254 (1981)
4.34 N.V. Kukhtarev: Pis'ma v Zh. Tech. Fiz. **2**, 1114-1119 (1976) [Engl. transl.: Sov. Tech. Phys. Lett. **2**, 438-440 (1976)]
4.35 S.I. Stepanov, M.P. Petrov: Nonstationary Holographic Recording for Efficient Amplification and Phase Conjugation, in *Photorefractive Materials and Applica-*

tions I, ed. by P. Günter, J.P. Huignard, Topics Appl. Phys., Vol.61 (Springer, Berlin, Heidelberg 1987) pp.263-289

4.36   G.C. Valley: J. Opt. Soc. Am. B1, 868-873 (1984)
4.37   G. Hamel de Montchenault, B. Loiseaux, J.P. Huignard: Electronics Lett. 22, 1030-1032 (1986)
4.38   M.C. Jonathan, R.W. Hellwarth, G. Roosen: IEEE J. QE-22, 1936-1941 (1986)
4.39   H. Rajbenbach, J.P. Huignard, B. Loiseaux: Opt. Commun. 48, 247-252 (1983)
4.40   J. Kumar, G. Albanese, W. Steier: Opt. Commun. 63, 191-193 (1987)
       B. Imbert, H. Rajbenbach, S. Mallick, J.P. Herriau, J.P. Huignard: Opt. Lett. 13, 327-329 (1988)
4.41   A.A. Kumshilin, M.G. Miteva, V.V. Kulikov, S.I. Stepanov: Patent of Bulgaria N 58343 (1982)
4.42   S.I. Stepanov, M.P. Petrov: Pis'ma v Zh. Tech. Fiz. 10, 1356-1360 (1984) [Engl. transl.: Sov. Tech. Phys. Lett. 10, 572-573 (1984)]
       S.I. Stepanov, M.P. Petrov: Opt. Commun. 53, 292-295 (1985)
4.43   S.L. Sochava, S.I. Stepanov: Pis'ma v Zh. Tech. Fiz. 15, 34-39 (1989) [Engl. transl.: Sov. Tech. Phys. Lett. 15, 594-595 (1989)]
4.44   J. Kumar, G. Albanese, W.H. Steier, M. Ziari: Opt. Lett. 12, 120-122 (1987)
       K. Walsh, T.J. Hall: Electron. Lett. 24, 477-478 (1988)
       M.B. Klein, S.W. McCahon, T.F. Boggess, G.C. Valley: J. Opt. Soc. of Am. B5, 2467-2472 (1988)
4.45   V.V. Bryksin, L.I. Korovin: Fiz. Tverd. Tela 24, 2030-2036 (1982) [Engl. transl.: Sov. Phys.-Solid State 24, 1159-1162 (1982)]
4.46   V.V. Bryksin, L.I. Korovin, M.P. Petrov: Zh. Tech. Fiz. 54, 1504-1511 (1984) [Engl. transl.: Sov. Phys.-Tech. Phys. 29, 878-882 (1984)]
4.47   V.V. Bryksin, L.I. Korovin, V.I. Marakhonov: Zh. Tech. Fiz. 53, 1133-1138 (1983) [Engl. transl.: Sov. Phys.-Tech. Phys. 28, 686-689 (1983)]
4.48   A.N. Astratov, A.V. Il'inskii: Fiz. Tverd. Tela 24, 108-115 (1982) [Engl. transl.: Sov. Phys.-Solid State 24, 61-64 (1982)]
4.49   V.V. Bryksin, L.I. Korovin, V.I. Marakhonov, A.V. Khomenko: Fiz. Tverd. Tela 24, 2978-2984 (1982) [Engl. transl.: Sov. Phys.-Solid State 24, 1686-1689 (1982)]
4.50   V.V. Bryksin, L.I. Korovin: Fiz. Tverd. Tela 25, 55-61 (1983) [Engl. transl.: Sov. Phys.-Solid State 25, 30-33 (1983)]
4.51   V.V. Bryksin, L.I. Korovin: Fiz. Tverd. Tela 26, 2456-2467 (1984) [Engl. transl.: Sov. Phys.-Solid State 26, 1487-1493 (1984)]; ibid. 26, 3415-3425 (1984) [Engl. transl.: Sov. Phys.-Solid State 26, 2051-2057 (1984)]; ibid. 26, 3651-3657 (1984) [Engl. transl.: Sov. Phys.-Solid State 26, 2195-2199 (1984)]
4.52   V.V. Bryksin, L.I. Korovin: Fiz. Tverd. Tela 25, 2346-2353 (1983) Engl. transl.: Sov. Phys.-Solid State 25, 1346-1351 (1983)
4.53   M.G. Shlyagin, A.V. Khomenko, V.V. Bryksin, L.I. Korovin, M.P. Petrov: Zh. Tekh. Fiz. 55, 119-126 (1985) [Engl. transl.: Sov. Phys. Tech. Phys. 30, 68-72 (1985)]
4.54   V.N. Astratov, A.V. Il'inskii, V.A. Kiselev, M.B. Mel'nikov: Fiz. Tverd. Tela 25, 2755-2758 (1983) [Engl. transl.: Sov. Phys.-Solid State 25, 1585-1587 (1983)]
4.55   V.N. Astratov, A.V. Il'inskii, V.A. Kiselev: Fiz. Tverd. Tela 26, 2843-2851 (1984) [Engl. transl.: Sov. Phys.-Solid State 26, 1720-1725 (1984)]
4.56   V.V. Bryksin, L.I. Korovin, V.I. Marakhonov, A.V. Khomenko: Pis'ma v Zh. Tekn. Fiz. 9, 385-390 (1983) [Engl. transl.: Sov. Tech. Phys. Lett. 9, 165-167 (1983)]
4.57   V.V. Bryksin, L.I. Korovin, Yu.I. Kuz'min: Fiz. Tverd. Tela 28, 2728-2736 (1986) [Engl. transl.: Sov. Phys.-Solid State 28, 1528-1533 (1986)]
4.58   G.C. Valley: IEEE J. QE-19, 1637-1645 (1983)
       G. Le Saux, A. Brun: IEEE J. of Quantum Electron. QE-23, 1680-1688 (1987)
       G.C. Valley, A.L. Smirl: IEEE J. of Quantum Electron. QE-24, 304-310 (1988)

4.59 S.I. Stepanov, V.V. Kulikov: Zh. Tech. Fiz. **53**, 2255-2257 (1983) [Engl. transl.: Sov. Phys.-Tech. Phys. **28**, 1384-1386 (1983)]
4.60 A.A. Kamshilin, M.G. Miteva: Opt. Commun. **36**, 429-433 (1981)
4.61 J.J. Amodei, D.L. Staebler: Appl. Phys. Lett. **18**, 540-542 (1971)
      D.L. Staebler, J.J. Amodei: Ferroelectrics **3**, 107-113 (1972)
4.62 V.V. Kulikov, S.I. Stepanov: Fiz. Tverd. Tela **21**, 3204-3208 (1979) [Engl. transl.: Sov. Phys.-Solid State **21**, 1849-1851 (1979)]
      V.V. Kulikov, M.P. Petrov, S.I. Stepanov: Avtometriya (USSR) No.1, 39-45 (1980)
4.63 W.D. Johnston, Jr.: Appl. Phys. **41**, 3279-3285 (1970)
4.64 A.P. Levanyuk, V.V. Osipov: Izv. Akad. Nauk SSSR, Ser. fiz. **41**, 752-770 (1977) Engl. transl.: Bull. Acad. Sci. USSR, Phys. Ser. **41**, N4, 83-98 (1977)
      A.P. Levanyuk, V.V. Osipov, A.S. Sigov: Ferroelectrics **18**, 147-151 (1978)
4.65 F. Micheron: Ferroelectrics **18**, 153-159 (1978)
4.66 J. Feinberg, D. Heiman, A.R. Tanguay, Jr., R.W. Hellwarth: J. Appl. Phys. **51**, 1297-1305 (1981)
4.67 A.R. Pogosyan, E.M. Uyukin, A.P. Levanyuk: Fiz. Tverd. Tela **22**, 3725-3727 (1980) [Engl. transl.: Sov. Phys.-Solid State **22**, 2182-2183 (1980)]
4.68 V.L. Vinetskii, N.V. Kukhtarev: Fiz. Tverd. Tela **16**, 3714-3716 (1974) [Engl. transl.: Sov. Phys.-Solid State **16**, 2414-2415 (1975)]
4.69 A. Krumins, P. Günter: Phys. Status Solidi A**63**, K111-K114 (1981)
4.70 G.S. Trofimov, S.I. Stepanov: Fiz. Tverd. Tela **30**, 919-921 (1988) [Engl. transl. Sov. Phys. Solid State **30**, 534-535 (1988)]
4.71 G.S. Trofimov, S.I. Stepanov: Fiz. Tverd. Tela **28**, 2785-2789 (1986) [Engl. transl.: Sov. Phys.-Solid State **28**, 1559-1562 (1986)]
      M.P. Petrov, S.I. Stepanov, G.S. Trofimov: Pis'ma v Zh. Tech. Fiz. **12**, 916-921 (1986) [Engl. transl.: Sov. Tech. Phys. Lett. **12**, 379-381 (1986)]
4.72 B.I. Sturman: Kvantovaya Elektron. **7**, 483-488 (1980) [Engl. transl.: Sov. J. Quantum Electron. **10**, 267-271 (1980)]
4.73 S.G. Odoulov: Pis'ma v Zh. Eksp. i Teor. Fiz. **35**, 10-12 (1982) [Engl. transl. JETP Lett. **35**, 10-12 (1982)]
4.74 S.G. Odoulov, O.I. Oleinik: Kvantovaya Elektron. **10**, 1498-1501 (1983) [Engl. transl. Sov. J. Quantum Electron. **13**, 980-982 (1983)]
4.75 I.N. Kiseleva, V.V. Obukhovski, S.G. Odoulov: Fiz. Tverd. Tela **28**, 2975-2980 (1986) [Engl. transl.: Sov. Phys. Solid State **28**, 1673-1676 (1986)]
4.76 A. Novikov, S. Odoulov, O. Oleinik, B. Sturman: Ferroelectrics **75**, 295-315 (1987)

Chapter 5

5.1 R.J. Collier, C.B. Burckhardt, L.H. Lin: *Optical Holography* (Academic, New York 1971)
5.2 J.M. Cowley: *Diffraction Physics* (North-Holland, Amsterdam 1975)
5.3 L. Solymar, D.J. Cooke: *Volume Holography and Volume Gratings* (Academic, London 1981)
5.4 M.P. Petrov, S.I. Stepanov, A.V. Khomenko: *Photosensitive Electro-optic Media in Holography and Optical Information Processing* (Nauka, Leningrad 1983)
5.5 L.D. Landau, E.M. Lifshitz: *Electrodynamic of Continuous Media* (Addison-Wesley, Reading, Mass., 1965).
5.6 Yu.N. Denisyuk: Dokl. Akad. Nauk SSSR **144**, 1275-1278 (1962) [Engl. transl.: Sov. Phys.-Dokl. **7**, 543-545 (1962)]; Optica i Spectrosk. **15**, 522-532 (1963) [Engl. transl.: Optics and Spectrosc. **15**, 279-286 (1963)]; Optica i Spektrosk. **18**, 275-283 (1965) [Engl. transl.: Optics and Spectrosc. **18**, 152-157 (1965)]

5.7 E.N. Leith, A. Kozma, J. Upatnieks, J. Marks, N. Massey: Appl. Opt. 5, 1303-1312 (1966)

5.8 V.V. Aristov, V.Sh. Shekhtman: Usp. Fiz. Nauk 104, 51-76 (1971) [Engl. transl.: Sov. Phys.-Usp. 14, 263-277 (1971)]

5.9 J.A. Ratcliffe: Reps. on Prog. in Phys. 19, 188-267 (1956)

5.10 H. Kogelnik: Bell Syst. Tech. J. 48, 2909-2947 (1969)

5.11 C. Elashi: Proc. IEEE 64, 1666-1698 (1976)

5.12 T.K. Gaylord, M.G. Moharam: Proc. IEEE 73, 894-937 (1985)

5.13 M. Born, E. Wolf: *Principles of Optics* (Pergamon, New York 1965)

5.14 A. Korpel: IEEE Proc. 69, 48-53 (1981)

5.15 A. Yariv, P. Yeh: *Optical Waves in Crystals* (Wiley, New York 1984)

5.16 S.I. Stepanov, M.P. Petrov, A.A. Kamshilin: Piz'ma v Zh. Tech. Fiz. 3, 849-854 (1977) [Engl. transl.: Sov. Tech. Phys. Lett. 3, 345-346 (1977)]

5.17 M.P. Petrov, S.I. Stepanov, A.A. Kamshilin: Opt. and Laser Techol. 149-151 June (1979)

5.18 M.P. Petrov, S.I. Stepanov, A.A. Kamshilin: Opt. Commun. 29, 44-48 (1979)

5.19 M.P. Petrov, T.G. Pencheva, S.I. Stepanov: J. Optics 12, 287-292 (1981)

5.20 T.G. Pencheva, M.P. Petrov, S.I. Stepanov: Opt. Commun. 40, 175-178 (1981)

5.21 S.V. Miridonov, M.P. Petrov, S.I. Stepanov: Piz'ma v Zh. Tech. Fiz. 4, 976-980 (1978) [Engl. transl.: Sov. Tech. Phys. Lett. 4, 393-394 (1978)]
M.P. Petrov, S.V. Miridonov, S.I. Stepanov, V.V. Kulikov: Opt. Commun. 31, 301-305 (1979)

5.22 N.V. Kukhtarev, S.G. Odulov: Piz'ma v Zh. Tech. Fiz. 6, 1176-1179 (1980) [Engl. transl.: Sov. Tech. Phys. Lett. 6, 503-505 (1980)]
L. Arizmendi, R.C. Powell: J. Appl. Phys. 61, 2128-2131 (1987)

5.23 N.V. Kukhtarev, E. Krätzig, M.C. Külich, R.A. Rupp, J. Albers: Appl. Phys. B35, 17-21 (1984)

5.24 E. Voit, P. Günter: Opt. Lett. 12, 769-771 (1987)
H.C. Külich, R.A. Rupp, H. Hesse, E. Krätzig: Opt. & Quantum Electron. 19, 93-107 (1987)

5.25 J.F. Nye: *Physical Properties of Crystals* (Clarendon, Oxford 1957)

5.26 J. Feinberg: Optical phase conjugation in photorefractive materials, in *Optical Phase Conjugation*, ed. by R.A. Fisher (Academic, New York 1983) pp.417-444

5.27 A.A. Izvanov, A.E. Mandel', N.D. Khat'kov, S.M. Shandarov: Avtometriya No.2, 79-84 (1986) [Engl. transl.: Autom. Monit. & Meas. No.2, 80-84 (1986)]
A. Mandel, N. Khat'kov, S. Shandarov: Ferroelectrics 83, 215-220 (1988)

5.28 T.G. Pencheva, M.P. Petrov, S.I. Stepanov: Avtometriya No.1, 122-126 (1980)

5.29 J. Feinberg, R.W. Hellwarth: Opt. Lett. 5, 519-521 (1980)

5.30 A. Marrakchi, R.V. Johnson, A.R. Tanguay, Jr.: J. Opt. Soc. Am. B3, 321-336 (1986)
A.G. Apostolidis: Polarization Properties of Phase Volume Gratings Recorded in a $Bi_{12}SiO_{20}$ Crystal for two Transverse Configurations, in *Electro-optic and Photorefractive Materials*, ed. by P. Günter, Springer Proc. Phys., Vol.18 (Springer, Berlin, Heidelberg 1987) pp.324-338
G. Pauliat, G. Roosen: Ferroelectrics 75, 281-294 (1987)

5.31 J.P. Huignard, F.M. Micheron: Appl. Phys. Lett. 29, 591-593 (1976)

5.32 S.I. Stepanov, S.M. Shandarov, N.D. Khat'kov: Fiz. Tverd. Tela 29, 3054-3058 (1987) [Engl. transl.: Sov. Phys.-Solid State 29, 1754-1756 (1987)]
S.I. Stepanov, M.P. Petrov, S.L. Sochava: Ferroelectrics 92, 199-204 (1989)

5.33 A. Marrakchi, J.P. Huignard, P. Günter: Appl. Phys. 24, 131-138 (1981)

5.34 J.P. Herriau, J.P. Huignard, P. Aubourg: Appl. Opt. 17, 1851-1852 (1978)

5.35 T.G. Pencheva, S.I. Stepanov: Fiz. Tverd. Tela 24, 1214-1216 (1982) [Engl. transl.: Sov. Phys.-Solid State 24, 687-688 (1982)≤

5.36 A. Partovi, E.M. Garmire, L.-J. Cheng: Appl. Phys. Lett. 51, 299-301 (1987)

5.37 E. Voit, C. Zaldo, P. Günter: Opt. Lett. 11, 309-311 (1986)

Chapter 6

6.1   F.S. Chen, J.T. La Macchia, D.B. Fraser: Appl. Phys. Lett. **13**, 223-225 (1968)
6.2   R.L. Townsend, J.T. La Macchia: J. Appl. Phys. **41**, 5188-5192 (1970)
6.3   T.K. Gaylord, T.A. Rabson, F.K. Tittel: Appl. Phys. Lett. **20**, 47-49 (1972)
6.4   R. Chiao, P.L. Kelly, E. Garmire: Phys. Rev. Lett. **17**, 1158-1161 (1966)
6.5   D.L. Staebler, J.J. Amodei: J. Appl. Phys. **43**, 1042-1049 (1972)
6.6   T.K. Gaylord, T.A. Rabson, F.K. Tittel, C.R. Quick: J. Appl. Phys. **44**, 896-897 (1973)
6.7   V. Kondilenko, V. Markov, S. Odulov, M. Soskin: Optica Acta **26**, 239-251 (1979)
6.8   Y. Ninomiya: J. Opt. Soc. Am. **63**, 1124-1130 (1973)
6.9   D.I. Stasel'ko, V.G. Sidorovich: Zh. Tech. Fiz. **44**, 580-587 (1974) [Engl. transl.: Sov. Phys.-Tech. Phys. **19**, 361-365 (1974)]
6.10   D.W. Vahey: J. Appl. Phys. **46**, 3510-3515 (1975)
6.11   B.I. Sturman: Zh. Tech. Fiz. **48**, 1010-1020 (1978) [Engl. transl.: Sov. Phys. Tech. Phys. **23**, 589-595 (1978)]
6.12   V.L. Vinetskii, N.V. Kukhtarev, S.G. Odulov, M.S. Soskin: Usp. Fiz. Nauk **129**, 113-137 (1979) [Engl. transl.: Sov. Phys.-Usp. **22**, 742 (1979)]
6.13   N.V. Kukhtarev, V.B. Markov, S.G. Odulov, M.S. Soskin, V.L. Vinetskii: Ferroelectrics **22**, 949-962 (1979)
6.14   R.J. Collier, C.B. Burckhardt, L.H. Lin: *Optical Holography* (Academic, New York 1971)
6.15   Yu.A. Anan'ev: Kvantovaya Elektron. **1**, No.7, 1669-1672 (1974) [Engl. transl.: Sov. J. Quantum. Electron. **4**, No.7, 929-931 (1975)]
6.16   J.P. Huignard, A. Marrakchi: Opt. Commun. **38**, 249-254 (1981)
6.17   N. Kukhtarev, V. Markov, S. Odulov: Opt. Commun. **23**, 338-343 (1977)
6.18   V.P. Kondilenko, V.B. Markov, S.G. Odulov, M.S. Soskin: Ukr. Fiz. Zh. **23**, 2039-2043 (1978)
6.19   N.V. Kukhtarev, V.B. Markov, S.G. Odulov: Zh. Tech. Fiz. **50**, 1905-1914 (1980) [Engl. transl.: Sov. Phys. Tech. Phys. **25**, 1109-1114 (1980)]
6.20   L. Solymar, J.M. Heaton: Opt. Commun. **51**, 76-78 (1984)
6.21   A. Yariv: *Quantum Electronics* (Wiley, New York 1975)
       Handbook of Lasers with Selected Data on Optical Technology, ed. by R.J. Pressley (Chemical Rubber Co., Cleveland 1971)
6.22   R.W. Hellwarth: J. Opt. Soc. Am. **67**, 1-3 (1977)
6.23   A. Yariv, D.M. Pepper: Opt. Lett. **1**, 16-18 (1977)
6.24   H. Kogelnik: Bell Syst. Techn. J. **44**, 2451-2455 (1965)
6.25   H. Kogelnik, K.S. Pennington: J. Opt. Soc. Am. **58**, 273-274 (1966)
6.26   E.N. Leith, J. Upatnieks: J. Opt. Soc. Am. **56**, 523 (1966)
6.27   M. Cronin-Golomb, J.O. White, B. Fisher, A. Yariv: Opt. Lett. **7**, 313-315 (1982)
6.28   M. Cronin-Golomb, B. Fisher, J.O. White, A. Yariv: IEEE J. QE-20, 12-30 (1984)
6.29   B. Fisher, M. Cronin-Golomb, J.O. White, A. Yariv: Opt. Lett. **6**, 519-521 (1981)
6.30   J.F. Lam: Appl. Phys. Lett. **42**, 155-157 (1983)
       K.R. MacDonald, J. Feinberg: Phys. Rev. Lett. **55**, 821-824 (1985)
6.31   J. Goltz, F. Laeri, T. Tschudi: Opt. Commun. **64**, 63-66 (1987)
6.32   S.I. Stepanov, M.P. Petrov, M.V. Krasin'kova: Zh. Tech. Fiz. **54**, 1223-1225 (1984) [Engl. transl.: Sov. Phys.-Tech. Phys. **29**, 703-705 (1984)]
       S.I. Stepanov, M.P. Petrov: Optica Acta **31**, 1335-1343 (1984)
6.33   S.I. Stepanov, M.P. Petrov: Opt. Commun. **53**, 64-68 (1985)
6.34   B. Fisher, Sh. Weiss: Appl. Phys. Lett. **53**, 257-259 (1988)

A. Erdmann, R. Kowarschik: IEEE J. Quantum Electron. **24**, 155-160 (1988)

A. Bledowski, W. Krolikowski: IEEE J. Quantum Electron. **24**, 652-659 (1988)

6.35 S.I. Stepanov, M.P. Petrov: Nonstationary holographic recording for efficient amplification and phase conjugation, in *Photorefractive Materials and Applications I*, ed. by P. Günter, J.P. Huignard (Springer, Berlin, Heidelberg 1987) pp.263-289

S.I. Stepanov, M.P. Petrov: Piz'ma v Zh. Tech. Fiz. **10**, 1356-1360 (1984) [Engl. transl.: Sov. Tech. Phys. Lett. **10**, 572-573 (1984)]

6.36 H. Rajbenbach, B. Imbert, J.P. Huignard, S. Mallick: Opt. Lett. **14**, 78-80 (1989)

6.37 H. Kong, C. Lin, A.M. Birnacki, M. Cronin-Golomb: Opt. Lett. **13**, 324-326 (1988)

6.38 J. Feinberg, R.W. Hellwarth: Opt. Lett. **5**, 519-521 (1980)

6.39 J.O. White, M. Cronin-Golomb, B. Fisher, A. Yariv: Appl. Phys. Lett. **40**, 450-452 (1982)

6.40 F. Laeri, T. Tschudi, J. Albers: Opt. Commun. **47**, 387-390 (1983)

6.41 J. Feinberg, G.D. Bacher: Opt. Lett. **9**, 420-422 (1984)

6.42 H. Rajbenbach, J.P. Huignard: Opt. Lett. **10**, 137-139 (1985)

J.P. Huignard, H. Rajbenbach, Ph. Refregier, L. Solymar: Opt. Engineering **24**, 586-592 (1985)

6.43 S.-K. Kwong, A. Yariv, M. Cronin-Golomb, I. Ury: Appl. Phys. Lett. **47**, 460-462 (1985)

S.-K. Kwong, M. Cronin-Golomb, A. Yariv: IEEE J. QE-22, 1508-1523 (1986)

6.44 W.E. Lamb, Jr.: Phys. Rev. **134**, A1429-A1450 (1964)

6.45 M. Cronin-Golomb, B. Fisher, J.O. White, A. Yariv: Appl. Phys. Lett. **41**, 689-691 (1982)

6.46 M. Cronin-Golomb, A. Yariv: Opt. Lett. **11**, 242-244 (1986)

6.47 S.G. Odulov, L.G. Sukhoverkhova: Kvantovaya Elektron. **11**, 575-581 (1984) [Engl. transl.: Sov. J. Quantum Electron. **14**, 390-393 (1984)]

6.48 S.G. Odulov, M.S. Soskin: Piz'ma v Zh. Eksp. i Teor. Fiz. **37**, 243-247 (1983) [Engl. transl.: JETP Lett. **37**, 289-293 (1983)]

6.49 M. Cronin-Golomb, B. Fisher, J. Nilsen, J.O. White, A. Yariv: Appl. Phys. Lett. **41**, 219-220 (1982)

6.50 M. Cronin-Golomb, B. Fisher, S.-K. Kwong, J.O. White, A. Yariv: Opt. Lett. **10**, 353-355 (1985)

6.51 J. Feinberg: Opt. Lett. **7**, 486-488 (1982)

K.R. MacDonald, J. Feinberg: J. Opt. Soc. Am. **73**, 548-554 (1983)

6.52 V.I. Odintsov, L.F. Rogacheva: Piz'ma v Zh. Eksp. i Teor. Fiz. **36**, 281-284 (1982) [Engl. transl.: JETP Lett. **36**, 344-347 (1982)]

6.53 M. Cronin-Golomb, B. Fisher, J.O. White, A. Yariv: Appl. Phys. Lett. **42**, 919-921 (1983)

6.54 I.M. Bel'dyugin, M.G. Galushkin, E.M. Zemskov: Kvantovaya Elektron. **11**, 887-893 (1984) [Engl. transl.: Sov. J. Quantum Electron. **14**, 602-605 (1984)]

6.55 B. Fisher, Sh. Sternklar: Appl. Phys. Lett. **47**, 1-3 (1985)

6.56 M. Cronin-Golomb, J. Paslaski, A. Yariv: Appl. Phys. Lett. **47**, 1131-1133 (1985)

6.57 O.M. Vokhnik, Yu.S. Kuz'minov, N.M. Polozkov: Kvantovaya Electron. **13**, 1633-1637 (1986) [Engl. transl.: Sov. J. Quantum Electron **16**, 1066-1069 (1986)]

6.58 S.L. Sochava, S.I. Stepanov, M.P. Petrov: Pis'ma v Zh. Tech. Fiz. **13**, 660-664 (1987) [Engl. transl.: Sov. Tech. Phys. Lett. **13**, 274-275 (1987)]

6.59 Sh. Sternklar, Sh. Weiss, M. Segev, B. Fisher: Opt. Lett. **11**, 528-530 (1986)

6.60 Sh. Weiss, Sh. Sternklar, B. Fisher: Opt. Lett. **12**, 114-116 (1987)

A.M.C. Smout, R.W. Eason: Opt. Lett. **12**, 498-500 (1987)

6.61 R.A. Rupp, J. Marotz, K. Ringhofer, S. Tveichel, S. Feng, E. Krätzig: IEEE J. Quantum Electron. QE-23, 2136-2141 (1987)

6.62 M. Segev, Sh. Weiss, B. Fisher: Appl. Phys. Lett. **50**, 1397-1399 (1987)
Sh. Weiss, M. Segev, B. Fisher: IEEE J. Quantum Electron. QE-24, 706-708 (1988)
6.63 Sh. Weiss, M. Segev, Sh. Sternklar, B. Fisher: Appl. Opt. **27**, 3422-3428 (1988)
6.64 M.P. Petrov, S.L. Sochava, S.I. Stepanov: Opt. Lett. **14**, 284-286 (1989)
6.65 M.D. Ewbank, P. Yeh, M. Khoshnevisan, J. Feinberg: Opt. Lett. **10**, 282-284 (1985)
P. Yeh, T.Y. Chang, M.D. Ewbank: J. Opt. Soc. of Am. B5, 1743-1749 (1988)
6.66 M. Cronin-Golomb, B. Fisher, S.-K. Kwong, J.O. White, A. Yariv: Opt. Lett. **10**, 353-355 (1985)
S.-K. Kwong, A. Yariv, M. Cronin-Golomb, B. Fisher: J. Opt. Soc. of Am. A3, 157-160 (1986)
6.67 A.D. Novikov, V.V. Obukhovski, S.G. Odoulov, B.I. Sturman: Piz'ma v Zh. Eksper. i Teor. Fiz. **44**, 418-421 (1986) [Engl. transl.: JETP Lett. **44**, 538-540 (1986)]
6.68 A. Novikov, S. Odoulov, O. Oleinik, B. Sturman: Ferroelectrics **74**, 295-315 (1987)
6.69 A. Novikov, V. Obukhovski, S. Odoulov, B. Sturman: Opt. Lett. **13**, 1017-1019 (1988)
6.70 T.Y. Chang, R.W. Hellwarth: Opt. Lett. **10**, 408-410 (1985)
6.71 A.V. Mamaev, V.V. Shkunov: Kvantovaya Elektron. **15**, 1317-1319 (1988); ibid. **16**, 1863-1869 (1989) [Engl. transl.: Sov. J. Quantum Electron. **18**, 829-830 (1988); ibid. **19**, 1199-1203 (1989)]
6.72 B. Ya. Zel'dovich, V.I. Popovichev, V.V. Ragulskii, F.S. Faizullov: Piz'ma v Zh. Eksper. **15**, 160-164 (1972) [Engl. transl.: JETP Lett. **15**, 109-111 (1972)]
6.73 G.C. Valley, G.J. Dunning: Opt. Lett. **9**, 513-515 (1984)
6.74 P. Günter, E. Voit, M.Z. Zha, A. Albers: Opt. Commun. **55**, 210-214 (1985)
6.75 J.P. Jiang, J. Feinberg: Opt. Lett. **12**, 266-268 (1987)
6.76 S.L. Sochava, S.I. Stepanov: Zh. Techn. Fiz. **58**, 1780-1079 (1988) [Engl. transl.: Sov. Phys. Tech. Phys. **33**, 1077-1079 (1988)]
6.77 P. Günter, J.P. Huignard (eds.): *Photrefractive Materials and their Applications II*, Topics Appl. Phys., Vol.62 (Springer, Berlin, Heidelberg 1989)
6.78 S.G. Odoulov, M.S. Soskin, A.I. Khyzhniak: *Optical Oscillators with OFWM (Dynamic Lasers)* (Harwood Academic, London 1990)

Chapter 7

7.1 V.V. Bryksin, L.I. Korovin: Fiz. Tverd. Tela **25**, 2346-2353 (1983) [Engl. transl.: Sov. Phys.-Solid State **25**, 1346-1351 (1983)]
7.2 M.G. Shlyagin, A.V. Khomenko, V.V. Bryksin, L.I. Korovin, M.P. Petrov: Zh. Tech. Fiz. **55**, 119-126 (1985) [Engl. transl.: Sov. Phys. Tech. Phys. **30**, 68-72 (1985)]
7.3 V.V. Bryksin, L.I. Korovin, V.I. Marakhonov, A.V. Khomenko: Pis'ma v Zh. Tekh. Fiz. **9**, 385-390 (1983) [Engl. transl.: Sov. Tech. Phys. Lett. **9**, 165-168 (1983)]
7.4 V.V. Bryksin, L.I. Korovin, V.I. Marakhonov: Zh. Tekh. Fiz- **53**, 1133-1138 (1983) [Engl. transl.: Sov. Phys. Tech. Phys. **28**, 686-689 (1983)]
7.5 A.V. Khomenko, M.P. Petrov, M.V. Krasin'kova: Pis'ma v Zh. Tekh. Fiz. **5**, 334-338 (1979) [Engl. transl.: Sov. Tech. Phys. Lett. **5**, 133-134 (1979)]
7.6 M.P. Petrov, A.V. Khomenko, M.V. Krasin'kova, V.I. Marakhonov, M.G. Shlyagin: Zh. Tekh. Fiz. **51**, 1422-1431 (1981) [Engl. transl.: Sov. Phys. Tech. Phys. **26**, 816-821 (1981)]
7.7 D.M. Shields, T.E. Luke: Opt. Commun. **55**, 391-392 (1985)

7.8 P.S. Theocaris, E.E. Gdoutos: *Matrix Theory of Photoelasticity*, Springer Ser. Opt. Sci., Vol.11 (Springer, Berlin, Heidelberg 1979)
7.9 W.J. Tabor, F.S. Chen: J. Apl. Phys. **40**, 2760-2765 (1969)
7.10 A. Yariv, P. Yeh: *Optical Waves in Crystals* (Wiley, New York 1983)
7.11 M.P. Petrov, A.V. Khomenko: Optik **67**, 247-256 (1984)
7.12 M.P. Petrov, S.I. Stepanov, A.V. Khomenko: *Photosensitive Electro-optic Media in Holography and Optical Information Processing* (Nauka, Leningrad 1983)
7.13 E.R. Mustel', V.N. Parygin: *Light Modulation and Deflection Techniques* (Nauka, Moscow 1970) (in Russian)
7.14 A.A. Berezhnoi, A.A. Bozhinskii, Yu.V. Popov: Zh. Tehn. Fiz. **54**, 1619-1622 (1984) [Engl. transl.: Sov. Phys.-Tech. Phys. **29**, 947-948 (1984)]
7.15 A.V. Khomenko, M.P. Petrov, M.V. Krasin'kova: Pis'ma v Zh. Tekh. Fiz. **5**, 334-338 (1979) [Engl. transl.: Sov. Tech. Phys. Lett. **5**, 133-134 (1979)]
7.16 W.R. Roach: IEEE Trans. ED-21, 453-459 (1974)
7.17 Y. Owechko, A.R. Tanguay, Jr.: J. Opt. Soc. Am. A1, 635-643 (1984)

Chapter 8

8.1 D.S. Oliver, P. Vohl, R.E. Aldrich, M.E. Behrndt, W.R. Buchan, R.C. Ellis, J.E. Genthe, J.R. Goff, S.L. Hou, G. McDaniel: Appl. Phys. Lett. **17**, 416-418 (1970)
8.2 D.S. Oliver, W.R. Buchan: IEEE Trans. ED-18, 769-773 (1971)
8.3 S.L. Hou, D.S. Oliver: Appl. Phys. Lett. **18**, 325-328 (1971)
8.4 J. Feinleib, D.S. Oliver: Appl. Opt. **11**, 2752-2759 (1972)
8.5 P. Vohl, P. Nisenson, D.S. Oliver: IEEE Trans. ED-20, 1032-1037 (1973)
8.6 S. Lipson, P. Nisenson: Appl. Opt. **13**, 2052-2060 (1974)
8.7 B.A. Horwitz, F.J. Corbett: Opt. Eng. **17**, 353-364 (1978)
8.8 Yu.N. Grehov, P.E. Kotl'ar, E.S. Nezhevenko, V.I. Fel'bush, N.I. Shadeev: Pis'ma v Zh. Tekh. Fiz. **2**, 457-461 (1976)
8.9 A.V. Khomenko, N.N. Kovalev, M.P. Petrov: Pis'ma v Zh. Tekh. Fiz. **2**, 1095-1098 (1976)
8.10 A.R. Tanguay, Jr.: Opt. Eng. **24**, 2-18 (1985)
8.11 A.A. Berezhnoi, V.Z. Gurevich, S.V. Morozov, Yu.V. Popov: Pis'ma v Zh. Tekh. Fiz. **2**, 198-200 (1976)
8.12 T. Minemoto, Y. Suemoto, S. Fujita: Jpn. J. Appl. Phys. **16**, 1683-1684 (1977)
8.13 T. Minemoto, S. Numata, K. Miyamoto: Appl. Optics **25**, 4046-4052 (1986)
8.14 A.S. Abrahams, C. Svensson, A.R. Tanguay, Jr.: Solid State Commun. **30**, 293-295 (1979)
8.15 M.P. Petrov, A.V. Khomenko, V.I. Berezkin, M.V. Krasin'kova: Mikroelektronika **8**, 20-23 (1979)
8.16 A.V. Khomenko, M.P. Petrov, M.V. Krasin'kova: Pis'ma v Zh. Tekh. Fiz. **5**, 334-338 (1979) [Engl. transl.: Sov. Tech. Phys. Lett. **5**, 133-134 (1979)]
8.17 D. Casasent, F. Caimi, A. Khomenko: Appl. Opt. **20**, 4215-4220 (1981)
8.18 D. Casasent: Appl. Opt. **18**, 2445-2453 (1979)
8.19 Y. Owechko, A.R. Tanguay, Jr.: SPIE **218**, 67-80 (1980)
8.20 R.A. Sprague: J. Appl. Phys. **46**, 1673-1678 (1975)
8.21 M.P. Petrov, A.V. Khomenko, M.V. Krasin'kova, V.I. Marachonov, M.G. Shlyagin: Ferroelectrics **28**, 407 (1980)
8.22 M.P. Petrov, A.V. Khomenko, M.G. Shlyagin: Applications of SLM PROM for optical information processing in real time, in *Proceedings of 10th Soviet Conference on Holography* (Leningrad 1978) pp.141-150
8.23 P. Nisenson, S. Iwasa: Appl. Opt. **11**, 2760-2767 (1972)
8.24 P. Nisenson, R.A. Sprague: Appl. Opt. **14**, 2602-2606 (1975)

8.25 S. Isawa: Appl. Opt. **15**, 1418-1424 (1976)

8.26 R.E. Brooks, R.F. Kemp: SPIE **218**, 119-125 (1980)

8.27 J.P. Benton, F. Corbett, T. Richard: SPIE **218**, 126-135 (1980)

8.28 J.C.H. Spencer, A. Olsen: SPIE **218**, 154-160 (1980)

8.29 T. Minemoto, K. Hata: Appl. Opt. **25**, 4065-4070 (1986)

8.30 Y. Imai, Y. Ohtsuka: Appl. Opt. **26**, 274-277 (1987)

8.31 M.P. Petrov, S.I. Stepanov, A.V. Khomenko: *Photosensitive Electro-optic Media in Holography and Optical Information Processing* (Nauka, Leningrad 1983)

8.32 M.P. Petrov, A.V. Khomenko, M.V. Krasin'kova, V.I. Marachonov, M.G. Shlyagin: Zh. Tekh. Fiz. **51**, 1422-1431 (1981) [Engl. transl.: Sov. Phys.-Tech. Phys. **26**, 816-821 (1981)]

8.33 M.P. Petrov, A.V. Khomenko: Optik **57**, 247-256 (1984)

8.34 D. Casasent, F. Caimi, M.P. Petrov, A.V. Khomenko: Appl. Opt. **21**, 3846-3854 (1982)

8.35 D.M. Shields, T.E. Luke: Opt. Commun. **55**, 391-392 (1985)

8.36 A.A. Berezhnoi, A.A. Bozhinskii, Yu.V. Popov: Optiko-Mechan. Prom (USSR), No.8, 24-27 (1985)

8.37 V.I. Fel'dbush: Avtometriya (USSR), No.6, 108-110 (1980)

8.38 M.P. Petrov, V.I. Marakhonov, V.I. Berezkin, M.V. Krasin'kova, A.V. Khomenko: Pis'ma v Zh. Tekh. Fiz. **11**, 260-263 (1985) [Engl. transl.: Sov. Tech. Phys. Lett. **11**, 106-107 (1985)]

8.39 V.V. Bryksin, L.I. Korovin, V.I. Marakhonov, M.P. Petrov, A.V. Khomenko: Pis'ma v Zh. Tekh. Fiz **9**, 1011-1015 (1983) [Engl. transl.: Sov. Techn. Phys. Lett. **9**, 434-436 (1983)]

8.40 M.P. Petrov, A.V. Khomenko: Fiz. Tverd. Tela **23**, 1350-1356 (1981) [Engl. transl.: Sov. Phys. - Solid State **23**, 789-792 (1981)]

8.41 A. Feldman, W.S. Brower, D. Horwitz: Appl. Phys. Lett. **15**, 201-202 (1970)

8.42 M.P. Petrov, A.V. Khomenko, V.I. Marakhonov: Pis'ma v Zh. Tekh. Fiz **9**, 193-197 (1983) [Engl. transl.: Sov. Tech. Phys. Lett. **9**, 85-86 (1983)]

8.43 M.P. Petrov, A.V. Khomenko, V.I. Marakhonov: Opt. Commun. **50**, 296-299 (1984)

8.44 L.M. Soroko: *Hilbert Optics* (Moscow 1981)

8.45 V.V. Bryksin, V.S. Voloshin, L.I. Korovin: Zh. Tekh. Fiz. **56**, 1040-1048 (1986) [Engl. transl.: Sov. Phys.-Tech. Phys. **31**, 609-614 (1986)]

8.46 V.I. Berezkin, A.V. Khomenko: Pis'ma v Zh. Tekh. Fiz. **6**, 1265-1268 (1980) [Engl. transl.: Sov. Tech. Phys. Lett. **6**, 542-543 (1980)]

8.47 V.V. Bryksin, L.I. Korovin, M.P. Petrov: Zh. Tekh. Fiz. **54**, 1504-1511 (1984) [Engl. transl.: Sov. Phys.-Tech. Phys. **29**, 878-882 (1984)]

8.48 M.G. Shlyagin, A.V. Khomenko: Zh. Tekh. Fiz. **57**, 2101-2104 (1987) [Engl. transl.: Sov. Phys. - Tech. Phys. **32**, 707-709 (1987)]

8.49 G. Maric: Ferroelectrics **18**, 9-14 (1976)

8.50 M.P. Petrov, A.V. Khomenko, V.I. Marakhonov, M.G. Shlyagin: Pis'ma v Zh. Tekh. Fiz. **6**, 385-388 (1980) [Engl. transl.: Sov. Tech. Phys. Lett. **6**, 165-166 (1980)]

8.51 A.A. Vasiliev, I.N. Kompanets, A.V. Parfenov: Optik **67**, 223-236 (1983)

8.52 D. Casasent, F. Caimi, A. Khomenko: Appl. Opt. **20**, 3090-3092 (1981)

8.53 A.M. Bliznetsov, V.V. Bryksin, L.I. Korovin, S.V. Miridonov, A.V. Khomenko: Zh. Tekh. Fiz. **57**, 1268-1275 (1987) [Engl. transl.: Sov. Phys. - Tech. Phys. **32**, 750-754 (1987)]

8.54 M.P. Petrov, M.G. Shlyagin, N.O. Shalaevskii, M.P. Petrov, A.V. Khomenko: Zh. Tekh. Fiz. **55**, 2247-2250 (1980) [Engl. transl.: Sov. Phys.-Tech. Phys. **30**, 1331-1332 (1980)]

8.55 V.N. Astratov, V.A. Il'inskii, M.B. Mel'nikov: Fiz. Tverd. Tela **25**, 2163-2168 (1983) [Engl. transl.: Sov. Phys. Solid State **25**, 1244-1247 (1983)]

8.56 V.I. Chmirev, V.M. Skorikov, M.I. Subotin: Neorganich. Mater. (USSR) 19, 269-273 (1983)

8.57 A.M. Bliznetsov, M.P. Petrov, A.V. Khomenko: Pis'ma v Zh. Tekh. Fiz. 10, 1094-1098 (1984) [Engl. transl.: Sov. Phys.-Tech. Lett. 10, 463-464 (1984)]

8.58 A.V. Khomenko, V.M. Petrov, M.G. Shlyagin, N.O. Shalaevskii: Proc. SPIE 1183, 309-316 (1990)

8.59 G. Matic: Ferroelectrics 10, 9-14 (1976)

8.60 D. Casasent: Opt. Eng. 17, 344-352 (1978)

8.61 G. Matic: Philips Pes. Repts. 22, 110-132 (1967)

8.62 M. Grenot, J. Pergrale, J. Donjon: Appl. Phys. Lett. 21, 83-85 (1972)

8.63 J. Donjon, F. Dument, M. Grenot: IEEE Trans. ED-20, 1037-1042 (1973)

8.64 D. Casasent: Opt. Eng. 17, 365-370 (1978)

8.65 D. Casasent: S. Natu, G. Lebreton, E. DeBazelaire: SPIE 202, 122-131 (1979)

8.66 D. Casasent: T.K. Luu: Appl. Opt. 18, 3307-3314 (1979)

8.67 A.S. Sonin, A.S. Vasilevskaya: *Electro-optical Crystals* (Atomizdat, Moscow 1971)

8.68 W.R. Roach: IEEE Trans. ED-21, 453-459 (1974)

8.69 V.G. Mal'shakov, S.K. Mankevich, A.I. Nagaev, V.N. Parygin, G.N. Stravrakov: Kvantovaya Electron. 6, 2393-2400 (1979) [Engl. transl.: Sov. J. Quantum Electron. 9, 1409-1412 (1979)]

8.70 M.M. Butusov, A.V. Ivanov, A.I. Kosarev, A.B. Krumin', A.R. Shternberg: Zh. Tekh. Fiz. 47, 2561-2565 (1977) [Engl. transl.: Sov. Phys.-Tech. Phys. 22, 1485-1487 (1977)]

8.71 K. Shinoda, Y. Suzuki: SPIE 613, 158-164 (1986)

8.72 A.V. Il'inskii, E.B. Shadrin: Pis'ma v Zh. Tekh. Fiz. 6, 520-523 (1980) [Engl. transl.: Sov. Tech. Phys. Lett. 6, 224-225 (1980)]

8.73 M.M. Butusov, A.Z. Dun, S.Yu. Merkin, R.Sh. Tukhvatulin: Zh. Tekh. Fiz. 51, 111-116 (1981) [Engl. transl.: Sov. Phys.-Tech. Phys. 26, 61-64 (1981)]

8.74 C. Warde, A.D. Fisher, D.M. Cocco, M.Y. Burmawi: Opt. Lett. 3, 196-198 (1978)

8.75 C. Warde, A.M. Weiss, A.D. Fesher: SPIE 218, 59-66 (1980)

8.76 T. Hara, K. Shinoda, T. Kato, M. Sugyiama, Y. Suzuki: Appl. Opt. 25, 2306-2310 (1986)

8.77 T. Hara, N. Mukohzaka, Y. Suzuki: SPIE 625, 30-34 (1986)

8.78 T. Hara, Y. Ooi, T. Kato, Y. Suzuki: SPIE 613, 153-157 (1986)

8.79 A.A. Kamshilin, M.P. Petrov: Pis'ma v Zh. Tekh. Fiz. 6, 337-341 (1980) [Engl. transl.: Sov. Tech. Phys. Lett. 6, 144-150 (1980)]

8.80 Y. Shi, D. Psaltis, A. Marrakchi, A.R. Tanguay, Jr.: Appl. Opt. 22, 3665 (1983)

8.81 A. Marrakchi, A.R. Tanguay, Jr., J.Yu., D. Psaltis: Opt. Eng. 24, 124-131 (1985)

8.82 A.A. Berezhnoi, A.A. Bozhinskii, Yu.V. Popov: Zh. Tekh. Fiz. 54, 1619-1622 (1984) [Engl. transl.: Sov. Phys. Tech. Phys. 29, 947-948 (1984)]; Opt. i Spektrosk. 60, 113-116 (1986) [Engl. transl.: Opt. and Spectrosc. 60, 93-96 (1986)]

8.83 A.A. Berezhnoi, A.A. Buzhinskii, Yu.V. Popov: Optiko-Mechanich. Prom. (USSR), No.3, 24-27 (1987)

8.84 V.M. Petrov, A.V. Khomenko, M.V. Krasin'kova: Zh. Tekh. Fiz. 58, 596-600 (1988) [Engl. transl.: Sov. Phys. - Tech. Phys. 33, 358-361 (1988)]

Chapter 9

9.1 R.J. Collier, C.B. Burckhardt, L.H. Lin: *Optical Holography* (Academic, New York 1971).

9.2 C.M. Vest: *Holographic Interferometry* (Wiley, New York 1979)

9.3 Yu.I. Ostrovsky, M.M. Butusov, G.V. Ostrovskaya: *Interferometry by Holography*, in Springer Ser. Opt. Sci., Vol.20 (Springer, Berlin, Heidelberg 1980)

9.4 G.B. Brandt: Holographic interferometry, in *Handbook of Optical Holography*, ed. by H.J. Caulfield (Academic, New York 1979) pp.463-502

9.5 Yu.I. Ostrovsky, V.P. Shchepinov, V.V. Yakovlev: *Holographic Interferometry in Experimental Mechanics*, Springer Ser. Opt. Sci., Vol.60 (Springer, Berlin, Heidelberg 1991)

9.6 J.P. Huignard, F. Micheron: Appl. Phys. Lett. **29**, 591-593 (1976)

9.7 J.P. Huignard, J.P. Herriau: Appl. Opt. **16**, 1807-1809 (1977)
J.P. Herriau, A. Marrakchi, J.P. Huignard: Rev. Tech. Thomson-CSF **13**, 501-520 (1981)

9.8 G.S. Trofimov, S.I. Stepanov: Pis'ma v Zh. Tech. Fiz. **11**, 615-621 (1985) [Engl. transl.: Sov. Tech. Phys. Lett. **11**, 256-257 (1985)]

9.9 T. Sato, T. Suzuki: Appl. Opt. **22**, 815-818 (1983)

9.10 F.M. Küchel, H.J. Tiziani: Opt. Commun. **38**, 17-20 (1981)

9.11 J.P. Huignard, J.P. Herriau, T. Valentin: Appl. Opt. **16**, 2796-2798 (1977)

9.12 A. Marrakchi, J.P. Huignard, J.P. Herriau: Opt. Commun. **34**, 15-18 (1980)

9.13 J.P. Huignard, A. Marrakchi: Opt. Lett. **6**, 622-624 (1981)

9.14 A.A. Kamshilin, M.P. Petrov: Opt. Commun. **53**, 23-26 (1985)
A.A. Kamshilin, E.V. Mokrushina, M.P. Petrov: Opt. Eng. **28**, 580-585 (1989)

9.15 H.J. Tiziani, K. Leonard, J. Klenk: Opt. Commun. **34**, 327-331 (1980)

9.16 T. Sato, M. Takehara, O. Ikeda: ICO-13, Conf. Digest (Sapporo, Jpn. 1984) pp.34-35

9.17 Y.H. Ja: Opt. and Quantum Electron. **14**, 367-369 (1982)

9.18 J.P. Huignard, J.P. Herriau, L. Pichon, A. Marrakchi: Opt. Lett. **5**, 436-437 (1980)

9.19 S.I. Stepanov: Photorefractive crystals for adaptive interferometry, in *Optical holography with recording in three-dimentional media*, ed. by Yu.N. Denis'juk (Nauka, Leningrad, 1989) pp.64-74
S.I. Stepanov, I.A. Sokolov: Proc. of 2nd. Int'l Conf. on Holographic Systems, Components and Applications (Bath, GB 1989) pp.95-100

9.20 A.V. Knyaz'kov, N.M. Kozhevnikov, Yu.S. Kuz'minov, N.M. Polozkov, A.S. Saikin, S.A. Sergushchenko: Zh. Tech. Fiz. **54**, 1737-1741 (1984) [Engl. transl.: Sov. Phys.-Tech. Phys. **29**, 1013-1015 (1984)]

9.21 S.I. Stepanov, S.M. Shandarov, N.D. Khat'kov: Fiz. Tverd. Tela, **29**, 3054-3058 (1987) [Engl. transl.: Sov. Phys. Solid State **29**, 1754-1756 (1987)]

9.22 P.A.M. DosSantos, L. Cescato, J. Frejlich: Opt. Lett. **13**, 1014-1016, (1988)
P.M. Garcia, L. Cescato, J. Frejlich: J. Appl. Phys. **66**, 47-49 (1989)

9.23 T.J. Hall, M.A. Fiddy, M.S. Ner: Opt. Lett. **5**, 485-487 (1980)

9.24 D.A. Jackson, R. Priest, A. Dandridge, A.B. Tveten: Appl. Opt. **19**, 2926-2929 (1980)

9.25 M.A. Nokes, B.C. Hill, A.E. Barelli: Rev. Sci. Instrum. **49**, 722-728 (1978)

9.26 Yu.O. Barmenkov, V.V. Zosimov, N.M. Kozhevnikov, L.M. Lyamshev, S.A. Sergushchenko: Dokl. Akad. Nauk SSSR **290**, 1095-1098 (1986) [Engl. transl.: Sov. Phys.-Dokl. **31**, 817-819 (1986)]

9.27 A.A. Kamshilin, E.V. Mokrushina: Pis'ma v Zh. Tech. Fiz. **12**, 363-369 (1986) [Engl. transl.: Sov. Tech. Phys. Lett. **12**, 149-151 (1986)]

9.28 A.A. Kamshilin, J. Frejlich, L.H.D. Cescato: Appl. Opt. **25**, 2375-2381 (1986)

9.29 R.W. Hellwarth: J. Opt. Soc. Am. **67**, 1-3 (1977)

9.30 A. Yariv, D.M. Pepper: Opt. Lett. 1, 16-18 (1977)

9.31 G.J. Dunning, R.C. Lind: Opt. Lett. 7, 558-560 (1982)

9.32 P. Yeh, M.D. Ewbank, M. Khoshevisan, J.M. Tracy: Opt. Lett. 9, 41-43 (1984)

9.33 B. Fisher, Sh. Sternklar: Appl. Phys. Lett. 46, 113-114 (1985)

9.34 G.S. Trofimov, S.I. Stepanov: Fiz. Tverd. Tela 28, 2785-2789 (1986); [Engl. transl.: Sov. Phys. Solid State 28, 1559-1562 (1986)]
M.P. Petrov, S.I. Stepanov, G.S. Trofimov: Pis'ma v Zh. Tech. Fiz. 12, 916-921 (1986) [Engl. transl.: Sov. Tech. Phys. Lett. 12, 379-381 (1986)]

9.35 M.P. Petrov, I.A. Sokolov, S.I. Stepanov, G.S. Trofimov: J. Appl. Phys. 68, 2216-2225 (1990)

9.36 H. Kogelnik: Bell Syst. Tech. J. 44, 2451-2455 (1965)

9.37 E.N. Leith, J. Upatnieks: J. Opt. Soc. Am. 56, 523 (1966)

9.38 A. Yariv: IEEE J. QE-14, 650-660 (1978)

9.39 C.R. Giuliano: Physics Today 4, 27-35 (1981)

9.40 B.Ya. Zel'dovich, N.F. Pilipetsky, V.V. Shkunov: *Principles of Phase Conjugation*, Springer Ser. Opt. Sci., Vol.42 (Springer, Berlin, Heidelberg 1985)

9.41 J.P. Huignard, J.P. Herriau, P. Aubourg, E. Spitz: Opt. Lett. 4, 21-23 (1979)

9.42 S. Odulov, M. Soskin, M. Vasnetsov: Opt. Commun. 32, 355-358 (1980)

9.43 M.D. Levenson, K.M. Johnson, V.C. Hanchett, K. Chiang: J. Opt. Soc. Am. 71, 737-743 (1981)

9.44 M.D. Levenson, K. Chiang: IBM J. Res. Develop. 26, 160-170 (1982)

9.45 A. Yariv: Appl. Phys. Lett. 28, 88-89 (1976)

9.46 B. Fischer, M. Cronin-Golomb, J.O. White, A. Yariv: Appl. Phys. Lett. 41, 141-143 (1982)

9.47 O. Ikeda, T. Suzuki, T. Sato: Appl. Opt. 22, 2192-2195 (1983)

9.48 J.W. Goodman, W.H. Huntley, Jr., D.W. Jackson, M. Lehmann: Appl. Phys. Lett. 8, 311-313 (1966)

9.49 H. Kogelnik, K.S. Pennington: J. Opt. Soc. Am. 58, 273-274 (1968)

9.50 Y.H. Ja: Opt. and Quantum Electron. 15, 457-459 (1983)

9.51 P.A. Belanger: Opt. Eng. 21, 266-270 (1982)

9.52 C.R. Giuliano, R.C. Lind, T.R. O'Meara, G.C. Valley: Laser Focus 19, 55-64 (February 1983)

9.53 M. Cronin-Golomb, B. Fischer, J. Nilsen, J.O. White, A. Yariv: Appl. Phys. Lett. 41, 219-220 (1982)

9.54 R.A. McFarlane, D.G. Steel: Opt. Lett. 8, 208-210 (1983)

9.55 J. Feinberg, G.D. Bacher: Opt. Lett. 9, 420-422 (1984)

9.56 W.B. Whitten, J.M. Ramsey: Opt. Lett. 9, 44-46 (1984)

9.57 M. Cronin-Golomb, K.Y. Lau, A. Yariv: Appl. Phys. Lett. 47, 567-569 (1985)

9.58 M. Cronin-Golomb, A. Yariv: Opt. Lett. 11, 455-457 (1986)

9.59 K. Vahala, K. Kyuma, A. Yariv, S.-K. Kwong, M. Cronin-Golomb, K.Y. Lau: Appl. Phys. Lett. 49, 1563-1566 (1986)

9.60 A. Litvinenko, S. Odulov: Opt. Lett. 9, 68-70 (1984)

9.61 R. Grousson, S. Mallick, S. Odulov: Opt. Commun. 51, 342-346 (1984)

9.62 J. Feinberg, G.D. Bacher: Appl. Phys. Lett. 48, 570-572 (1986)

9.63 M. Cronin-Golomb, A. Yariv, I. Ury: Appl. Phys. Lett. 48, 1240-1242 (1986)

9.64 M.A. Kramer, S. Sifuentes, C.M. Clayton: Appl. Opt. 27, 1371-1374 (1988)

9.65 J.O. White, G.C. Valley: Appl. Opt. 27, 5026-5030 (1988)

9.66 Sh. Sternklar, Sh. Weiss, M. Segev, B. Fisher: Opt. Lett. 11, 528-530 (1986)

9.67 M. Segev, Sh. Weiss, B. Fisher: Appl. Phys. Lett. 50, 1397-1399 (1987)

9.68 Sh. Weiss, M. Segev, B. Fisher: IEEE J. QE-24, 706-708 (1988)

9.69 F. Aronowitz: The laser gyro, in *Laser Applications*, ed. by M. Ross (Academic, New York 1971) pp.131-200

9.70 E. Udd: Laser Focus/Electro-Optics No.12, 64-74 (1985)

9.71 S. Ezekiel, H.J. Arditty (eds.): *Fiber-Optic Rotation Sensors*, Springer Ser. Opt. Sci., Vol.32 (Springer, Berlin, Heidelberg 1982)

9.71 M.M. Teherany: Proc. SPIE 412, 186-191 (1983)
9.72 B. Fischer, Sh. Sternklar: Appl. Phys. Lett. 47, 1-3 (1985)
      S.K. Kwong, A. Yariv, M. Cronin-Golomb, I. Ury: Appl. Phys. Lett. 47, 460-462 (1985)
9.74 P. Yeh, I. McMichael, M. Khoshnevisan: Appl. Opt. 25, 1029-1030 (1986)
9.75 P. Yeh: Opt. Commun. 51, 195-197 (1984)
9.76 I. McMichael, M. Khoshnevisan, P. Yeh: Opt. Lett. 11, 525-527 (1986)
9.77 M.P. Petrov, S.I. Stepanov, A.A. Kamshilin: Opt. Commun. 21, 297-300 (1977)
9.78 Sh. Sternklar, Sh. Weiss, B. Fisher: Appl. Opt. 24, 3121-3122 (1985)
9.79 P.J. Van Heerden: Appl. Opt. 2, 392-400 (1963)
9.80 D.L. Staebler, W.J. Burke, W. Phillips, J.J. Amodei: Appl. Phys. Lett. 26, 182-184 (1975)
9.81 V.V. Aristov: Opt. Commun. 3, 194-196 (1971)
9.82 R. Güther, S. Kusch: Optica Applicata 4, No.4, 121-129 (1976)
9.83 K. Blotekjaer: Appl. Opt. 18, 57-67 (1979)
9.84 J.J. Amodei, D.L. Staebler: Appl. Phys. Lett. 18, 540-542 (1971)
9.85 V.I. Bobrinev, Z.G. Vasil'eva, E.Kh. Gulanyan, A.L. Mikaelyan: Pis'ma v Zh. Eksp. Teor. Fiz 18, 267-269 (1973) [Engl. transl.: JETP Lett. 18, 159-160 (1973)]
9.86 H. Kurz: Optica Acta, 24, 463-473 (1977)
9.87 F. Micheron, C. Mayeux, J.C. Trotier: Appl. Opt. 13, 784-787 (1974)
9.88 E.Kh. Gulanyan, I.R. Dorosh, V.D. Iskin, A.L. Mikaelyan, M.A. Maiorchuk: Kvantovaya Elektron., 6, 1097-1100 (1979) [Engl. transl.: Sov. J. Quantum Electron. 9, 647-649 (1979)]
9.89 D. Von der Linde, A.M. Glass, K.F. Rodgers: Appl. Phys. 25, 155-157 (1974)
9.90 D. Von der Linde, A.M. Glass, K.F. Rodgers: J. Appl. Phys. 47, 217-220 (1976)
9.91 M.P. Petrov, S.I. Stepanov, A.A. Kamshilin: Opt. Commun. 29, 44-48 (1979)
9.92 M.P. Petrov, S.I. Stepanov, A.A. Kamshilin: Opt. and Laser Techn. 11, No6, 149-151 (1979)
9.93 A.A. Kamshilin, M.P. Petrov, S.I. Stepanov: Pis'ma v Zh. Tech. Fiz. 5, 374-377 (1979) [Engl. transl.: Sov. Tech. Phys. Lett. 5, 150-152 (1979)]
9.94 M.P. Petrov, S.V. Miridonov, S.I. Stepanov, V.V. Kulikov: Opt. Commun. 31, 301-305 (1979)
9.95 M.P. Petrov, S.I. Stepanov, S.V. Miridonov, V.V. Kulikov: Proc. SPIE 213, 44-49 (1979)
9.96 L. D'Auria, J.P. Huignard, C. Slezak, E. Spitz: Appl. Opt. 13, 808-818 (1974)
9.97 F.M. Smits, L.E. Gallaher: Bell Syst. Techn. J. 46, 1267-1278 (1967)
      K.L. Anderson: Bell Lab. Rec. 46, 318-325 (1968)
9.98 Di Chen, J.D. Zook: IEEE Proc. 63, 1207-1230 (1975)
      T. Sugaya, M. Ishikawa, I. Hoshino: Appl. Opt. 20, 3104-3118 (1981)
      B.V. Vanyushev, A.V. Volkov, I.S. Gibin, V.A. Dumbrovskii, S.A. Dumbrovskii, T.L. Mantush, E.F. Pen, V.I. Pechurkin, V.A. Polivanov, A.N. Potapov, P.E. Tverdokhleb, A.I. Chernyshev, L.F. Chernyshev: Avtometriya No.3, 19-26 (1984) [Engl. transl.: Autom. Monit. & Meas. No.3, 16-22 (1984)]
9.99 B.H. Soffer, G.J. Dunning, Y. Owechko, E. Marom: Opt. Lett. 11, 118-120 (1986)
9.100 A. Yariv, S.K. Kwong: Opt. Lett. 11, 186-188 (1986)
      A. Yariv, S.K. Kwong, K. Kyuma: Appl. Phys. Lett. 48, 1114-1116 (1986)
9.101 D.K. Anderson, M.C. Erie: Opt. Eng. 26, 434-444 (1987)
      H.J. White, N.B. Aldridge, I. Lindsay: Opt. Eng. 27, 30-37 (1988)
9.102 M.P. Petrov, S.I. Stepanov, A.V. Khomenko: Photosensitive Electro-optic Media in Holography and Optical Information Processing (Nauka, Leningrad 1983) (in Russian)
9.103 R.P. Kenan, C.M. Verber, E. Van Wood: Appl. Phys. Lett. 24, 428-430 (1974)
9.104 O. Mikami: Opt. Commun. 19, 42-44 (1976)

9.105 S.I. Stepanov, A.A. Kamshilin, M.P. Petrov: Pis'ma v Zh. Tech. Fiz. 3, 89–93 (1977) [Engl. transl.: Sov. Tech. Phys. Lett. 3, 36–38 (1977)]
9.106 T. Yasuhira, Y. Mitsuhashi, T. Morikawa, J. Shimada, T. Kamijo: Appl. Opt. 16, 2532–2534 (1977)
9.107 M.P. Petrov, S.I. Stepanov, A.A. Kamshilin: Ferroelectrics 22, 631–634 (1978)
9.108 P.D. Henshaw: Appl. Opt. 21, 2323–2325 (1982)
9.109 G.T. Sincerbox, G. Roosen: Appl. Opt. 22, 690–697 (1983)
9.110 G. Pauliat, J.P. Herriau, A. Delboulbe, G. Roosen, J.P. Huignard: J. Opt. Soc. Am. B3, 306–313 (1986)
9.111 A. Yariv, P. Yeh: Optical Waves in Crystals (Wiley, New York 1984)
9.112 E.G.H. Lean, C.F. Quate, H.J. Shaw: Appl. Phys. Lett. 10, 48–50 (1967)
9.113 E. Voit, C. Zaldo, P. Günter: Opt. Lett. 11, 309–311 (1986)
9.114 T.G. Pencheva, M.P. Petrov, S.I. Stepanov: Opt. Commun. 40, 175–178 (1981)
9.115 J.P. Huignard, B. Ledu: Opt. Lett. 7, 310–312 (1982)
9.116 D. Rak, I. Ledoux, J.P. Huignard: Opt. Commun. 49, 302–306 (1984)
9.117 A. Vander Lught: IEEE Trans. IT-10, 139–145 (1964)
9.118 P. Nisenson, S. Iwasa: Appl. Opt. 11, 2760–2767 (1972)
9.119 A.V. Khomenko, V.M. Petrov, M.G. Shlyagin, N.O. Shalaevskii: Proc. SPIE 1183, 309–316 (1990)
9.120 M.P. Petrov, V.I. Marakhonov, M.G. Shlyagin, A.V. Khomenko, M.V. Krasin'kova: Zh. Tekh. Fiz. 50, 1311–1314 (1980) [Engl. transl.: Sov. Phys. Tech. Phys. 25, 752–753 (1980)]
9.121 R.A. Sprague: Appl. Opt. 17, 2762–2767 (1978)
9.122 S.G. Lipson, P. Nisenson: Appl. Opt. 13, 2052–2060 (1974)
9.123 K.A. Boyarchuk, K.I. Volyak, A.I. Malyarovskii, S.V. Miridonov: Kratkie Soobzhen. po Fiz. No.7, 5–6 (1986)
9.124 D. Casasent, T.K. Luu: Appl. Opt. 18, 3307–3314 (1979)
9.125 D. Pepper, J. Auyeung, D. Fekete: Opt. Lett. 3, 7–9 (1978)
9.126 S.G. Odulov, M.S. Soskin: Dokl. Akad. Nauk SSSR 252, No.2, 336–339 (1980) [Engl. transl.: Sov. Phys.-Dokl. 25, No.5, 380–381 (1980)]
9.127 J.O. White, A. Yariv: Appl. Phys. Lett. 37, 5–7 (1980)
9.128 J.O. White, A. Yariv: Opt. Eng. 21, 224–230 (1982)
9.129 T.R. O'Meara, D.M. Pepper, J.O. White: Application of nonlinear optical phase conjugation, in Optical Phase Conjugation, ed. by R.A. Fisher (Academic, New York 1978) pp.537–595
9.130 B. Loiseaux, G. Illiaquer, J.P. Huignard: Opt. Eng. 24, 144–149 (1985)
     L. Pichon, J.P. Huignard: Opt. Commun. 37, 277–280 (1981)
9.131 M.G. Nicholson, G.G. Gibbons, L.C. Laycock, C.R. Petts: Electron Lett. 22, 1200–1202 (1986)
9.132 G. Gheen, L.J. Cheng: Appl. Opt. 27, 2756–2761 (1988)
9.133 S.I. Stepanov, V.D. Gural'nik: Pis'ma v Zh. Tech. Fiz. 8, 114–118 (1982) [Engl. transl.: Sov. Tech. Phys. Lett. 8, 49–51 (1982)]
9.134 T.J. Wang: White light real time optical correlator using BSO crystal. ICO-13, Conf. Digest (Sapporo, Jpn. 1984) pp.166–167
9.135 J.P. Huignard, J.P. Herriau: Appl. Opt. 17, 2671–2672 (1978)
9.136 J. Feinberg: Opt. Lett. 5, 330–332 (1980)
9.137 J. Feinberg: Optical phase conjugation in photorefractive materials, in Optical Phase Conjugation, ed. by R.A. Fisher (Academic, New York 1989) pp.417–443
9.138 E. Ochoa, L. Hesselink, J.W. Goodman: Appl. Opt. 24, 1826–1832 (1985)
9.139 Y.H. Ja: Opt. Commun. 44, 24–28 (1982)
9.140 D. Gabor, G.W. Stroke, R. Restrick, A. Funkhouser, D. Brumm: Phys. Lett. 18, 116–118 (1965)
9.141 J.P. Huignard, J.P. Herriau, F. Micheron: Appl. Phys. Lett. 26, 256–258 (1975)
9.142 J.P. Huignard, J.P. Herriau, F. Micheron: Ferroelectrics 11, 393–396 (1076)

9.143 T.K. Gaylord, R. Magnusson, J.E. Weaver: Opt. Commun. **20**, 365-366 (1977)

9.144 Y.H. Ja: Opt. Commun. **42**, 377-380 (1982)

9.145 L.M. Bernardo, O.D.D. Soares: Appl. Opt. **25**, 592-593 (1986)

9.146 A.E. Chiou, P. Yeh: Opt. Lett. **11**, 306-308 (1986)

9.147 V. Markov, S. Odulov, M. Soskin: Opt. and Laser Technol. **11**, 95-99 (1979)

9.148 N.V. Kukhtarev, V.B. Markov, S.G. Odulov, M. Soskin, V.L. Vinetskii: Ferroelectrics **22**, 961-964 (1979)

9.149 S.K. Kwong, Y.-H. Chung, M. Cronin-Golomb, A. Yariv: Opt. Lett. **10**, 359-361 (1985)

9.150 S.L. Sochava, S.I. Stepanov: Zh. Tek. Fiz. **57**, 1763-1766 (1987) [Engl. transl.: Sov. Phys. - Tech. Phys. **32**, 1054-1056 (1987)]

9.151 Y. Fainman, C.C. Guest, S.H. Lee: Appl. Opt. **25**, 1598-1603 (1986)

9.152 J.W. Goodman: *Introduction to Fourier Optics* (McGraw-Hill, New York, 1968)

9.153 D. Casasent, S. Notu, G. Lebreton: Proc. SPIE **202**, 122-131 (1979)

9.154 E. Ochoa, J.W. Goodman, L. Hesselink: Opt. Lett. **10**, 430-432 (1985)

9.155 S.-K. Kwong, A. Yariv: Appl. Phys. Lett. **48**, 564-566 (1986)

9.156 A.E. Chiou, P. Yeh: Opt. Lett. **11**, 461-463 (1986)

9.157 M. Cronin-Golomb, D.Z. Anderson: Appl. Phys. Lett. **47**, 346-348 (1985)

9.158 M. Cronin-Golomb, A. Yariv: J. Appl. Phys. **57**, 4906-4910 (1985)

9.159 S.K. Kwong, M. Cronin-Golomb, A. Yariv: Appl. Phys. Lett. **45**, 1016-1018 (1984)

9.160 S.K. Kwong, A. Yariv: Opt. Lett. **11**, 377-379 (1986)

9.161 M.P. Petrov, A.V. Khomenko, V.V. Marakhonov et al.: Pis'ma v Zh. Tech. Fiz. **6**, 386-388 (1980) [Engl. transl.: Sov. Tech. Phys. Lett. **6**, 165-166 (1980)]

9.162 D. Casasent, F. Caimi, A. Khomenko: Appl. Opt. **20**, 3090-3092 (1981)

9.163 S.I. Stepanov, V.V. Kulikov: Zh. Tech. Fiz. **53**, 2255-2257 (1983); [Engl. transl. Sov. Phys. - Tech. Phys. **28**, 1384-1386 (1983)]

9.164 D.Z. Anderson, D.M. Lininger, J. Feinberg: Opt. Lett. **12**, 123-125 (1987)
    R.S. Cudney, R.M. Pierce, J. Feinberg: Nature, **332**, 424-426 (1988)
    D.Z. Anderson, J. Feinberg: IEEE J. QE-**25**, 635-647 (1989)

9.165 M. Cronin-Golomb, A.M. Biernacki, C. Lin, H. Kong: Opt. Lett. **12**, 1029-1031 (1987)

9.166 J.E. Ford, Y. Fainman, S.H. Lee: Opt. Lett. **13**, 856-858 (1988)

9.167 D. Psaltis, J. Yu, J. Hong: Appl. Opt. **24**, 3860-3865 (1985)

9.168 D. Psaltis, D. Brady, K. Wagner: Appl. Opt. **27**, 1752-1759 (1988)
    P. Yeh, A.E.T. Chiou, J. Hong: Appl. Opt. **27**, 2093-2096 (1988)

Appendix A

A.1   A.A. Ballman: J. Amer. Ceram. Soc. **48**, 112-113 (1965)

A.2   P. Lerner, C. Legras, J.P. Dumas: J. Cryst. Growth **3**, 231-235 (1968)

A.3   Yu.S. Kuz'minov: *Lithium Niobate and Lithium Tantalate Materials for Nonlinear Optics* (Nauka, Moscow, 1975)

A.4   S.C. Abrahams, J.M. Reddy, J.L. Bernstein: J. Phys. Chem. Solids **27**, 997 (1966)

A.5   A.W. Warner, M. Onoe, G.A. Coquin: J. Acoust. Soc. Am. **42**, 1223-1231 (1967)

A.6   D. Redfield, W.J. Burke: J. Appl. Phys. **45**, 4566-4571 (1974)

A.7   M.G. Clark, F.J. DiSalvo, A.M. Glass, G.E. Peterson: J. Chem. Phys. **59**, 6209-6219 (1973)

A.8   R.R. Shah, D.M. Kim, T.A. Rabson, F.K. Tittel: J. Appl. Phys. **47**, 5421-5431 (1976)

A.9   G.D. Boyd, W.L. Bond, H.L. Carter: J. Appl. Phys. **38**, 1941-1943 (1967)

A.10   E.H. Turner: Appl. Phys. Lett. **8**, 303-304 (1966)
A.11   I.P. Kaminov, E.H. Turner: IEEE Proc. **54**, 1374-1390 (1966)
A.12   E.R. Mustel', V.N. Parygin: *Light Modulation and Deflection Techniques* (Nauka, Moscow 1970)
A.13   A. Yariv, P. Yeh: *Optical Waves in Crystals* (Wiley, New York 1984)
A.14   Y. Ohmori, Y. Yasojima, H. Adachi, Y. Inuishi: Technology Rep. of Osaka Univ. **24**, 105-114 (1974)
A.15   Y. Ohmori, Y. Yasojima, Y. Inuishi: Jpn. J. Appl. Phys. **14**, 1291-1300 (1975)
A.16   E. Krätzig, H. Kurz: Optica Acta **24**, 475-482 (1977)
A.17   V.A. Pashkov, N.M. Solov'eva, E.M. Uyukin: Fiz. Tverd. Tela **21**, 1879-1882 (1979) [Engl. transl.: Sov. Phys.-Solid State **21**, 1079-1081 (1979)]
A.18   V.V. Kulikov, S.I. Stepanov: Fiz. Tverd. Tela **21**, 3204-3208 (1979) [Engl. transl.: Sov. Phys.-Solid State **21**, 1849-1851 (1979)]
A.19   A.M. Glass, D. Von der Linde, T.J. Negran: Appl. Phys. Lett. **25**, 233-235 (1974)
A.20   E. Krätzig, H. Kurz: Ferroelectrics **13**, 295-296 (1976)
A.21   K.G. Belabaev, V.B. Markov, S.G. Odulov: Ukr. Fiz. Zh. **24**, 366-371 (1979)
A.22   A. Ashkin, G.D. Boyd, J.M. Dziedzic, R.G. Smith, A.A. Ballman, J.J. Levinstein, K. Nassau: Appl. Phys. Lett. **9**, 72-74 (1976)
A.23   F.S. Chen: J. Appl. Phys. **40**, 3389-3396 (1969)
A.24   B. Dischler, J.P. Herrington, A. Räuber: Sol. St. Commun. **14**, 1233-1236 (1974)
A.25   H. Kurz: Wavelength dependence of the photorefractive process in doped LiNbO$_3$. *Photonics*, ed. by M. Balkanski, P. Lallemand (Gauthier-Villars, Paris, 1975) pp.193-198
A.26   H. Wang, G. Shi, Z. Wu: Phys. St. Sol.(a) **89**, K211-K213 (1985)
        D.A. Bryan, R. Gerson, H.E. Tomaschke: Appl. Phys. Lett. **44**, 847-849 (1984)
A.27   J. Wen, L. Wang, Y. Tang, H. Wang: Appl. Phys. Lett. **53**, 260-261 (1988)
A.28   F.S. Chen, J.T. LaMacchia, D.F. Fraser: Appl. Phys. Lett. **13**, 223-225 (1968)
A.29   J.J. Amodei, W. Phillips, D.L. Staebler: IEEE J. QE-7, 63 (1971)
A.30   J.J. Amodei, W. Phillips, D.L. Staebler: Appl. Opt. **11**, 390-396 (1972)
A.31   W. Phillips, J.J. Amodei, D.L. Staebler: RCA Rev. **33**, 94-109 (1972)
A.32   O. Mikami, A. Ishida: Opt. Commun. **9**, 354-356 (1973)
A.33   E. Okamoto, M. Ikeo, K. Mato: Appl. Opt. **14**, 2453-2455 (1975)
A.34   A.L. Mikaelyan, E. Kh. Gulanyan, E.I. Dmitrieva, I.R. Dorosh: Kvantovaya Elektron., **5**, 440-442 (1978) [Engl. transl.: Sov. J. Quantum Electron. **8**, 257-258 (1978)]
A.35   E. Kh. Gulanyan, I.R. Dorosh, A.I. Zhmurko: Vopr. Radioelektron. (Ser. Obstcheteckn.) 95-105 (1979)
A.36   S.I. Stepanov, M.P. Petrov, A.A. Kamshilin: Pis'ma v Zh. Tech. Fiz. **3**, 849-854 (1977) [Engl. transl.: Sov. Tech. Phys. Lett. **3**, 345-346 (1977)]
A.37   N.V. Kukhtarev, S.G. Odulov: Pis'ma v Zh. Tech. Fiz. **6**, 1176-1179 (1980) [Engl. transl.: Sov. Tech. Phys. Lett. **6**, 503-505 (1980)]
A.38   L. Arizmendi, R.C. Powell: J. Appl. Phys. **61**, 2128-2131 (1987)
A.39   L. Arizmendi: J. Appl. Phys. **64**, 4654-4656 (1988)
A.40   T.G. Pencheva, M.P. Petrov, S.I. Stepanov: Avtometriya N.1, 122-126 (1980)
A.41   M.P. Petrov, T.G. Pencheva, S.I. Stepanov: J. Optics **12**, 287-292 (1981)
A.42   A. Mandel', N. Khat'kov, S. Shandarov: Ferroelectrics **83**, 215-220 (1988)
A.43   J.J. Amodei: RCA Rev. **32**, 185-198 (1971)
A.44   J.J. Amodei, D.L. Staebler: RCA Rev. **33**, 71-93 (1972)
A.45   D.L. Staebler, J.J. Amodei: J. Appl. Phys. **43**, 1042-1049 (1972)
A.46   V.L. Vinetskii, N.V. Kukhtarev, V.B. Markov, S.G. Odulov, M.S. Soskin: Izv. Akad. Nauk SSSR, Ser. Fiz. **41**, 811-820 (1977) [Engl. transl.: Bull Acad. Sci. USSR, Phys. Ser. **41**, 135-143 (1977)]

A.47 D. Von der Linde, A.M. Glass: Appl. Phys. **8**, 85-100 (1975)
A.48 G.A. Alphonse, W. Phillips: RCA Rev. **37**, 184-205 (1976)
A.49 D.M. Kim, R.R. Shah, T.A. Rabson, F.K. Tittel: Apl. Phys. Lett. **29**, 84-86 (1976)
A.50 G.E. Peterson, A.M. Glass, T.J. Negran: Appl. Phys. Lett. **19**, 130-132 (1971)
A.51 W.D. Cornish, M.G. Moharam, L. Young: J. Appl. Phys. **47**, 1479-1484 (1976)
A.52 R. Orlowski, E. Krätzig, H. Kurz: Opt. Commun. **20**, 171-174 (1977)
A.53 S.G. Odulov, O.I. Oleinik: Kvantovaya Elektron **10**, 1498-1502 (1983) [Engl. transl.: Sov. J. Quantum Electron. **13**, 980-982 (1983)]
A.54 S.G. Odulov: Pis'ma v Zh. Eksp. i Teor. Fiz. **35**, 10-12 (1982) [Engl. transl.: JETP Lett. **35**, 10-14 (1982)]
A.55 D.L. Staebler, W. Phillips: Appl. Opt. **13**, 788-794 (1974)
A.56 J.J. Amodei, D.L. Staebler: Appl. Phys. Lett. **18**, 540-542 (1971)
A.57 D.L. Staebler, J.J. Amodei: Ferroelectrics 3, 107-113 (1972)
A.58 D.L. Staebler, W.J. Burke, W. Phillips, J.J. Amodei: Appl. Phys. Lett. **26**, 182-184 (1975)
A.59 W. Bollmann, H.J. Stöhr: Phys. St. Solidi (a) **39**, 477-484 (1977)
A.60 V.V. Kulikov, M.P. Petrov, S.I. Stepanov: Avtometriya No.1, 39-45 (1980)
A.61 I.B. Barkan, M.V. Entin, S.I. Marennikov: Phys. St. Solidi (a) **44**, K91-K94 (1977)
A.62 K.G. Belabaev, V.B. Markov, S.G. Odulov: Ukr. Fiz. Zh. **21**, 1550-1554 (1976)
A.63 V.I. Bobrinev, Z.G. Vasil'eva, E.Kh. Gulanyan, A.L. Mikaelyan: Pis'ma v Zh. Eksp. i Teor. Fiz. **18**, 267-269 (1973) [Engl. transl.: JETP Lett. **18**, 159-160 (1973)]
A.64 H. Kurz: Optica Acta **24**, 463-473 (1977)
A.65 D. Von der Linde, A.M. Glass, K.F. Rodgers: Appl. Phys. Lett. **25**, 155-157 (1974)
A.66 D. Von der Linde, A.M. Glass, K.F. Rodgers: J. Appl. Phys. **47**, 217-220 (1976)
A.67 M.P. Petrov, S.I. Stepanov, A.A. Kamshilin: Opt. Commun. **29**, 44-48 (1979)
A.68 M.P. Petrov, S.I. Stepanov, A.A. Kamshilin: Opt. and Laser Techn. **11**, 149-151 (1979)
A.69 A.A. Kamshilin, M.P. Petrov, S.I. Stepanov: Pis'ma v Zh. Tech. Fiz. **5**, 374-377 (1979) [Engl. transl.: Sov. Tech. Phys. Lett. **5**, 150-152 (1979)]
A.70 T.K. Gaylord, T.A. Rabson, F.K. Tittel, C.R. Quick: J. Appl. Phys. **44**, 896-897 (1973)
A.71 V.L. Vinetskii, N.V. Kukhtarev, V.B. Markov, S.G. Odulov, M.S. Soskin: Preprint No.15, Inst. of Phys. of Ukr. Acad. of Sci., Kiev (1976)
A.72 N. Kukhtarev, V. Markov, S. Odulov: Opt. Commun. **23**, 338-343 (1977)
A.73 V.P. Kondilenko, V.B. Markov, S.G. Odulov, M.S. Soskin: Ukr. Fiz. Zh. **23**, 2039-2043 (1978)
A.74 I. Kanaev, V. Malinovski, B. Sturman: Opt. Commun. **34**, 95-100 (1980)
A.75 M.P. Petrov, S.I. Stepanov, A.V. Khomenko: *Photosensitive Electro-optic Materials in Holography and Optical Information Processing* (Nauka, Moscow, 1983)
A.76 R. Magnussen, T.K. Gaylord: Appl. Opt. **13**, 1545-1548 (1974)
A.77 N.D. Khat'kov, S.M. Shandarov: Avtometriya No.2, 61-65 (1983)
A.78 R. Grousson, S. Mallick, S. Odulov: Opt. Commun. **51**, 342-346 (1984)
A.79 N. Kukhtarev, S.G. Odulov: Pis'ma v Zh. Eksp. i Teor. Fiz. **30**, 6-11, (1979) [Engl. transl.: JETP Lett. **30**, 4-8 (1979)]
A.80 N.V. Kukhtarev, S.G. Odulov: SPIE Proc. **213**, 2-9 (1979)
A.81 N.V. Kukhtarev, S.G. Odulov: Opt. Commun. **32**, 183-186 (1980)
A.82 A.D. Novikov, V.V. Obuchovskii, S.G. Odulov, B.I. Sturman: Pis'ma v Zh. Eksper. i Teor. Fiz. **44**, 418-421 (1986) [Engl. transl.: JETP Lett. **44**, 538-542 (1986)]

261

A.83   S.G. Odulov, M.S. Soskin: Pis'ma v Zh. Eksper. i. Teor. Fiz. 37, 243-247 (1983) [Engl. transl.: JETP Lett. 37, 289-293 (1983)]

A.84   S.G. Odulov: Kvantovaya Elektron. 11, 529-536 (1984) [Engl. transl.: Sov. J. Quantum Electron 14, 360-364 (1984)]

A.85   T.K. Gaylord, T.A. Rabson, F.K. Tittel, C.R. Quick: Appl. Opt. 12, 414-415 (1973)

A.86   P. Shah, T.A. Rabson, F.K. Tittel, T.K. Gaylord: Appl. Phys. Lett. 24, 130-131 (1974)

A.87   I.F. Kanaev, V.K. Malinovskii: Fiz. Tverd. Tela 16, 3694-3696 (1974) [Engl. transl.: Sov. Phys.-Solid State 16, 2398-2399 (1975)]

A.88   I.B. Barkan, E.V. Pestryakov, M.V. Entin: Avtometriya No.4, 18-22 (1976)

A.89   C.T. Chen, D.M. Kim, D. Von der Linde: Appl. Phys. Lett. 34, 321-324 (1979)

A.90   A.M. Glass, I.P. Kaminov, A.A. Ballman, D.H. Olson: Appl. Opt. 19, 276-281 (1980)

A.91   O. Mikami: Opt. Commun. 11, 30-32 (1974)

A.92   O.V. Kandidova, V.V. Lemanov, B.V. Sukharev: Pis'ma v Zh. Tech. Fiz. 9, 777-781 (1983) [Engl. transl.: Sov. Tech. Phys. Lett. 9, 335-336 (1983)]

A.93   O.V. Kandidova, V.V. Lemanov, B.V. Sukharev: Zh. Tech. Fiz. 54, 1748-1754 (1984) [Engl. transl.: Sov. Phys.-Tech. Phys. 29, 1019-1022 (1984)]

A.94   J.P. Nisius, E. Krätzig: Solid St. Commun. 53, 743-746 (1985)

A.95   Z.I. Shapiro, S.A. Fedulov, Yu.N. Venevtsev: Fiz. Tverd. Tela 16, 316-317 (1964) [Engl. transl.: Sov. Phys.-Solid State 6, 254-255 (1964)

A.96   E.G. Spencer, P.V. Lenzo, A.A. Ballman: IEEE Proc. 55, 2074-2108 (1967)

A.97   W.L. Bond: J. Appl. Phys. 36, 1674 (1965)

A.98   P.V. Lenzo, E.H. Turner, E.G. Spencer, A.A. Ballman: Appl. Phys. Lett. 8, 81-82 (1966)

A.99   J.M. Spinhirne, D. Aug, C.S. Joiner, T.L. Estle: Appl. Phys. Lett. 30, 89-91 (1977)

A.100   P.A. Augustov, K.K. Shwarts, E. Krätzig: Phys. Status Solidi (a) 87, K73-K76 (1985)

A.101   J.M. Spinhirne, T.L. Estle: Appl. Phys. Lett. 25, 38-39 (1974)

A.102   E. Krätzig, R. Orlowski: Appl. Phys. 15, 133-139 (1978)

A.103   E. Krätzig, R. Orlowski, V. Doorman, M. Rosenkranz: SPIE Proc. 164, 33-37 (1978)

A.104   H. Tsuya: J. Appl. Opt. 46, 4323-4333 (1975)

A.105   H. Vormann, E. Krätzig: Solid State Commun. 49, 843-847 (1984)

A.106   E.M. Avakyan, K.G. Balabaev, I.N. Kiseleva, S.G. Odulov, E.I. Renkachish-skaya: Ukr. Fiz. Zh. 29, 790-793 (1984)

A.107   S. Odulov, K. Belabaev, I. Kiseleva: Opt. Lett. 10, 31-33 (1985)

A.108   S. Odulov, M. Soskin, M. Vasnetsov: Opt. Commun. 32, 355-358 (1980)

A.109   A. Khizhnyk, V. Kondilenko, V. Kremenitski, S. Odulov, M. Soskin: SPIE Proc. 213, 18-25 (1979)

A.110   V.P. Kondilenko, S.G. Odulov, M.S. Soskin: Izv. Akad. Nauk SSSR. Ser. Fiz. 45, 958-962 (1981) [Engl. transl.: Bull. Acad. Sci. USSR, Phys. Ser. 45, 55-58 (1981)]

A.111   V. Belrus, J. Kalnajs, A. Linz, R.C. Folweiler: Mat. Res. Bull. 6, 899-905 (1971)

A.112   Acousto-optical Crystals, ed. by M.P. Shaskol'skaya (Nauka, Moscow 1982)

A.113   F. Jona, G. Shirane: Ferroelectric Crystals (Pergamon, Oxford 1962)

A.114   M.B. Klein: Physics of the photorefractive effect in BaTiO₃, in Photorefractive Crystals and Applications I, ed. by P. Günter, J.P. Huignard, Topics Appl. Phys., Vol.61 (Springer, Berlin, Heidelberg 1988) pp.195-236

A.115   S. Ducharme, J. Feinberg: J. Opt. Soc. Am. B3, 283-292 (1986)

A.116 N.V. Kukhtarev, E. Krätzig, H.C. Külich, R.A. Rupp, J. Albers: Appl. Phys. B35, 17-21 (1984)
A.117 J. Feinberg, D. Heiman, A.R. Tanguay, Jr., R.W. Hellwarth: J. Appl. Phys. 51, 1297-1305 (1980)
A.118 M.B. Klein, R.N. Schwartz: J. Opt. Soc. Am. B3, 293-305 (1986)
A.119 S.H. Wemple, M. DiDomenico, Jr., I. Camlibel: J. Phys. Chem. Solids 29, 1797-1803 (1968)
A.120 M.B. Klein, G.C. Valley: J. Appl. Phys. 57, 4901-4905 (1985)
A.121 B.A. Wechsler, M.B. Klein: J. Opt. Soc. Am. B 5, 1711-1723 (1988)
A.122 R.L. Townsend, J.T. LaMacchia: J. Appl. Phys. 41, 5188-5192 (1970)
A.123 J. Feinberg, R.W. Hellwarth: Opt. Lett. 5, 519-521 (1980)
A.124 F. Micheron, G. Bismuth: Appl. Phys. Lett. 20, 79-81 (1972)
A.125 F. Laeri, T. Tschudi, J. Albers: Opt. Commun. 47, 387-390 (1983)
A.126 M. Cronin-Golomb, K.Y. Lau, A. Yariv: Appl. Phys. Lett. 47, 567-569 (1985)
A.127 Yu.B. Afanas'ev, A.A. Petrov, M.P. Petrov, S.I. Stepanov, G.S. Trofimov: Pis'ma v Zh. Tech. Fiz. 13, 1161-1164 (1987) [Engl. transl.: Sov. Phys. Tech. Lett. 13, 486-487 (1987)]
A.128 S. Ducharme, J. Feinberg: J. Appl. Phys. 56, 838-842 (1984)
A.129 D. Rak, I. Ledoux, J.P. Huignard: Opt. Commun. 49, 302-306 (1984)
A.130 T. Tschudi, A. Herden, J. Golts, H. Klumb, J. Albers: IEEE J. QE-22, 1493-1502 (1986)
A.131 Y. Fainman, E. Klancnik, S.H. Lee: Opt. Eng. 25, 228-234 (1986)
A.132 S.K. Kwong, Y.H. Chung, M. Cronin-Golomb, A. Yariv: Opt. Lett. 10, 359-361 (1985)
A.133 J.O. White, M. Cronin-Golomb, B. Fisher, A. Yariv: Appl. Phys. Lett. 40, 450-452 (1982)
A.134 S.K. Kwong, A. Yariv: Appl. Phys. Lett. 48, 564-566 (1986)
A.135 S.K. Kwong, A. Yariv, M. Cronin-Golomb, I. Ury: Appl. Phys. Lett. 47, 460-462 (1985)
A.136 J. Feinberg: J. Opt. Soc. Am. 72, 46-51 (1982);
R.A. Rupp, F.W. Drees: Appl. Phys. B 39, 223-229 (1986);
J.F. Ford, Y. Fainman, S.H. Lee: Opt. Lett. 13, 856-858 (1988)
A.137 H. Kong, C Lin, A.M. Birnaki, M. Cronin-Golomb: Opt. Lett. 13, 324-326 (1988)
A.138 J. Goltz, C. Denz, H. Klumb, T. Tschudi: Opt. Lett. 13, 321-323 (1988)
A.139 M. Cronin-Golomb, B. Fisher, S.K. Kwong, J.O. White, A. Yariv: Opt. Lett. 10, 353-355 (1985)
A.140 M. Cronin-Golomb, B. Fischer, J. Nilsen, J.O. White, A. Yariv: Appl. Phys. Lett. 41, 219-220 (1982)
A.141 M. Cronin-Golomb, B. Fischer, J.O. White, A. Yariv: Appl. Phys. Lett. 41, 689-691 (1982)
A.142 M. Cronin-Golomb, S.K. Kwong, A. Yariv: Appl. Phys. Lett. 44, 727-729 (1984)
A.143 J. Feinberg: Opt. Lett. 7, 486-488 (1982)
J. Feinberg, G.D. Bacher: Opt. Lett. 9, 420-422 (1984)
A.144 R.A. McFarlane, D.G. Steel: Opt. Lett. 8, 208-210 (1983)
W.B. Whitten, J.M. Ramsey: Opt. Lett. 9, 44-46 (1984)
A.145 M.C. Gower: Opt. Lett. 11, 458-460 (1986)
M.C. Gower, P. Hribek: J. Opt. Soc. Am. B 5, 1750-1757 (1988)
A.146 M. Cronin-Golomb, B. Fischer, J.O. White, A. Yariv: Appl. Phys. Lett. 42, 919-921 (1983)
A.147 B. Fischer, Sh. Sternklar: Appl. Phys. Lett. 47, 1-3 (1985)
A.148 M. Cronin-Golomb, J. Paslaski, A. Yariv: Appl. Phys. Lett. 47, 1131-1133 (1985)

A.149 M. Cronin-Golomb, A. Yariv: Opt. Lett. 11, 455-457 (1986)
A.150 K. Vahala, K. Kyuma, A. Yariv, S.K. Kwong, M. Cronin-Golomb, K.Y. Lau: Appl. Phys. Lett. 49, 1563-1565 (1986)
A.151 Sh. Sternklar, Sh. Weiss, M. Segev, B. Fisher: Opt. Lett. 11, 528-530 (1986)
Sh. Weiss, Sh. Sternklar, B. Fisher: Opt. Lett. 12, 114-116 (1987);
Sh. Sternklar, B. Fisher: Opt. Lett. 12, 711-713 (1987)
A.152 M. Segev, Sh. Weiss, B. Fisher: Appl. Phys. Lett. 50, 1397-1399 (1987)
Sh. Weiss, M. Segev, B. Fisher: IEEE J. QE-24, 706-708 (1988)
A.153 Sh. Sternklar, Sh. Weiss, B. Fisher: Opt. Eng. 26, 423-427 (1987)
Sh. Weiss, M. Segev, Sh. Sternklar, B. Fisher: Appl. Opt. 27, 3422-3428 (1988)
A.154 M.D. Ewbank: Opt. Lett. 13, 47-49 (1988)
A.155 J. Feinberg: Physics Today 41, 46-52 (1988)
A.156 R.W. Eason, A.M.C. Smout: Opt. Lett. 12, 51-53 (1987)
A.M.C. Smout, R.W. Eason: Opt. Lett. 12, 498-500 (1987)
A.157 M.A. Kramer, S. Sifuentes, C.M. Clayton: Appl. Phys. 27, 1371-1374 (1988)
A.158 L.K. Lam, T.Y. Chang, J. Feinberg, R.W. Hellwarth. Opt. Lett. 6, 475-477 (1981)
A.159 A.L. Smirl, G.C. Valley, R.A. Mullen, K. Bohnert, C.D. Mire, T.F. Boggess: Opt. Lett. 12, 501-502 (1987)
A.160 A.A. Ballman, H. Brown: J. Cryst. Growth 1, 311-314 (1967)
A.161 R.R. Neorgaonkar, W.K. Cory, J.R. Oliver, M.D. Ewbank, W.F. Hall: Opt. Eng. 26, 392-405 (1987)
A.162 Yu.S. Kuz'minov: *Ferroelectric Crystals for Laser Beam Modulation* (Nauka, Moscow 1982)
A.163 P.V. Lenzo, E.G. Spencer, A.A. Ballman: Appl. Phys. Lett. 11, 23-24 (1967)
A.164 V.V. Voronov, Yu.S. Kuz'minov, I.G. Lukina: Fiz. Tverd. Tela 18, 1047-1050 (1976) [Engl. transl.: Sov. Phys.-Solid State 18, 598-600 (1976)]
A.165 E.L. Venturini, E.G. Spencer, P.L. Lenzo, A.A. Ballman: J. Appl. Phys. 39, 343-344 (1968)
A.166 A.V. Guinzberg, K.D. Kochev, Yu.S. Kuz'minov, T.R. Volk: Ferroelectrics 18, 71-73 (1978)
A.167 J.B. Thaxter: Appl. Phys. Lett. 15, 210-212 (1969)
A.168 J.B. Thaxter, M. Kestigian: Appl. Opt. 13, 913-924 (1974)
A.169 K. Megumi, H. Kozuka, M. Kobayashi: Appl. Phys. Lett. 12, 631-633 (1977)
A.170 V.V. Voronov, E. Kh. Gulanyan, I.R. Dorosh, Yu.S. Kuz'minov, A.L. Mikaelyan, V.V. Osiko, N.M. Polozkov, A.M. Prokhorov: Kvantovaya Elektron. 6, 1993-1999 (1979) [Engl. transl.: Sov. J. Quantum Electron 9, 1172-1175 (1979)]
A.171 I.R. Dorosh, Yu.S. Kuz'minov, N.M. Poloskov, A.M. Prokhorov, V.V. Osiko, N.V. Tkachenko, V.V. Voronov, D.Kh. Nurligareev: Phys. Status Solidi (a) 65, 513-522 (1981)
A.172 I.R. Dorosh, Yu.S. Kuz'minov, V.V. Osiko, N.V. Tkachenko: Fiz. Tverd. Tela 23, 609-611 (1981) [Engl. transl.: Sov. Phys.-Solid State 23, 345-346 (1981)]
A.173 A.V. Knyaz'kov, N.M. Kozhevnikov, Yu.S. Kuz'minov, N.M. Polozkov, A.S. Saikin, S.A. Sergushchenko: Pis'ma v Zh. Tech. Fiz. 9, 399-401 (1983) [Engl. transl.: Sov. Tech. Phys. Lett. 9, 171 (1983)]
A.174 A.V. Knyaz'kov, N.M. Kozhevnikov, Yu.S. Kuz'minov, V.V. Kulikov, N.M. Polozkov, S.A. Sergushchenko: Zh. Tech. Fiz. 54, 1379-1381 (1984) [Engl. transl.: Sov. Phys.-Tech. Phys. 29, 801-802 (1984)]
A.175 F. Micheron, C. Mayeux, J.C. Trotier: Appl. Opt. 13, 784-787 (1974)
A.176 S. Redfield, L. Hesselink: Opt. Lett. 13, 380-382 (1988)
A.177 Ya. Seglins, S. Odulov, A. Krumins: Coherent light oscillator by induced holographic gratings in SBN:Ce crystals. Abstract book of VI Int'l Meeting on Ferroelectricity (Kobe, Jpn. 1985) p.58
A.178 A.E.T. Chiou, P. Yeh: Opt. Lett. 10, 621-623 (1985)

A.179 A.V. Knyaz'kov, N.M. Kozhevnikov, Yu.S. Kuz'minov, N.M. Polozkov, A.S. Saikin, S.A. Sergushchenko: Zh. Tech. Fiz. **54**, 1737-1740 (1984) [Engl. transl.: Sov. Phys.-Tech. Phys. **29**, 1013-1015 (1984)]

A.180 Yu.O. Barmenkov, V.V. Zosimov, N.M. Kozhevnikov, L.M. Lyamshev, S.A. Sergushchenko: Dokl. Akad. Nauk SSSR **290**, 1095-1098 (1986) [Engl. transl.: Sov. Phys. Dokl. **31**, 817-819 (1986)]

A.181 G.A. Rakuljic, K. Sayano, A. Agranat, A. Yariv: Appl. Phys. Lett. **53**, 1465-1467 (1988)

A.182 V.V. Voronov, I.R. Dorosh, Yu.S. Kuz'minov, N.V. Tkachenko: Kvantovaya Elektron. **7**, 2313-2318 (1980) [Engl. transl.: Sov. J. Quantum Electron. **10**, 1346-1349 (1980)]

A.183 B. Fischer, M. Cronin-Golomb, J.O. White, A. Yariv, R. Neurgaonkar: Appl. Phys. Lett. **40**, 863-865 (1982)

A.184 O.M. Vokhnik, Yu.S. Kuz'minov, N.M. Polozkov: Kvantovaya Elektron. **13**, 1633-1637 (1986) [Engl. transl.: Sov. J. Quantum Electron **16**, 1066-1069 (1986)]

A.185 G. Salamo, M.J. Miller, W.W. Clark III, G.L. Wood, E.J. Sharp: Opt. Commun. **59**, 417-422 (1986)

A.186 M.J. Miller, E.J. Sharp, G.L. Wood, W.W. Clark III, G.J. Salamo, R.R. Neurgaonkar: Opt. Lett. **12**, 340-342 (1987)

A.187 S.R. Montgomery, J. Yarrison-Rice, D.O. Pederson, G.J. Salamo, M.J. Miller, W.W. Clark III, G.L. Wood, E.J. Sharp, R.R. Neurgaonkar: Techn. Digest of Topical Meeting on Photorefractive Materials, Effects and Devices (August 1987, Los Angeles) pp.171-174

A.188 L. Hesselink, S. Redfield: Opt. Lett. **13**, 877-879 (1988)

A.189 S. Singh, D.A. Draegert, J.E. Geusic: Phys. Rev. B2, 2709-2724 (1970)

A.190 J.J. Amodei, D.L. Staebler, W. Stephens: Appl. Phys. Lett. **18**, 507-509 (1971)

A.191 S.G. Odulov, O.I. Oleinik: Fiz. Tverd. Tela 27, 3470-3473 (1985) [Engl. transl.: Sov. Phys.-Solid State **27**, 2093-2094 (1985)]

A.192 K.I. Zemskov, M.A. Kazaryan, S.F. Lyuksyutov, S.G. Odoulov, N.G. Orlova, G.G. Petrasch, M.S. Soskin: Pis'ma v Zh. Eksp. i Teor. Fiz. **48**, 187-189 (1988) [Engl. transl. JETP Lett. **48**, 202-205 (1988)]

A.193 S.G. Odoulov, O.I. Oleinik: Kvantovaija Electron. **14**, 886-889 (1987) [Engl. transl. Sov. J. Quantum Electron. **17**, 562-564 (1987)]

A.194 T. Fukuda, H. Hirano, Y. Uematsu: Jpn. J. Appl. Phys. **13**, 1021-1022 (1974)

A.195 P. Günter: Opt. Commun. **11**, 285-290 (1974)

A.196 C. Medrano, E. Voit, P. Amrhein, P. Günter: J. Appl. Phys. **64**, 4668-4673 (1988)

A.197 P. Günter, V. Flückiger, J.P. Huignard, F. Micheron: Ferroelectrics 13, 297-299 (1976)

A.198 P. Günter, F. Micheron: Ferroelectrics **18**, 27-38 (1978)

A.199 A. Krumins, P. Günter: Appl. Phys. **19**, 153-163 (1979)

A.200 P. Günter, A. Krumins: Appl. Phys. **23**, 199-209 (1983)

A.201 E. Voit, M.Z. Zha, P.Amrhein, P. Günter: Techn. Digest of Topical Meeting on Photorefractive Materials, Effects and Applications (Los Angeles, CA 1987) pp.2-4

A.202 P. Günter: Ferroelectrics 40, 43-47 (1982)

A.203 P. Günter: Coherent light amplification and optical phase conjugation with photorefractive materials. J. Physique, C2, Suppl. No.3, **44**, 141-147 (1983)

A.204 P. Günter: Opt. Lett. **7**, 10-12 (1982)

A.205 E. Voit: In *Electro-optic and Photorefractive Materials*, ed. by P. Günter, Springer, Proc. Phys. **18**, 246-265 (Springer, Berlin, Heidelberg 1987)

A.206 E. Voit, C. Zaldo, P. Günter: Opt. Lett. **11**, 309-311 (1987)

A.207 E. Voit, P. Günter: Opt. Lett. **12**, 769-771 (1987)

A.208   A. Krumins, P. Günter: Phys. Status Solidi (a) 63, K111–K114 (1981)

A.209   A.A. Ballman: J. Cryst. Growth 1, 37–40 (1967)

A.210   E.L. Venturini, E.G. Spencer, A.A. Ballman: J. Appl. Phys. 40, 1622–1624 (1969)

A.211   S.C. Abrahams, J.L. Bernstein, C. Svensson: J. Chem. Phys. 71, 788–792 (1979)

A.212   R.E. Aldrich, S.L. Hou, M.L. Harwill: J. Appl. Phys. 42, 493–494 (1971)

A.213   S.L. Hou, R.B. Lauer, R.E. Aldrich: J. Appl. Phys. 44, 2652–2658 (1973)

A.214   O.A. Gudaev, V.A. Detinenko, V.K. Malinovskii: Fiz. Tverd. Tela 23, 195–201 (1981) [Engl. transl.: Sov. Phys.-Solid State 23, 109–113 (1981)]

A.215   A. Feldman, W.S. Brower, Jr., D. Horowitz: Appl. Phys. Lett. 16, 201–202 (1970)

A.216   S.C. Abrahams, C. Svensson, A.R. Tanguay, Jr.: Sol. St. Commun. 30, 293–295 (1979)

A.217   S.I.Stepanov, M.P. Petrov: Nonstationary holographic recording for efficient amplification and phase conjugation. *Photorefractive Materials and Applications I*, ed. by P. Günter, J.P. Huignard, Topics Appl. Phys., Vol.61 (Springer, Berlin, Heidelberg 1988) pp.263–289

A.218   T.G. Pencheva, S.I. Stepanov: Fiz. Tverd. Tela 24, 1214–1216 (1982) [Engl. transl.: Sov. Phys.-Solid State 24, 687–688 (1982)]

A.219   R.A. Sprague: J. Appl. Phys. 46, 1673–1678 (1975)

A.220   R. Grousson, M. Henry, S. Mallick: J. Appl. Phys. 56, 224–229 (1984)

A.221   J.P. Huignard, F. Micheron: Appl. Phys. Lett. 29, 591–593 (1976)

A.222   R.B. Lauer: J. Appl. Phys. 45, 1794–1797 (1974)

A.223   V.I. Berezkin: Fiz. Tverd. Tela 23, 3482–3484 (1981) [Engl. transl.: Sov. Phys.-Solid State 23, 2025–2026 (1981)]

A.224   B.Kh. Kostyuk, A.Yu. Kudzin, G.Kh. Sokolyanskii: Fiz. Tverd. Tela 22, 2454–2459 (1980) [Engl. transl.: Sov. Phys.-Solid State 22, 1429–1432 (1980)]

A.225   A.A. Kamshilin, M.P. Petrov: Fiz. Tverd. Tela 23, 3110–3116 (1981) [Engl. transl.: Sov. Phys.-Solid State 23, 1811–1814 (1981)]

A.226   M.P. Petrov, A.I. Grachev: Pis'ma v Zh. Eksp. i Teor. Fiz. 30, 18–21 (1979) [Engl. transl.: JETP Lett. 30, 15–18 (1979)]

A.227   A.I. Grachev, M.P. Petrov: Ferroelectrics 43, 181–184 (1982)

A.228   S.L. Hou, D.S. Oliver, Appl. Phys. Lett. 18, 325–328 (1971)

A.229   J.P. Herriau, J.P. Huignard, P. Aubourg: Appl. Opt. 17, 1851–1852 (1978)

A.230   S.V. Miridonov, M.P. Petrov, S.I. Stepanov: Pis'ma v Zh. Tech. Fiz. 4, 976–981 (1978) [Engl. transl.: Sov. Tech. Phys. Lett. 4, 393–394 (1978)]

A.231   M.P. Petrov, S.V. Miridonov, S.I. Stepanov, V.V. Kulikov: Opt. Commun. 31, 301–305 (1979)

A.232   M.P. Petrov, S.I. Stepanov, T.G. Pencheva, V.V. Kulikov: Opt. i Spektrosk. 55, 326–330 (1983) [Engl. transl.: Opt. & Spectrosc. 55, 192–195 (1983)]

A.233   J.P. Herriau, J.P. Huignard, A.G. Apostolidis, S. Mallick: Opt. Commun. 56, 141–144 (1985)

A.234   A. Marrakchi, R.V. Johnson, A.R. Tanguay, Jr.: J. Opt. Soc. Am. B 3, 321–336 (1986)

A.235   A.G. Apostolidis: Polarization properties of phase volume gratings recorded in a $Bi_{12}SiO_{20}$ crystal for two transverse configurations. *Electro-Optic and Photorefractive Materials*, ed. by P. Günter, Springer Proc. Phys. 18, 324–356 (Springer, Berlin, Heidelberg 1987)

A.236   T.G. Pencheva, S.I. Stepanov, S.V. Miridonov: Zh. Tech. Fiz. 53, 114–117 (1983) [Engl. transl.: Sov. Phys.-Tech. Phys. 28, 66–68 (1983)]

A.237   T.G. Pencheva, M.P. Petrov, S.I. Stepanov: Opt. Commun. 40, 175–178 (1981)

A.238   S.I. Stepanov, S.M. Shandarov, N.D. Khat'kov: Fiz. Tverd. Tela 29, 3054–3058 (1987) [Engl. transl.: Sov. Phys. Solid State 29, 1754–1756 (1987)]

A.239   M. Peltier, F. Micheron: J. Appl. Phys. 48, 3683–3690 (1977)

A.240 J.P. Huignard, J.P. Herriau, G. Rivet, P. Günter: Opt. Lett. 5, 102-104 (1980)

A.241 A. Marrakchi, J.P. Huignard, P. Günter: Appl. Phys. 24, 131-138 (1981)

A.242 G.S. Trofimov, S.I. Stepanov: Fiz. Tverd. Tela 30, 919-921 (1988) [Engl. transl. 30, 534-535 (1988)]

A.243 S.I. Stepanov, V.V. Kulikov, M.P. Petrov: Pis'ma v Zh. Tech. Fiz. 8, 527-531 (1982) [Engl. transl.: Sov. Tech. Phys. Lett. 8, 229-230 (1982)]

A.244 S.I. Stepanov, V.V. Kulikov, M.P. Petrov: Opt. Commun. 44, 19-23 (1982)

A.245 G. Hamel de Montchenault, B. Loiseaux, J.P. Huignard: Electron. Lett. 22, 1030-1032 (1986)

A.246 A.A. Kamshilin, M.G. Miteva: Opt. Commun. 36, 429-433 (1981)

A.247 S.I. Stepanov, V.V. Kulikov: Zh. Tech. Fiz. 53, 2255-2257 (1983) [Engl. transl.: Sov. Phys.-Tech. Phys. 28, 1384-1386 (1983)]

A.248 M.A. Powell, C.R. Petts: Opt. Lett. 11, 36-38 (1986)

A.249 N.I. Katsavets, E.I. Leonov, V.M. Orlov, E.B. Shadrin: Pis'ma v Zh. Tech. Fiz. 9, 424-428 (1983) [Engl. transl.: Sov. Tech. Phys. Lett. 9, 183-184 (1983)]

A.250 J.-M.C. Jonathan, R.W. Hellwarth, G. Roosen: IEEE J. QE-22, 1936-1941 (1986)

A.251 M.P. Petrov, S.I. Stepanov, S.V. Miridonov, V.V. Kulikov: SPIE Proc. 213, 44-49 (1979)

A.252 J.P. Huignard, B. Ledu: Opt. Lett. 7, 310-312 (1982)
J.P. Herriau, A. Delboulbe, B. Loiseaux, J.P. Huignard: J. Optics 15, 314-318 (1984)
G. Pauliat, J.P. Herriau, A. Delboulbe, G. Roosen, J.P. Huignard: J. Opt. Soc. Am. B3, 306-313 (1986)

A.253 G.S. Trofimov, S.I. Stepanov: Pis'ma v Zh. Tech. Fiz. 10, 669-673 (1984) [Engl. transl.: Sov. Tech. Phys. Lett. 10, 282-283 (1984)]
S.I. Stepanov, G.S. Trofimov: Zh. Tech. Fiz. 55, 559-566 (1985) [Engl. transl.: Sov. Phys.-Tech. Phys. 30, 331-334 (1985)]

A.254 J.P. Herriau, J.P. Huignard: Appl. Phys. Lett. 49, 1140-1142 (1986)

A.255 L. Arizmendi: J. Appl. Phys. 65, 423-427 (1989)

A.256 J.P. Huignard, A. Marrakchi: Opt. Commun. 38, 249-254 (1981)

A.257 H. Rajbenbach, J.P. Huignard, B. Loiseaux: Opt. Commun. 48, 247-252 (1983)

A.258 Ph. Refregier, L. Solimar, H. Rajbenbach, J.P. Huignard: Electron. Lett. 20, 656-657 (1984)

A.259 J.P. Huignard, J.P. Herriau, P. Aubourg, E. Spitz: Opt. Lett. 4, 21-23 (1978)

A.260 H. Rajbenbach, J.P. Huignard, Ph. Refregier: Opt. Lett. 9, 558-560 (1984)

A.261 H. Rajbenbach, J.P. Huignard: Opt. Lett. 10, 137-139 (1985)

A.262 J.P. Huignard, H. Rajbenbach, Ph. Refregier, L. Solymar: Opt. Eng. 24, 586-592 (1985)

A.263 P. Pellat-Finet, J.L. de Bougrenet de la Tochnaye: Opt. Commun. 55, 305-310 (1985)

A.264 J.P. Hermann, J.P. Herriau, J.P. Huignard: Appl. Opt. 20, 2173-2175 (1981)

A.265 G. LeSaux, G. Roosen, A. Brun: Opt. Commun. 56, 374-378 (1986)

A.266 G. LeSaux, G. Roosen, A. Brun: Opt. Commun. 58, 238-240 (1986)

A.267 G. LeSaux, A. Brun: IEEE J. QE-23, 1680-1688 (1987)

A.268 J.L. Ferrier, J. Gazengel, X. Nguyen Phy, G. Rivoire: Opt. Commun. 58, 343-348 (1986)

A.269 S.C. Abrahams, P.B. Jamieson, J.L. Bernstain: J. Chem. Phys. 47, 4034-4041 (1967)

A.270 G.G. Douglas, R.N. Zitter: J. Appl. Phys. 39, 2133-2135 (1968)

A.271 P.V. Lenzo, E.G. Spencer, A.A. Ballman: Appl. Opt. 5, 1688-1689 (1966)

A.272 G. Pauliat, J.M. Cohen-Jonathan, M. Allain, J.C. Launay, G. Roosen: Opt. Commun. 59, 266-271 (1986)
G. Pauliat, M. Allain, J.C. Launay, G. Roosen: Opt. Commun. 61, 321-323 (1987)

267

A.273 J.P. Herriau, D. Rojas, J.P. Huignard, J.M. Bassat, J.C. Launay: Ferroelectrics 75, 271-279 (1987)

A.274 P. Günter: Opt. Commun. 41, 83-88 (1982)

A.275 Y.H. Ja: Electron. Lett. 17, 488-489 (1981)

A.276 Y.H. Ja: Opt. Commun. 41, 159-163 (1982); Opt. and Quantum Electron. 14, 363-365 (1982); ibid. 16, 355-358 (1984)

A.277 A.A. Petrov, V.V. Prokof'ev, M.V. Krasin'kova, I.V. Micheeva, G.I. Dolivo-Dobrovol'skaya, B.B. Zhdanova: Digests of VI Int'l Conf. on Crystal Growth (ICCG-6, Moscow 1980) Vol.3, 117-118

A.278 A.A. Ballman, H. Brown, P.K. Tien, R.J. Martin: J. Cryst. Growth 20, 251-255 (1973)

A.279 A.J. Fox, T.M. Bruton: Appl. Phys. Lett. 27, 360-363 (1975)

A.280 Sh.M. Efendiev, V.E. Bagiev, A.Sh. Zeinally, V.A. Balashov, V.A. Lomonov, A.A. Majer: Phys. Status Solidi (a) 63, K19-K22 (1981)

A.281 G.S. Trofimov, S.I. Stepanov: Pis'ma v Zh. Tech. Fiz. 11, 615-621 (1985) [Engl. transl.: Sov. Tech. Phys. Lett. 11, 256-257 (1985)]

A.282 S.I. Stepanov, M.P. Petrov: Opt. Commun. 53, 292-295 (1985)

A.283 S.I. Stepanov, M.P. Petrov, S.L. Sochava: Ferroelectrics 92, 199-204 (1989)

A.284 M.S. Brodin, V.I. Volkov, N.V. Kukhtarev, A.V. Privalko: Ukr. Fiz. Zh. 33, 1781-1783 (1988)

A.285 A.A. Kamshilin, M.P. Petrov: Opt. Commun. 53, 23-26 (1985)

A.286 A.A. Kamshilin, S.V. Miridonov, M.G. Miteva, E.V. Mokrushina: Zh. Tech. Fiz. 59, 113-117 (1989) [Engl. transl.: Sov. Phys. Tech. Phys. 34, 66-68 (1989)]

A.287 S.L. Sochava, S.I. Stepanov: Zh. Tech. Fiz. (USSR) 57, 1763-1766 (1987) [Engl. transl.: Sov. Phys.-Tech. Phys. 32, 1054-1056 (1987)]

A.288 S.I. Stepanov, M.P. Petrov, M.V. Krasin'kova: Zh. Tech. Fiz. 54, 1223-1225 (1984) [Engl. transl.: Sov. Phys.-Tech. Phys. 29, 703-705 (1984)]
S.I. Stepanov, M.P. Petrov: Optica Acta 31, 1335-1343 (1984)

A.289 S.I. Stepanov, M.P. Petrov: Pis'ma v Zh. Tech. Fiz. 10, 1356-1360 (1984) [Engl. transl.: Sov. Tech. Phys. Lett. 10, 572-573 (1984)]
M.P. Petrov, S.I. Stepanov: New mechanisms of holographic recording and wavefront conjugation in cubic photorefractive crystals. ICO-13 (Sapporo, Jpn. 1984) Digest pp.430-431

A.290 S.L. Sochava, S.I. Stepanov, M.P. Petrov: Pis'ma v Zh. Tech. Fiz. 13, 660-665 (1987) [Engl. transl.: Sov. Tech. Phys. Lett. 13, 274-275 (1987)]

A.291 S.L. Sochava, S.I. Stepanov: Zh. Tech. Fiz. 58, 1780-1783 (1988) [Engl. transl. Sov. Phys. Tech. Phys. 33, 1077-1079 (1988)]

A.292 M.P. Petrov, S.L. Sochava, S.I. Stepanov: Opt. Lett. 14, 284-286 (1989)

A.293 N.V. Kukhtarev, M.S. Brodin, V.I. Volkov: Fiz. Tverd. Tela 30, 2757-2760 (1988) [Engl. transl. Sov. Phys. Solid State 30, 1588-1590 (1988)]

A.294 A.M. Glass, A.M. Johnson, D.H. Olson, W. Simpson, A.A. Ballman: Appl. Phys. Lett. 44, 948-950 (1984)

A.295 A.M. Glass, J. Strait: Photorefractive four-wave mixing in electro-optic semi-conductors. VI Int'l Meeting on Ferroelectricity (Jpn. 1985) Abstract book, p.58

A.296 A.M. Glass, J. Strait: The photorefractive effect in semiconductors. *Photorefractive Materials and their Applications I*, ed. by P. Günter, J.P. Huignard, Topics Appl. Phys., Vol.61 (Springer, Berlin, Heidelberg 1988) pp.237-262

A.297 K. Walsh, T.J. Hall, R.E. Burge: Opt. Lett. 12, 1026-1028 (1987)

A.298 S.G. Odoulov, S.S. Slusarenko, K.V. Scherbin: Pis'ma v Zh. Tech. Fiz. 15, 10-14 (1989) [Engl. transl. Sov. Tech. Phys. Lett. 15, 417-418 (1989)]

A.299 L.J. Cheng, P. Yeh: Opt. Lett. 13, 50-52 (1988)

A.300 A. Partovi, E.M. Garmire, L.J. Cheng: Appl. Phys. Lett. 51, 299-301 (1987)
L.J. Cheng, G. Gheen, T.H. Chao, H.K. Liu, A. Partovi, J. Katz, E.M. Garmire: Opt. Lett. 12, 705-707 (1987)

A.301   R.B. Bylsma, D.H. Olson, A.M. Glass: Opt. Lett. 13, 853-855 (1988)
A.302   M.B. Klein: Opt. Lett. 9, 350-352 (1984)
A.303   A.M. Glass, M.B. Klein, G.C. Valley: Electron. Lett. 21, 220-221 (1985)
A.304   R.B. Bylsma, D.H. Olson, A.M. Glass: Appl. Phys. Lett. 52, 1083-1085 (1988)
A.305   L.J. Cheng, A. Partovi: Appl. Opt. 27, 1760-1763 (1988)
A.306   D.T.H. Liu, L.J. Cheng, M.F. Rau, F.C. Wang: Appl. Phys. Lett. 53, 1369-1371 (1988)
A.307   B. Mainguet: Opt. Lett. 13, 657-659 (1988)
A.308   J. Kumar, G. Albanese, W. Steier: Opt. Commun. 63, 191-193 (1987)
A.309   B. Imbert, H. Rajbenbach, S. Mallick, J.P. Herriau, J.P. Huignard: Opt. Lett. 13, 327-329 (1988)
A.310   H. Rajbenbach, B. Imbert, J.P. Huignard, S. Mallick: Opt. Lett. 14, 78-80 (1989)
A.311   J. Kumar, G. Albanese, W.H. Steier, M. Ziari: Opt. Lett. 12, 120-122 (1987)
A.312   M.B. Klein, S.W. McCahon, T.F. Boggess, G.C. Valley: J. Opt. Soc. Am. B 5, 2467-2472 (1988)
A.313   G.S. Trofimov, S.I. Stepanov, M.P. Petrov, M.V. Krasin'kova: Pis'ma v Zh. Tech. Fiz. 13, 265-269 (1987) [Engl. transl.: Sov. Tech. Phys. Lett. 13, 108-109 (1987)]
        I.A. Sokolov, G.S. Trofimov, S.I. Stepanov: Zh. Tech. Phys. 58, 429-431 (1988) [Engl. transl.: Sov. Phys. Tech. Phys. 33, 261-262 (1988)]
A.314   G.C. Valley, A.L. Smirl, M.B. Klein, K. Bohnert, T. Boggess: Opt. Lett. 11, 647-649 (1986)
A.315   A.L. Smirl, G.C. Valley, K.M. Bohnert, T.F. Boggess, Jr.: IEEE J. QE-24, 289-303 (1988)
A.316   H. Mao, Y. Liu, F. Li, W. Yu, X. Zhu, X. Deng: Opt. Commun. 69, 166-168 (1988)
A.317   J.C. Fabre, J.M.C. Jonathan, G. Roosen: J. Opt. Soc. Am. B 5, 1730-1736 (1988)
A.318   S.R. King, T.S. Hartwick, A.B. Chase: Appl. Phys. Lett. 21, 312-314 (1972)
A.319   D. Von der Linde, A.M. Glass, K.F. Rodgers: Appl. Phys. Lett. 26, 22-24 (1975)
A.320   L.A. Boatler, E. Krätzig, R. Orlowski: Ferroelectrics 27, 247-250 (1980)
A.321   R. Orlowski, L.A. Boatler, E. Krätzig: Opt. Commun. 35, 45-48 (1980)
A.322   F. Micheron, C. Mayeux, A. Hermasin, J. Nicolas: J. Am. Cer. Soc. 57, 306-308 (1974)
A.323   B. Houlier, F. Micheron: J. Appl. Phys. 50, 343-345 (1979)
A.324   J.W. Burgess: Appl. Opt. 15, 1550-1557 (1976)
A.325   M.M. Butusov, A.V. Knyaz'kov, A.E. Krumins, N.V. Kukhtarev, A.S. Saikin: Pis'ma v Zh. Tech. Fiz. 7, 914-917 (1981) [Engl. transl.: Sov. Tech. Phys. Lett. 7, 393-394 (1981)]
A.326   M.M. Butusov, A.V. Knyaz'kov, A.S. Saikin, N.V. Kukhtarev, A.E. Krumins: Ferroelectrics 45, 63-70 (1982)
A.327   A.E. Krumins, A.V. Knyaz'kov, A.S. Saikin, Ya.A. Seglin'sh: Fiz. Tverd. Tela 25, 1570-1572 (1983) [Engl. transl.: Sov. Phys.-Solid State 25, 908-909 (1983)]
A.328   A.V. Alekseev-Popov, A.V. Knyaz'kov, A.S. Saikin: Pis'ma v Zh. Tech. Fiz. 9, 1108-1112 (1983) [Engl. transl.: Sov. Tech. Phys. Lett. 9, 475-477 (1983)]
A.329   A.V. Knyaz'kov, M.N. Lobanov, A. Krumins, J. Seglin'sh: Ferroelectrics 69, 81-87 (1986)
A.330   A. Krumins, R.A. Rupp, K. Kerperin: Ferroelectrics 80, 281-284 (1988)

Subject Index

Phase distortions 133
Phase distortion compensation via phase
  conjugation 186,187
– one-way geometries 188
Phase homogeneity 25,102,187
Phase locking (synchronization)
  of lasers 194,195
Phase synchronism 55,57
Phase transfer 92
Photocarrier lifetime 44
Photoconductivity 40
– bipolar 52,53
– electron 44,234,237,239
– hole 55,228
Photocurrent 44,45
Photoexcitation rate 44,52
Photoexcited charge transport 2
Photogalvanic
– effect 3,44,45
– recording mechanism 47,225,227
Photoinduced noise 59,226
Photolitography 187
Photorefractive
– crystal 1
– effect 1
Phototitus 158
Photovoltaic
– current 44
– effect 1,2
– voltage 45
Photosensitive centers 1
Pixel 24
PLZT ceramics 241
Picosecond pulses 230,241
Pockels effect 9
Pockels readout optical modulator
  (PROM) 137,211
Poisson's equation 43,49,123
Polarization of light
– circular 70,72,81
– linear 10,68,70,77,82,83
– rotation 70,77,83,84,86,87,181
Poling 223,228
PRIZ spatial light modulator 144,213
Pump wave 95
– misalignment 98,102

Quantum efficiency 5,44,48,234
Quasi-neutrality condition
  violation 49-51,73
Quasi-stationary photoelectron
  distribution 44

Readout 139
Real-time operation of SLM 210
Relaxation (erasure, decay) characteristic
  time 47,52
Recombination 44
Recording (buildup) characteristic
  time 47
Recording mechanism 4
Reduction 52,225,228
Reflectivity of four-wave mixing
  geometry 97,98,100-102
Refractive index change 68,202
Resolution 49-51,127,136,141,147,
  163,167
Resonance
– hologram amplification 55
– speed of moving pattern 54
Response time 47
Rh 224
Ring
– photorefractive oscillator 103,104,197
– passive phase-conjugate geome-
  try 106,107,191,193,197-199

Saturation of impurities 49-51
Scattering
– light-induced 59,226,239
Screening effect 42
Selectivity of volume hologram
– angular 21,22,83,84
– wavelength 21,22,83-85
Self-diffraction (two-wave
  mixing) 22,88,92-94
– anisotropic 22,70,71
Self-pumped (passive) phase conjuga-
  tion 102-108,185,189-194,198,221
Semiconductor photorefractive 79,80,
  239-241
Sensitivity 131,140,150
– holographic 38
– of photorefractive crystal 39,48
Sign of photocarriers 52,55,237
Signal-to-noise ratio (SNR) 87,134,136,
  151,182
Sillenites 233-239
Space charge 1
Space-charge field 6,43,46,50,52,56
– maximum 49
Spatial
– frequency 3
– harmonics 43,206
Spatial light modulator
  (SLM) 24,109,136,210

Springer Series in Optical Sciences

Editorial Board: D.L. MacAdam A.L. Schawlow K. Shimoda A.E. Siegman T. Tamir